Earth Features and Their Meaning

An Introduction to Geology for the Student and the General Reader

William Herbert Hobbs

Writat

This edition published in 2022

ISBN : 9789356391291

Contents

PREFACE

THE series of readings contained in the present volume give in somewhat expanded form the substance of a course of illustrated lectures which has now for several years been delivered each semester at the University of Michigan. The keynote of the course may be found in the dominant characteristics of the different earth features and the geological processes which have been betrayed in the shaping of them. Such a geological examination of landscape is replete with fascinating revelations, and it lends to the study of Nature a deep meaning which cannot but enhance the enjoyment of her varied aspects.

That there is a real place for such a cultural study of geology within the University is believed to be shown by the increasing number of students who have elected the work. Even more than in former years the American travels afar by car or steamship, and the earth's surface features in all their manifold diversity are thus one after the other unrolled before him. The thousands who each year cross the Atlantic to roam over European countries may by historical, literary, or artistic studies prepare themselves to derive an exquisite pleasure as they visit places identified with past achievement of one form or another. Yet the Channel coast, the gorge of the Rhine, the glaciers of Switzerland, and the wild scenery of Norway or Scotland have each their fascinating story to tell of a history far more remote and varied. To read this history, the runic characters in which it is written must first of all be mastered; for in every landscape there are strong individual lines of character such as the pen artist would skillfully extract for an outline sketch. Such *character profiles* are often many times repeated in each landscape, and in them we have a key to the historical record.

An object of the present readings has thus been to enable the student to himself pick out in each landscape these more significant lines and so read directly from Nature. In the landscapes which[ix] have been represented, the aim has been to draw as far as possible upon localities well known to travelers and likely to be visited, either because of their historical interest or their purely scenic attractions. It should thus be possible for a tourist in America or Europe to pursue his landscape studies whenever he sets out upon his travels. The better to aid him in this endeavor, some suggestions concerning the itinerary of journeys have been supplied in an appendix.

Regarded as a textbook of geology, the present work offers some departures from existing examples. Though it has been customary to combine in a single text historical with dynamical and structural geology, a tendency has already become apparent to treat the historical division apart from the others. Again, a desire to treat the science of geology comprehensively has led some authors into including so many subjects as to render their texts unnecessarily encyclopedic and correspondingly uninteresting to the general reader. It is the author's belief that there is a real need for a book which may be read intelligently by the general public, and it must be recognized that the beginner in the subject cannot cover the entire field by a single course of readings. The present work has, therefore, been prepared with a view to selecting for study those dominant geological processes which are best illustrated by features in northern North America and Europe. It is this desire to illustrate the readings by travels afield, which accounts for the prominence given to the subject of glaciation; for the larger number of colleges and universities in both America and Europe are surrounded by the heavy accumulations that have resulted from former glaciations.

Emphasis has also been placed upon the dependence of the dominant geological processes of any region upon existing climatic conditions, a fact to which too little attention has generally been given. This explains the rather full treatment of desert regions, of which, in our own country particularly, much may be illustrated upon the transcontinental railway journeys.

More than in most texts the attempt has here been made to teach directly through the eye with the efficient aid of apt illustrations intimately interwoven with the text. For such success as has been reached in this endeavor, the author is greatly indebted to two students of the University of Michigan,—Mr. James H. Meier, who has prepared the line drawings of landscapes, and Mr. Hugh M.[x] Pierce, who has draughted the diagrams. Though credit has in most cases been given where illustrations have been made from another's photographs, yet especial mention should here be made of the debt to Dr. H. W. Fairbanks of Berkeley, California, whose beautiful and instructive photographs are reproduced upon many a page.

As given at the University of Michigan, the lectures reflected in the present volume are supplemented by excursions and by so much laboratory practice as is necessary to become familiar with the more common minerals and rocks, and to read intelligently the usual topographical and geological maps. In the appendices the means for carrying out such studies, in part with newly devised apparatus, have been indicated.

The scope of the book precludes the possibility of furnishing the reader with the sources for the body of fact and theory which is presented, although much may be inferred from the names which appear beneath the illustrations, and more definite knowledge will be found in the references to literature supplied at the ends of chapters. A large amount of original and unpublished material is for a similar reason unlabeled, and it has been left for the professional geologist to detect these new strands which have been drawn into the web.

WILLIAM HERBERT HOBBS.

ANN ARBOR,
MICHIGAN,
October 25, 1911.

CHAPTER I

THE COMPILATION OF EARTH HISTORY

The sources of the history.—The science which deals with the chapters of earth history that antedate the earliest human writings is geology. The pages of the record are the layers of rock which make up the outer shell of our world. Here as in old manuscripts pages are sometimes found to be missing, and on others the writing is largely effaced so as to be indistinct or even illegible. An intelligent interpretation of this record requires a knowledge of the materials and the structure of the earth, as well as a proper conception of the agencies which have caused change and so developed the history. These agencies in operation are physical and chemical processes, and so the sciences of physics and chemistry are fundamental in any extended study of geology. Not only is geology, so to speak, founded upon chemistry and physics, but its field overlaps that of many other important sciences. The earliest earth history has to do with the form, size, and physical condition of a minor planet in the solar system. The earliest portion of the story belongs therefore to astronomy, and no sharp line can be drawn to separate this chapter from those later ones which are more clearly within the domain of geology.

Subdivisions of geology.—The terms "cosmic geology" and "astronomic geology" have sometimes been used to cover the astronomy of the earth planet. The later earth history develops, among other things, the varied forms of animal and vegetable life which have had a definite order of appearance. Their study is to a large extent zoölogy and botany, though here considered from an essentially different viewpoint. This subdivision of our science is called paleontological geology or paleontology, which in common usage includes the plant as well as the animal world, or what is sometimes called paleobotany. In order to fix the order of events in geological history, these biological studies are necessary, for the pages of the record have many of them been misplaced as a result of the vicissitudes of earth history, and the remains of life in the rock layers supply a pagination from which it is possible to correctly rearrange the misplaced pages. As compiled into a consecutive history of the earth since life appeared upon it, we have the division of historical geology; though this differs but little from stratigraphical geology, the emphasis in the case of the former being placed on the history itself and in the latter upon the arrangement of events—the pagination of the record.

So far as they are known to us, the materials of which the earth is composed are minerals grouped into various characteristic aggregates known as rocks. Here the science is founded upon mineralogy as well as chemistry, and a study of the rock materials of the earth is designated petrographical geology or petrography. The various rocks which enter into the composition of the earth's outer shell—the only portion known to us from direct observation—are built into it in an architecture which, when carefully studied, discloses important events in the earth's history. The division of the science which is concerned with earth architecture is geotectonic or structural geology.

The study of earth features and their significance.—The features upon the surface of the earth have all their deep significance, and if properly understood, a flood of light is thrown, not only upon present conditions, but upon many chapters of the earth's earlier history. Here the relation of our study to topography and geography is very close, so that the lines of separation are but ill defined. The terms "physiographical geology", "physiography", and "geomorphology" are concerned with the configuration of the earth's surface—its physiognomy—and with the genesis of its individual surface features. It is this genetical side of physiography which separates it from topography and lends it an absorbing interest, though it causes it to largely overlap the division of dynamical geology or the study of geological processes. In fact, the difference between dynamical geology and physiography is largely one of emphasis, the stress being laid upon the processes in the former and upon the resultant features in the latter.

Under dynamical geology are included important subdivisions, such as seismic geology, or the study of earthquakes, and vulcanology, or the study of volcanoes. Another large subject, glacial geology, belongs within the broad frontier common to both dynamical geology and physiography. A relatively new subdivision of geological science is orientational geology, which is concerned with the trend of earth features, and is closely related both to physiography and to dynamical and structural geology.

Tabular recapitulation.—In a slightly different arrangement from the above order of mention, the subdivisions of geology are as follows:—

Subdivisions of Geology

Petrographical Geology.	Materials of the earth.
Geotectonic Geology.	Architecture of the earth's outer shell.
Dynamical Geology.	Earth processes.
Seismic Geology—earthquakes. Vulcanology—volcanoes. Glacial Geology—glaciers, etc.	
Physiographical Geology.	Earth physiognomy and its genesis.
Orientational Geology.	The arrangement and the trend of earth features.

In one way or another all of the above subdivisions of geology are in some way concerned in the genesis of earth physiognomy, and they must therefore be given consideration in a work which is devoted to a study of the meaning of earth features. The compiled record of the rocks is, however, something quite apart and without pertinence to the present work. As already indicated its subdivisions are:—

Astronomic Geology.	Planetary history of the earth.
Statigraphic Geology.	The pagination of earth records.
Historical Geology.	The compiled record and its interpretation.
Paleontological Geology.	The evolution of life upon the earth.

In every attempt at systematic arrangement difficulties are encountered, usually because no one consideration can be used throughout as the basis of classification. Such terms as "economic geology" and "mining geology" have either a pedagogical or a commercial significance, and so would hardly fit into the system which we have outlined.

Geological processes not universal.—It is inevitable that the geology of regions which are easily accessible for study should have absorbed the larger measure of attention; but it should not be forgotten that geology is concerned with the history of the entire world, and that perspective will be lost and erroneous conclusions drawn if local conditions are kept too often before the eyes. To illustrate by a single instance, the best studied regions of the globe are those in which fairly abundant precipitation in the form of rain has fitted the land for easy conditions of life, and has thus permitted the development of a high civilization. In degree, and to some extent also in kind, geologic processes are markedly different within those widely extended regions which, because either arid or cold, have been but ill fitted for human habitation. Yet in the historical development of the earth, those geologic processes which obtain in desert or polar regions are none the less important because less often and less carefully observed.

Change, and not stability, the order of nature.—Man is ever prone to emphasize the importance of apparent facts to the disadvantage of those less clearly revealed though equally potent. The ancient notion of the *terra firma*, the safe and solid ground, arose because of its contrast with the far more mobile bodies of water; but this illusion is quickly dispelled with the sudden quaking of the ground. Experience has clearly shown that, both upon and beneath the earth's surface, chemical and physical changes are going on, subject to but little interruption. "The hills rock-ribbed and ancient as the sun" is a poetical metaphor; for the Himalayas, the loftiest mountains upon the globe, were, to speak in geological terms, raised from the sea but yesterday. Even to-day they are pushing up their heads, only to be relentlessly planed down through the action of the atmosphere, of ice, and of running water. Even more than has generally been supposed, the earth suffers change. Often within the space of a few seconds, to the accompaniment of a heavy earthquake, many square miles of territory are bodily uplifted, while neighboring areas may be relatively depressed. Thus change, and not stability, is the order of nature.

Observational geology *versus* speculative philosophy.—There appears to be a more or less prevalent notion that the views which are held by scientists in one generation are abandoned by those of the next; and this is apt to lead to the belief that little is really known and that much is largely guessed. Some ground there undoubtedly is for such skepticism, though much of it may be accounted for by a general failure among scientists, as well as others, to clearly differentiate that which is essentially speculative from what is based broadly upon observed facts. Even with extended observation, the possibility of explaining the facts in more than one way is not excluded; but the line is nevertheless a broad one which separates this entire field of observation from what is essentially speculative philosophy. To illustrate: the mechanics of the action which goes on within volcanic craters is now fairly well understood as a result of many and extended observations, and it is little likely that future generations of geologists will discredit the main conclusions which have been reached. The cause of the rise of the lava to the earth's surface is, on the other hand, much less clearly demonstrated, and the views which are held express rather the differing opinions than any clear deductions from observation. Again, and similarly, the physical history of the great continental glaciers of the so-called "ice age" is far more thoroughly known than that of any existing glacier of the same type; but the cause of the climatic changes which brought on the glaciation is still largely a matter for speculation.

In the present work, the attempt will be, so far as possible, to give an exposition of geologic processes and the earth features which result from them, with hints only at those ultimate causes which lie hidden in the background.

The scientific attitude and temper.—The student of science should make it his aim, not only clearly to separate in his studies the proximate from the ultimate causes of observed phenomena, but he should keep his mind always open for reaching individual conclusions. No doctrines should be accepted finally upon faith merely, but subject rather to his own reasoning processes. This should not be interpreted to mean that concerning matters of which he knows little or nothing he should not pay respect to the recognized authorities; but his acceptance of any theory should be subject to review so soon as his own horizon has been sufficiently enlarged. False theories could hardly have endured so long in the past, had not too great respect been given to authorities, and individual reasoning processes been held too long in subjection.

The value of the hypothesis.—Because all the facts necessary for a full interpretation of observed phenomena are not at one's hand, this should not be made to stand in the way of provisional explanations. If science is to advance, the use of hypothesis is absolutely essential; but the particular hypothesis adopted should be regarded as temporary and as indicating a line of observation or of experimentation which is to be followed in testing it. Thus regarded with an open mind, inadequate hypotheses are eventually found to be untenable, whereas correct explanations of the facts by the same process are confirmed. Most hypotheses of science are but partially correct, for we now "see through a glass darkly"; but even so, if properly tested, the false elements in the hypothesis are one after the other eliminated as the embodied truth is confirmed and enlarged. Thus "working hypothesis" passes into theory and becomes an integral part of science.

Reading References for Chapter I

The most comprehensive of general geological texts written in English is Chamberlin and Salisbury's "Geology" in three volumes (Henry Holt, 1904-1906), the first volume of which is devoted exclusively to geological processes and their results. An abridged one-volume edition of the work intended for use as a college text was issued in 1906 (College Geology, Henry Holt). Other standard texts are:—

SIR ARCHIBALD GEIKIE. Text-book of Geology, 4th ed. 2 vols. London, 1902, pp. 1472.

W. B. SCOTT. An Introduction to Geology. 2d ed. Macmillan, 1907, pp. 816.

J. D. DANA. Manual of Geology. New edition. American Book Company, 1895, pp. 1087.

JOSEPH LECONTE. Elements of Geology. (Revised by Fairchild.) Appleton, 1905, pp. 667.

A very valuable guide to the recent literature of dynamical and structural geology is Branner's "Syllabus of a Course of Lectures on Elementary Geology" (Stanford University, 1908).

On the relation of geology to landscape, a number of interesting books have been written:—

JAMES GEIKIE. Earth Sculpture or the Origin of Land-Forms. New York and London, 1896, pp. 397.

JOHN E. MARR. The Scientific Study of Scenery. Methuen, London, 1900, pp. 368.

SIR A. GEIKIE. The Scenery of Scotland. 3d ed. Macmillan, London, 1901, pp. 540.

SIR JOHN LUBBOCK. The Scenery of Switzerland and the Causes to which it is Due. Macmillan, London, 1896, pp. 480.

LORD AVEBURY. The Scenery of England. Macmillan, London, 1902, pp. 534.

SIR A. GEIKIE. Landscape in History, and Other Essays. Macmillan, London, 1905, pp. 352.

N. S. SHALER. Aspects of the Earth. Scribners, New York, 1889, pp. 344.

G. DE LA NOE ET EMM. DE MARGERIE. Les Formes du Terrain, Service Géographique de l'Armée. Paris, 1888, pp. 205, pls. 48.

W. M. DAVIS. Practical Exercises in Physical Geography, with Accompanying Atlas. Ginn and Co., Boston, 1908, pp. 148, pls. 45.

JOHN MUIR. The Mountains of California. Unwin, London, 1894, pp. 381.

Upon the use and interpretation of topographic maps in illustration of characteristic earth features, the following are recommended:—

R. D. SALISBURY and W. W. ATWOOD. The Interpretation of Topographic Maps, Prof. Pap., 60 U.S. Geol. Surv., pp. 84, pls. 170.

D. W. JOHNSON and F. E. MATTHES. The Relation of Geology to Topography, in Breed and Hosmer's Principles and Practice of Surveying, vol. 2. Wiley, New York, 1908.

GÉNÉRAL BERTHAUT. Topologie, Étude du Terrain, Service Géographique de l'Armée. Paris, 1909, 2 vols., pp. 330 and 674, pls. 265.

The United States Geological Survey issues free of charge a list of 100 topographic atlas sheets which illustrate the more important physiographic types. In his "Traité de Géographie Physique", Professor E. de Martonne has given at the end of each chapter the important foreign maps which illustrate the physiographic types there described.

"The Principles of Geology", by Sir Charles Lyell, published first in three volumes, appeared in the years 1830-1833, and may be said to mark the beginning of modern geology. Later reduced to two volumes, an eleventh edition of the work was issued in 1872 (Appleton) and may be profitably read and studied to-day by all students of geology. Those familiar with the German language will derive both pleasure and profit from a perusal of Neumayr's "Erdgeschichte" (2d ed. revised by Uhlig. Leipzig and Vienna, 2 vols., 1895-1897), and especially the first volume, "Allgemeine Geologie." A recent French work to be recommended is Haug's "Traité de Géologie" (Paris, 1907).

Some texts of physical geography may well be consulted, especially Emm. de Martonne's "Traité de Géographie Physique." Colin, Paris, 1909, pp. 910, pls. 48, and figs. 396.

NOTE. An explanatory list of abbreviations used in the reading references follows the List of Illustrations.

CHAPTER II

THE FIGURE OF THE EARTH

The lithosphere and its envelopes.—The stony part of the earth is known as the *lithosphere*, of which only a thin surface shell is known to us from direct observation. The relatively unknown central portion, or "core", is sometimes referred to as the centrosphere. Inclosing the lithosphere is a water envelope, the *hydrosphere*, which comprises the oceans and inland bodies of water, and has a mass $^{1}/_{4540}$ that of the lithosphere. If uniformly distributed, the hydrosphere would cover the lithosphere to the depth of about two miles, instead of being collected in basins as it now is. Though apparently not continuous, if we take into account the zone of underground water upon the continents, the hydrosphere may properly be considered as a continuous film about the lithosphere. It is a fact of much significance that all the ocean basins are connected, so that the levels are adjusted to furnish a common record of deposits over the entire surface that is sea-covered.

Enveloping the hydrosphere is the gaseous envelope, the *atmosphere*, with a mass $^{1}/_{1200000}$ that of the lithosphere. The atmosphere is a mixture of the gases oxygen and nitrogen in parts by volume of one of the former to four of the latter, with a relatively small percentage of carbon dioxide. Locally, and at special seasons, the atmosphere may be charged with relatively large percentages of water vapor; and we shall see that both the carbon dioxide and the vapor contents are of the utmost importance in geological processes and in the influence upon climate. Unlike the water which composes the hydrosphere, the gases of the atmosphere are compressible. Forced down by the weight of superincumbent gas, the layers of the atmosphere at the level of the sea sustain a pressure of about fifteen pounds to the square inch; but this pressure steadily decreases in ascending to higher levels. From direct instrumental observation, the air has now been investigated to a height of more than twelve miles from the earth's surface.

The evolution of ideas concerning the earth's figure.—The ideas which in all ages have been promulgated concerning the figure of the earth have been many and varied. Though among them are not wanting the purely speculative and fantastic, it will be interesting to pass in review such theories as have grown directly out of observation.

The ancient Hebrews and the Babylonians were dwellers of the desert, and in the mountains which bounded their horizon they saw the confines of the earth. Pushing at last westward beyond the mountains, they found the Mediterranean, and thus arrived at the view that the earth was a disk with a rim of mountains which was floated upon water. The rare but violent rainfalls to which they were accustomed—the desert cloudburst—further led them to the belief that the mountain rim was continued upward in a dome or firmament of transparent crystal upon which the heavenly bodies were hung and from which out of "windows of heaven" the water "which is above the earth" was poured out upon the earth's surface. Fantastic as this theory may seem to-day, it was founded upon observation, and it well illustrates the dangers of reasoning from observation within too limited a field.

As soon as men began to sail the sea, it was noticed that the water surface is convex, for the masts of ships were found to remain visible long after their hulls had disappeared below the horizon. It is difficult to say how soon the idea of the earth's rotundity was acquired, but it is certainly of great antiquity. The Dominican monk Vincentius of Beauvais, in a work completed in 1244, declared that the surfaces of the earth and the sea were both spherical. The poet Dante made it clear that these surfaces were one, and in his famous address upon "The Water and the Land", which was delivered in Verona on the 20th of January, 1320, he added a statement that the continents rise higher than the ocean. His explanation of this was that the continents are pulled up by the attraction of the fixed stars after the manner of attraction of magnets, thus giving an early hint of the force of gravitation.

The earth's rotundity may be said to have been first proven when Magellan's ships in 1521 had accomplished the circumnavigation of the globe. Circumnavigation, soon after again carried out by Sir Francis Drake, proved that the earth is a closed body bounded by curving surfaces in part enveloped by the oceans and everywhere by the atmosphere. The great discovery of Copernicus in 1530 that the earth, like Venus, Mars, and the other planets, revolves about the sun as a part of a system, left little room for doubt that the figure of the earth was essentially that of a sphere.

The oblateness of the earth.—Every schoolboy is to-day familiar with the fact that the earth departs from a perfect spherical figure by being flattened at the ends of its axis of rotation. The polar diameter is usually given as $^1/_{299}$ shorter than the equatorial one. This oblateness of the spheroid was proven by geodesists when they came to compare the lengths of measured degrees of arc upon meridians in high and in low latitudes.

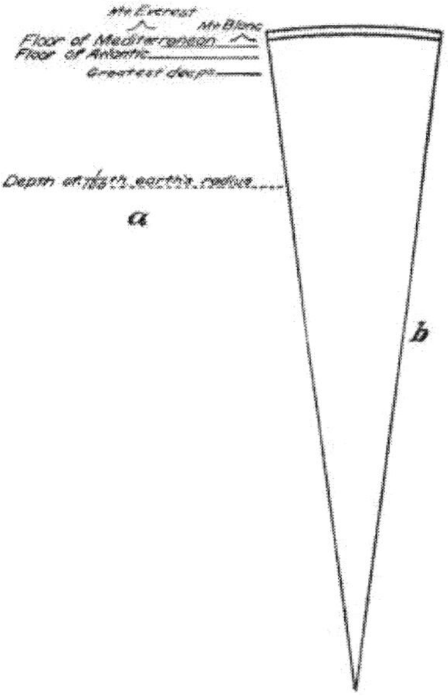

FIG. 1.—Diagrams to afford a correct impression of the measure of the inequalities upon the earth's surface compared to the earth's radius. The shell represented in *b* is $^1/_{100}$ of the earth's radius, and in *a* this zone is magnified for comparison with surface inequalities.

The oblateness of the geoid is well understood from accepted hypotheses to be the result of the once more rapid rotation of the planet when its materials were more plastic, and hence more responsive to deformation. An elastic hoop rotating rapidly about an axis in its plane appears to the eye as a solid, and becomes flattened at the ends of its axis in proportion as the velocity of rotation is increased. Like the earth, the other planets in the solar system are similarly oblate and by amounts dependent on the relative velocities of rotation.

The departure of the geoid from the spherical surface, owing to its oblateness, is so small that in the figures which we shall use for illustration it would be less than the thickness of a line. Since it is well recognized and not important in our present consideration, we shall for the time being speak of the figure of the earth in terms of departures from a standard spherical surface.

The arrangement of oceans and continents.—There are other departures from a spherical surface than the oblateness just referred to, and these departures, while not large, are believed to be full of significance. Lest the reader should gain a wrong impression of their magnitude, it may be well to introduce a diagram drawn to scale and representing prominent elevations and depressions of the earth (Fig. 1).

Wrong impressions concerning the figure of the lithosphere are sometimes gained because its depressions are obliterated by the oceans. The oceans are, indeed, useful to us in showing where the depressions are located, but the figure of the earth which we are considering is the naked surface of the rock. In a broad way, the earth's shape will be given by the arrangement of the oceans and the continents. As soon as we take up the study of this arrangement, we find that quite significant facts of distribution are disclosed.

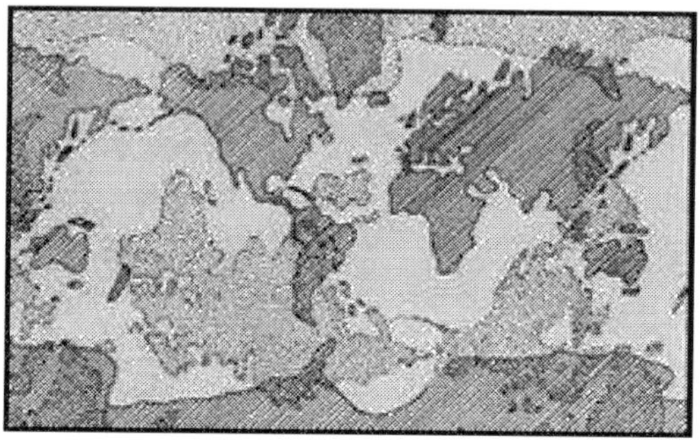

FIG. 2.—Map on Mercator's projection to show the reciprocal relation of the land and sea areas (after Gregory and Arldt).

One of the most significant facts involved in the distribution of land and sea, is a concentration of the land areas within the northern and the seas within the southern hemisphere. The noteworthy exception is the occurrence of the great and high Antarctic continent centered near the earth's south pole; and there are extensions of the northern continent as narrowing land masses to the southward of the equator. Hardly less significant than the existence of land and water hemispheres is the reciprocal or antipodal distribution of land and sea (Fig. 2). A third fact of significance is a dovetailing together of sea and land along an east-and-west direction. While the seas are generally A-shaped and narrow northward, the land masses are V-shaped and narrow southward, *but this occurs mainly in the southern hemisphere.* Lastly, there is some indication of a belt of sea dividing the land masses into northern and southern portions along the course of a great circle which makes a small angle with the earth's equator. Thus the western continent is nearly divided by a mediterranean sea,—the Caribbean,—and the eastern is in part so divided by the separation of Europe from Africa.

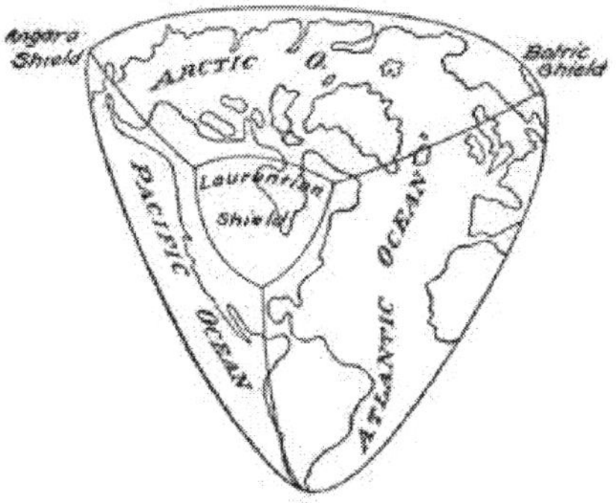

FIG. 3.—The form toward which the figure of the earth is tending, a tetrahedron with symmetrically truncated angles.

The figure toward which the earth is tending.—Thus far in our discussion of the earth's figure we have been guided entirely by the present distribution of land and water. There are, however, depressions upon the surface of the land, in some cases extending below the level of the sea, which are not to-day occupied by water. By far the most notable of these is the great Caspian Depression, which with its extension divides central and eastern Asia upon the east from Africa and Europe upon the west. This depression was quite recently occupied by the sea, and when added to the present ocean basins to indicate depressions of the lithosphere, it shows that the earth's figure departs from the standard spheroid *in the direction* of the form represented in Fig. 3. This form approximates to a tetrahedron, a figure bounded by four equal triangular faces, here with symmetrically truncated angles. Of all regular figures with plane surfaces the tetrahedron has the smallest volume for a given surface, and it presents moreover a reciprocal relation of projection to depression. Every line passing through its center thus finds the surface nearer than the average distance upon one side and correspondingly farther upon the other (Fig. 4).

Astronomical *versus* geodetic observations.—Confirmation of the conclusions arrived at from the arrangement of oceans and continents has been secured in other fields. It was pointed out that the earth's oblateness was proven by comparison of the measured degrees of latitude upon the earth's surface in lower and higher latitudes, the degree being found longer as the pole is approached. Any variation from the spherical surface must obviously increase the size of the measured degree of latitude in proportion to the departure from the standard form, and so the tetrahedral figure with one of its angles at the south pole will require that the degrees of latitude be longer in the southern than they are in the northern hemisphere. This has been found by measurement to be the case, and the result is further confirmed by pendulum studies upon the distribution of the earth's attraction or gravity. If less of the mass of the earth is concentrated in the southern hemisphere, its attraction as measured in vibrations of the pendulum should be correspondingly smaller.

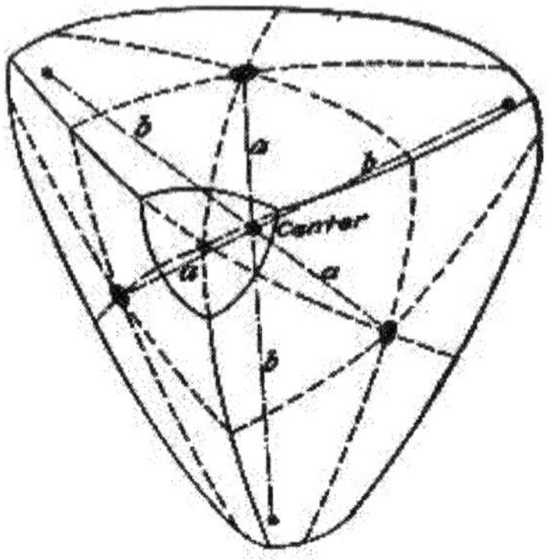

FIG. 4.—A truncated tetrahedron, showing how the depression upon one side of the center is balanced by the opposite projection.

Other confirmations of the tetrahedral figure of the earth have been derived from a comparison of astronomical data, which assume the earth to be a perfect spheroid, with geodetic measurements, which are based upon direct measurements. Thus the arc measured in an east-and-west direction across Europe revealed a different curvature near the angle of the tetrahedral figure from what was found farther to the eastward.

Changes of figure during contraction of a spherical body.—If we inquire why the earth in cooling should tend to approach the tetrahedral figure, an answer is easily found. When formed, the earth appears to have been a but slightly oblate spheroid, or practically a sphere—the shape which of all incloses the most space for a given surface. Cooled and solidified at the surface to the temperature of the surrounding air, and the core still hot and continuing to lose heat, the core must continue to contract though the outer shell is no longer able to do so. The superficial area being thus maintained constant while the volume continues to diminish, the figure must change from the initial one of greatest bulk to others of smaller volume, and ultimately, if the process should continue indefinitely, to the tetrahedron, which of all regular figures has the minimum volume for a given surface.

That a contracting sphere does indeed pass through such a series of changes has been shown by the behavior of contracting soap bubbles and of rubber balloons, as well as by experiments upon the exhaustion of air contained in hollow metal spheres of only moderate strength. In all these instances, the ultimate form produced indicates an indenting of four sides of the sphere which have the positions of the faces of a tetrahedron. The late

Professor Prinz of Brussels secured some extremely interesting results in which he obtained intermediate forms with six angles, but unfortunately these studies were not prepared for publication at the time of his death.

The earth's departure from the spheroid in the direction of the modified tetrahedron is, as we have seen, no hypothesis, but observed fact revealed in (1) the concentration of the land about a central ocean in the northern hemisphere; in (2) the antipodal relation of the land to the water areas, and in (3) the threefold subdivision of the surface into north and south belts by the two greater oceans and the Caspian Depression.

The earlier figures of the earth.—The manner in which continent and ocean are dovetailed into each other in an east-and-west direction has been generally adduced as additional evidence for the tetrahedral figure as above described. Closer examination shows that instead of being in harmony with this figure, it indicates a departure from it, and, as we shall see, a significant departure which undoubtedly has its origin in the earlier history of the planet. The mediterranean seas of both the eastern and the western hemispheres likewise interfere with the perfection of the tetrahedral figure and require an explanation.

Let us then examine in outline the past history of the world with reference especially to the evolution of the continents and to the times and the manners of surface change. It is now well known that there have been three major periods of great deformation of the earth's shell. The first of these of which we have record came at the end of the first great era of geologic history, the so-called Eozoic era; a second great transformation came at the close of the second or Paleozoic era; and a third began at the end of the next or Mesozoic era, an adjustment which is apparently continuing to-day. Each of these great surface deformations was accompanied by great volcanic eruptions of which we have the evidence in the lavas remaining for our inspection, and each was followed by the formation of great glaciers which spread over large areas of the existing continents.

Before the earliest of these great changes, the earth appears to have approximated in its figure somewhat closely to the ideal spheroid, for it was everywhere enveloped in the hydrosphere as a universal ocean. Toward the close of this period came the adjustments which brought the lithosphere to protrude through the hydrosphere in shield-like continents whose distribution, as shown by the rocks of this period, is of great significance. Within the northern hemisphere rose three land shields spaced at nearly equal intervals and at nearly equal distances from the northern pole. One of these was centered where now is Hudson Bay, another about the present Baltic Sea, and the relics of the third are found in northeastern Siberia. These earliest continents have been referred to as the Laurentian, Baltic, and Angara shields. Within the southern hemisphere shields appear to have developed in somewhat similar grouping, namely, in South America, in South Africa, and in Australia (Figs. 3 and 5).

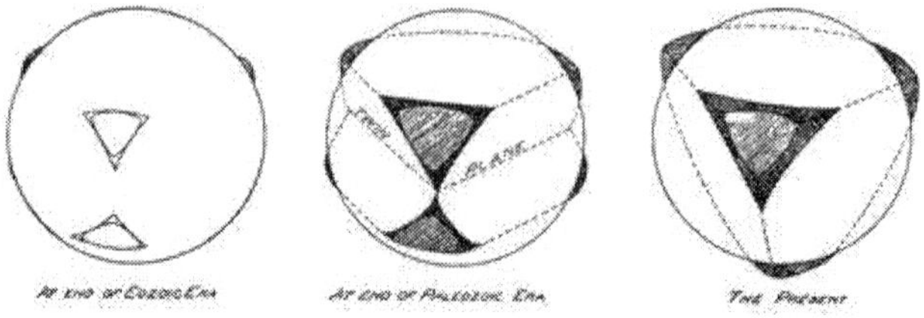

FIG. 5.—Approximations to earlier and present figures of the earth.

These coigns or angles of a form into which the earlier spheroid of the earth was being transformed have persisted through the greater part of subsequent geologic time, and have been enlarged by the growth of sediments about them as well as by the later elevation and wrinkling of these deposits into marginal mountain ranges.

The continents and oceans which arose at the close of the Paleozoic era.—At the close of the second great era in the recorded history of the earth, the now somewhat enlarged continents were profoundly altered during a series of convulsive movements within the surface shell of the lithosphere. When these convulsions were over, there was a new disposition of land and sea, but one quite different from the present arrangement. Instead of being extended in north-south belts, as they are at present, the continents stretched out in broad east-west zones, one in the northern and the other in the southern hemisphere. To the broad southern continent of which so little now remains, the name "Gondwana Land" has been given, and to the sea which divided the northern from the southern continent the name "Ocean of Tethys." The northern continent stretched across the site of the present Atlantic Ocean as the "North Atlantis", its northern shore to the westward being somewhat farther south than the present northern coast of North America, since life forms migrated in the northern ocean from the site of Behring Sea to that of the present North Atlantic.

This arrangement of land and water during the middle period of the earth's recorded history, when considered with reference both to its earlier and to its later evolution, may perhaps be best accounted for by the assumption that the lithosphere was then shaped like Fig. 5 (middle). In this figure two truncated tetrahedrons are joined in a common plane of contact which may be described as the twin plane. This medial depression upon the lithosphere was occupied by the intercontinental sea, the Ocean of Tethys.

Near the close of this second great era of the earth's continental history, crustal convulsions, which were perhaps the most remarkable in the entire record, resulted in the almost complete disappearance of the southern continent and a concentration of the land within the northern hemisphere as a somewhat interrupted belt surrounding a central polar ocean (Figs. 3 and 5).

Upon the assumption of twin tetrahedrons in the intermediate era of continental evolution, both the Ocean of Tethys of that time and its present remnants, the Caribbean and Mediterranean seas, are accounted for. The V-shaped continent extensions and the A-shaped oceans of the southern hemisphere (Fig. 2) may likewise be considered as relics of the now largely submerged tetrahedron of the southern hemisphere, since this had its apex to the northward (Fig. 6).

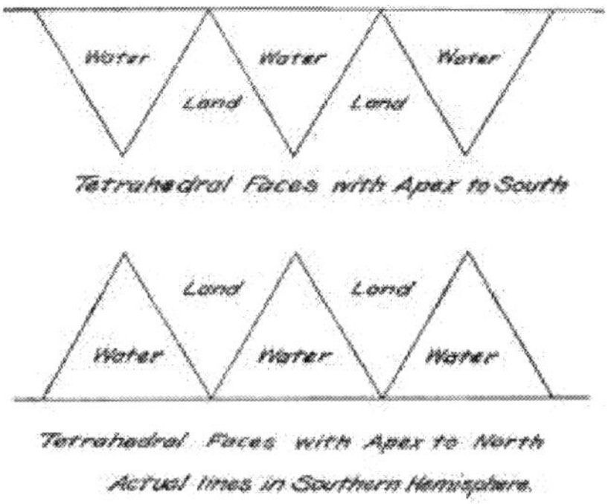

FIG. 6.—Diagrams for comparison of shore lines upon tetrahedrons which have an angle, the first at the south and the second at the north.

Thus we see that the lithosphere can scarcely be regarded as a perfect spheroid, since in the course of geologic ages it has undergone successive departures from this original form. In its present state it has been described as tetrahedral, though we must keep in mind that the sharp angles of that figure are deeply truncated. The soundings first by Nansen and more recently by Peary in the Arctic basin, far to the north of the continental border, showed that this depression is characterized by profound depths, and so have afforded confirmation of the tetrahedral figure. To match this depression at the northern extremity of the earth's axis, a high continent reaching to elevations in excess of 10,000 feet has been penetrated by Sir Ernest Shackleton at the opposite extremity of this polar diameter. Considering its size and its elevation, the Antarctic continent with its glacier mantle is the largest protuberance upon the surface of the lithosphere.

In our study of the departures of the earth from the standard spheroidal surface, we might even go a step farther and show how the tetrahedron, which best represents the symmetry of the present figure, is somewhat deformed by a flattening perpendicular to the Pacific Ocean. To draw attention to this flattening of the earth, it has sometimes been described as "potato-shaped", since the outline perpendicular to this face is imperfectly heart-shaped or like a flattened "peg top."

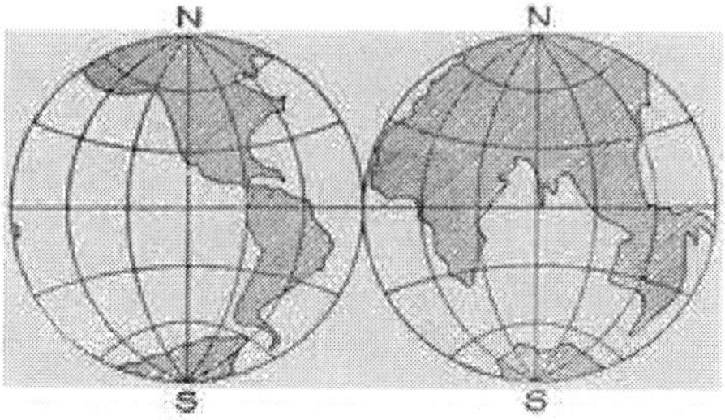

FIG. 7.—The continents with submerged portions added (after Gilbert).

The flooded portions of the present continents.—We are accustomed to think of the continents as ending at the shores of the oceans. If, however, we are to regard them as platforms which rise from the ocean depressions, their margins should be considerably extended, for a submerged shelf now practically surrounds all the continents to a nearly uniform depth of 100 fathoms or 600 feet. The oceans thus more than fill their basins and may be thought of as spilling over upon the continents. In Fig. 7, the submerged portions of the continents have been joined to those usually represented, thus adding about 10,000,000 square miles to their area, and giving them one third, instead of one fourth, of the lithosphere surface.

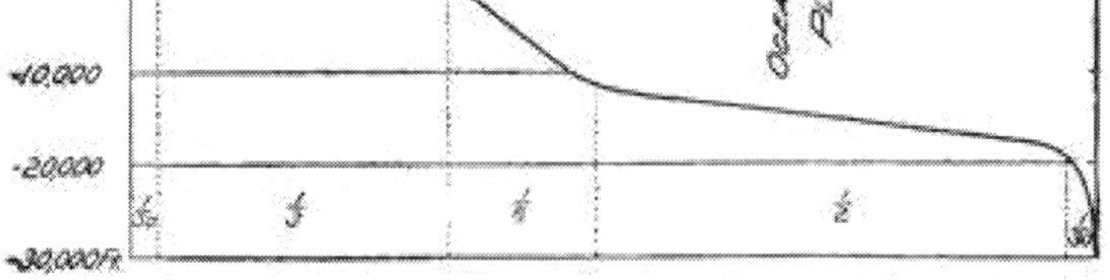

FIG. 8.—Diagram to indicate the altitude of different parts of the lithosphere surface.

The floors of the hydrosphere and atmosphere.—The highest altitudes upon the continents and the profoundest deeps of the ocean are each removed about 30,000 feet, or nearly 6 miles, from the level of the sea. In comparison with the entire surface of the lithosphere, these extremes of elevation represent such small areas as to be almost inappreciable. Only about $^1/_{80}$ of the lithosphere surface rises more than 6000 feet above sea level, and about the same proportion lies deeper than 18,000 feet below the same datum plane (Fig. 8). Almost the entire area of the lithosphere is included either in the so-called continental plateau or platform, in the oceanic platform, or in the slope which separates the two. The continental platform includes the continental shelf above referred to, and represents about one third of the entire area of the planet. This platform has a range of elevation from 6000 feet above to 600 feet below sea level and has an average altitude of about 2300 feet. The oceanic platform slopes more steeply, ranges in depth from 12,000 to 18,000 feet below sea level, and comprises about one half the lithosphere surface. The remaining portion of the surface, something less than one eighth of all, is included in the steep slopes between the two platforms, between 600 and 12,000 feet below sea. The two platforms and the slope between them must not, however, be thought of as continuous features upon the surface, but merely as representing the average elevations of portions of the lithosphere.

Reading References for Chapter II

On the evolution of ideas concerning the earth's figure:—

SUESS. The Face of the Earth (Clarendon Press, 1906), vol. 2, Chapter 1.

V. ZITTEL. History of Geology and Paleontology (Walter Scott, London, 1901), Chapters 1-2.

The departure of the spheroid toward the tetrahedron:—

W. LOWTHIAN GREEN. Vestiges of the Molten Globe, Part 1. London, 1875. (Now a rare work, but it contains the original statement of the idea.)

J. W. GREGORY. The Plan of the Earth and Its Causes, Geogr. Jour., vol. 13, 1899, pp. 225-251 (the best general statement of the arguments for a tetrahedral form).

W. PRINZ. L'échelle reduite des expériences géologiques, Bull. Soc. Belge d'Astronomie, 1899.

B. K. EMERSON. The Tetrahedral Earth and Zone of the Intercontinental Seas, Bull. Geol. Soc. Am., vol. 11, 1911, pp. 61-106, pls. 9-14.

M. P. RUDSKI. Physik der Erde (Tauchnitz, Leipzig, 1911), Chapters 1-3 (the best discussion of the geoid from the purely mathematical standpoint, so far as the spheroid is concerned).

The earlier figures of the earth:—

TH. ARLDT. Die Entwicklung der Kontinente und ihrer Lebewelt. Engelmann, Leipzig, 1907. (Contains a valuable series of map plates, showing the probable boundaries of the continents in the different geological periods).

CHAPTER III

THE NATURE OF THE MATERIALS IN THE LITHOSPHERE

The rigid quality of our planet.—For a long time it was supposed that the solid earth constituted a crust only which was floated upon a liquid interior. This notion was clearly an outgrowth of the then generally accepted Laplacian hypothesis of the origin of the universe, which assumed fluid interiors for the planets, the crust being suggested by the winter crust of frozen water upon the surface of our inland lakes. To-day the nebular hypothesis in the Laplacian form is fast giving place to quite different conceptions, in which solid particles, and not gaseous ones, are conceived to have built up the lithosphere. The analogy with frozen water has likewise been abandoned with the discovery that frozen rock, instead of floating, sinks in its molten equivalent.

Yet even more cogent arguments have been brought forward to show that whatever may be the state of aggregation within the earth's core—and it may be different from any now known to us—it nevertheless has many of the properties recognized as belonging to solid and rigid bodies. Provisionally, therefore, we may regard the earth's core as rigid and essentially solid. It was long ago pointed out by the late Lord Kelvin that if our lithosphere were not more rigid than a ball of glass of the same size, it would be constantly passing through periodic six-hourly distortions of great amplitude in response to the varying attractions of the moon. An equally striking argument emanating from the same high authority is furnished by the well-known egg-spinning demonstration. For illustration, Kelvin was accustomed to take two eggs, one boiled and the other raw, and attempt to spin them upon their ends. For the boiled, and essentially solid, egg this is easily accomplished, but internal friction of the liquid contents of the raw egg quickly stops any rotary motion which is imparted to it. Upon the same grounds it is argued that had the earth's interior possessed the properties of a liquid, rotation must long since have ceased.

A stronger proof of earth rigidity than either of these has been lately furnished by the instrumental study of earthquakes. With the delicate apparatus which is now installed for the purpose, heavy earthquakes may be sensed which have occurred anywhere upon the earth's surface, the earth movement sending its own message by the shortest route through the core of the earth to the observing station. A heavy shock which occurs in New Zealand is recorded in England, almost diametrically opposite, in about twenty-one minutes after its occurrence. The laws of wave propagation and their relation to the properties of the transmitting medium are well known, and in order to explain such extraordinary velocity it is necessary to assume that for such impulses the earth's interior is much more rigid than the finest tool steel.

Probable composition of the earth's core.—In deriving views concerning the nature of the earth's interior we are greatly aided by astronomical studies. The common origin long ago indicated for the planets of the solar system and the sun has been confirmed by the analysis of light with the aid of the spectroscope. It has thus been found that the same chemical elements which we find in the earth are present also in the sun and in the other stellar bodies. Again, the group of planets of the solar system which are nearest to the sun—Mercury, Venus, the Earth, and Mars—have each a high density, all except Mars, the most distant, having specific gravities very closely $5\frac{1}{2}$, that of Mars being about 4. This average specific gravity is also that of the solid bodies, the so-called meteorites, which reach the surface of our planet from the surrounding space. Yet though the earth as a whole is thus found to have a specific gravity five and a half times that of water, its surface shell has an average density of less than half this value, or 2.7.

The study of meteorites has given us a possible clew to the nature of the earth's interior; for when both terrestrial and celestial rock types are classified and placed in orderly arrangement, it is found that the chemical elements which compose the two groups are identical, and that these are united according to the same physical and chemical laws. No new element has been discovered in the one group that has not been found in the other, and though some compounds of these elements, the minerals, occur in the earth's crust that have not been found in meteorites, and though some occur in meteorites which are not known from the earth, yet of those which are common to both bodies there is agreement, even to the minor details (Fig. 9). It is found, however, that the commonest of the minerals in the earth's shell are absent from meteorites, as the commoner constituents of meteorites are wanting in the earth's crust. This observation would go far to show that we may in the two cases be examining different portions of quite similar bodies; and this view is strikingly confirmed when the rocks of the two groups are arranged in the order of their densities (Fig. 9).

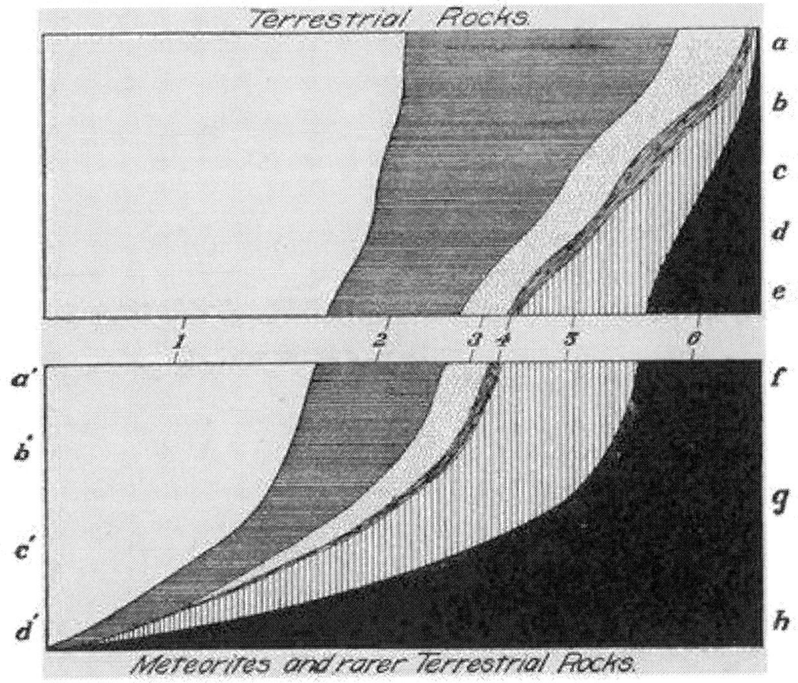

FIG. 9.—**Diagram to show how terrestrial rocks grade into those of the meteorites. 1, oxygen; 2, silicon; 3, aluminium; 4, alkali metals; 5, alkaline earth metals; 6, iron, nickel, cobalt, etc.; *a*, granites and rhyolites; *b*, syenites and trachytes; *c*, diorites and andesites; *d*, gabbros and basalts; *e*, ultra-basic rocks; *f*, basic inclosures in basalt, etc.; *g*, iron basalts of west Greenland; *h*, iron masses of Ovifak, west Greenland; *a'*-*d'*, meteorites in order of density (after Judd).**

In a broad way, density, structure, and chemical composition are all similarly involved in the gradations illustrated by the diagram; and it is significant that while there are terrestrial rocks not represented by meteorites, the densest and the most unusual of the terrestrial rocks are chemically almost identical with the less dense of the celestial bodies.

The earth a magnet.—The denser, and likewise the more common, of the meteorite rocks—the so-called meteoric irons—are composed almost entirely of the elements iron, nickel, and cobalt. Such aggregates are not known as yet from terrestrial sources, although transitional types appear to exist upon the island of Disco off the west coast of Greenland. If it were possible to explore the earth's interior, would such combinations of the iron minerals be encountered? Apart from the surprising velocity of transmission of earthquake waves, the strongest argument for an iron core to the lithosphere is found in the magnetic property of the earth as a whole. The only magnetic elements known to us are those of the heavy meteorites—iron, nickel, and cobalt,—and the earth is, as we know, a great magnet whose northern pole in British America and whose southern pole in Antarctica have at last been visited by Amundsen and David, respectively. The specific gravity of iron is 7.15, and those of nickel and cobalt, which in the meteorites are present in relatively small amounts, are 7.8 and 7.5, respectively. Considering that the surface shell of the earth has a specific gravity of 2.7, these values must be regarded as agreeing well with the determined density of the earth (5.6) and the other planets of its group (Mercury 5.7, Venus 5.4, Mars 4).

The chemical constitution of the earth's surface shell.—The number of the so-called chemical elements which enter into the earth's composition is more than eighty, but few of these figure as important constituents of the portion known to us. Nearly one half of the mass of this shell is oxygen, and more than a quarter is silicon. The remaining quarter is largely made up of aluminium, iron, calcium, magnesium, and the alkalies sodium and potassium, in the order named. These eight constituent elements are thus the only ones which play any important rôle in the composition of the earth's surface shell. They are not found there in the free condition, but combined in the definite proportions characteristic of chemical compounds, and as such they are known as *minerals*.

The essential nature of crystals.—A crystal we are accustomed to think of as something transparent bounded by sharp edges and angles, our ideas having been obtained largely from the gem minerals. This outward symmetry of form is, however, but an expression of a power which resides, so to speak, in the heart or soul of the crystal individual—it has its own structural make-up, its individuality. No more correct estimates of the comparison of crystal individualities would be obtained by the study of outward forms alone of two minerals than would be gained by a judgment of persons from the cut of their clothing. Too often this outward dress tells only of the conditions by which both men and crystals have been surrounded, and but little of the power inherent in the individual. Many a battered mineral fragment with little beauty to recommend it, when placed under suitable conditions for its development, has grown into a marvel of beauty. Few minerals are so mean that they have not within them this inherent power of individuality which lifts them above the world of the amorphous and shapeless.

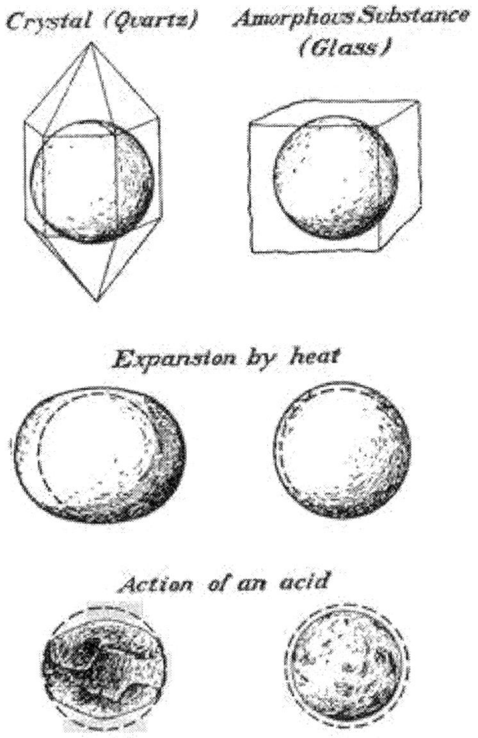

FIG. 10.—Comparison of a crystalline with an amorphous substance when expanded by heat and when attacked by acid.

Just as the real nature of a person is first disclosed by his behavior under trying circumstances, so of a crystal it is its conduct under stress of one sort or another which brings out its real character. By way of illustration let us prepare a sphere from the mineral quartz—it matters not whether we destroy the beautiful outlines of the crystal or employ a battered fragment—and then prepare a sphere of similar size and shape from a noncrystalline or amorphous substance like glass. If now these two spheres be introduced into a bath of oil and raised to a higher temperature, the glass globe undergoes an enlargement without change of its form; but the crystal ball reveals its individuality by expanding into a spheroid in which each new dimension is nicely adjusted to this more complex figure (Fig. 10).

We may, instead of submitting the two balls to the "trial by fire", allow each to be attacked by the powerful reagent, hydrofluoric acid. The common glass under the attack of the acid remains as it was before, a sphere, but with shrunken dimensions. The crystal, on the other hand, is able to control the action of the solvent, and in so doing its individuality is again revealed in a beautifully etched figure having many curving outlines—it is as though the crystal had possessed a soul which under this trial has been revealed. This glimpse into the nature of the crystal, so as to reveal its structural beauty, is still more surprising when the crystal is taken from the acid in the early stages of the action and held close beneath the eye. Now the little etchings upon the surface display each the individuality of the substance, and joining with their neighbors they send out a beautifully symmetrical and entirely characteristic picture (Fig. 11).

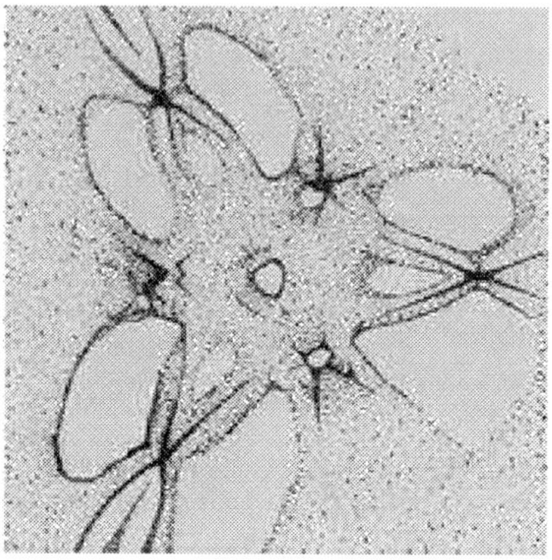

FIG. 11.—"Light figure" seen upon an etched surface of a crystal of calcite (after Goldschmidt and Wright).

The lithosphere a complex of interlocking crystals.—To the layman the crystal is something rare and expensive, to be obtained from a jeweler or to be seen displayed in the show cases of the great museums. Yet the one most striking quality of the lithosphere which separates it from the hydrosphere and the atmosphere is its crystalline structure,—a structure belonging also to the meteorite, and with little doubt to all the planets of the earth group. A snowflake caught during its fall from the sky reveals all the delicate tracery of crystal boundary; collected from a thick layer lying upon the ground, it appears as an intricate aggregate of broken fragments more or less firmly cemented together. And so it is of the lithosphere, for the myriads of individuals are either the ruins of former crystals, or they are grown together in such a manner that crystal facets had no opportunity to develop.

Such mineral individuals as once possessed the crystal form may have been broken and their surfaces ground away by mutual attrition under the rhythmic beating of the waves upon a shore or in the continuous rolling of pebbles on a stream bed, until as battered relics they are piled away together in a bed of sand. Yet no amount of such rough handling is sufficient to destroy the crystal individuality, and if they are now surrounded with conditions which are suitable for their growth, their individual nature again becomes revealed in new crystal outlines. Many of our sandstones when turned in the bright sunlight send out flashes of light to rival a bank of snow in early spring. These bright flashes proceed from the facets of minute crystals formed about each rounded grain of the sand, and if we examine them under a lens, we may note the beauty of line formed with such exactness that the most delicate instruments can detect no difference between the similar angles of neighboring crystals (Fig. 12).

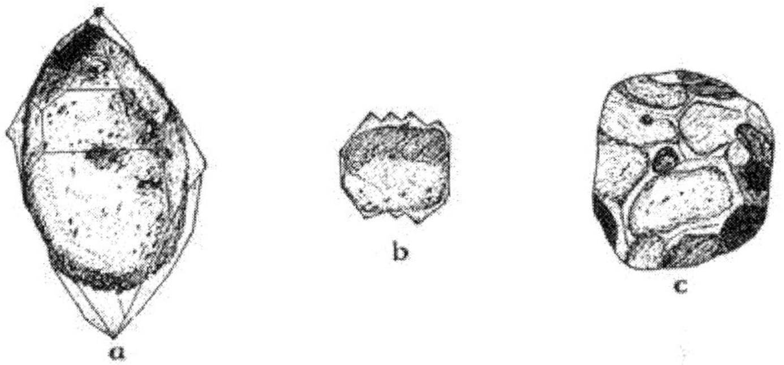

FIG. 12.—Battered sand grains which have taken on a new lease of life and have developed a crystal form. *a*, a single grain grown into an individual crystal; *b*, a parallel growth about a single grain; *c*, growth of neighboring grains until they have mutually interfered and so destroyed the crystal facets—the common condition within the mass of a rock (after Irving and Van Hise).

This individual nature of the crystal is believed to reside in a symmetrical grouping of the chemical molecules of the substance into larger and so-called "crystal molecules." The crystal quality belongs to the chemical elements and to their compounds in the solid condition, but not to ordinary mixtures of them.

Some properties of natural crystals, minerals.—No two mineral species appear in crystals of the same appearance, any more than two animal species have been given the same form; and so minerals may be recognized by the individual peculiarities of their crystals. Yet for the reason that crystals have so generally been prevented from developing or retaining their characteristic faces, in the vast number of instances it is the behavior, and not the appearance, of the mineral substance which is made use of for identification.

When a mineral is broken under the blow of a hammer, instead of yielding an irregular fracture, like that of glass, it generally tends to part along one or more directions so as to leave plane surfaces. This property of *cleavage* is strikingly illustrated for a single direction in the mineral mica, for two directions in feldspar, and for three directions in calcite or Iceland spar. Other properties of minerals, such as hardness, specific gravity, luster, color, fusibility, etc., are all made use of in rough determinations of the minerals. Far more delicate methods depend upon the behavior of minerals when observed in polarized light, and such behavior is the basis of those branches of geological science known as optical mineralogy and as microscopical petrography. An outline description of some of the common minerals and the means for identifying them will be found in appendix A.

The alterations of minerals.—By far the larger number of minerals have been formed in the cooling and consequent consolidation of molten rock material such as during a volcanic eruption reaches the earth's surface as lava. Beginning their growth at many points within the viscous mass, the individual crystals eventually may grow together and so prevent a development of their crystal faces.

Another class of minerals are deposited from solution in water within the cavities and fissures of the rocks; and if this process ceases before the cavities have been completely closed, the minerals are found projecting from the walls in a beautiful lining of crystal—the *Krystallkeller* or "crystal cellar." It is from such pockets or veins within the rocks that the valuable ores are obtained, as are the crystals which are displayed in our mineral cabinets.

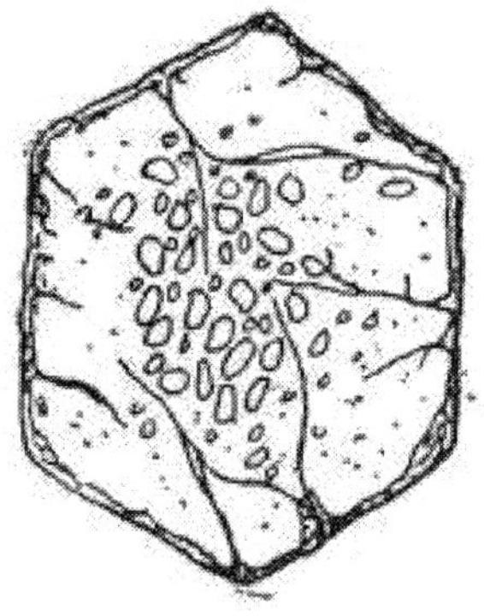

FIG. 13.—Crystal of garnet developed in a schist with grains of quartz included because not assimilated.

There is, however, a third process by which minerals are formed, and minerals of this class are produced within the solid rock as a product of the alteration of preëxisting minerals. Under the enormous pressures of the rocks deep below the earth's surface, they are as permeable to the percolating waters as is a sponge at the surface. Under these conditions certain minerals are dissolved and their material redeposited after traveling in the solution, or solution and redeposition of mineral matter may go on together within the mass of the same rock. One new mineral may have been produced from the dissolved materials of a number of earlier species, or several new minerals may be the result of the alteration of a preëxisting mineral with a more complex chemical structure. Where the new mineral has been formed "in place", it has sometimes been able to utilize the materials of all the minerals which before existed there, or it may have been obliged to inclose within itself those earlier constituents which it could not assimilate in its own structure (Fig. 13).

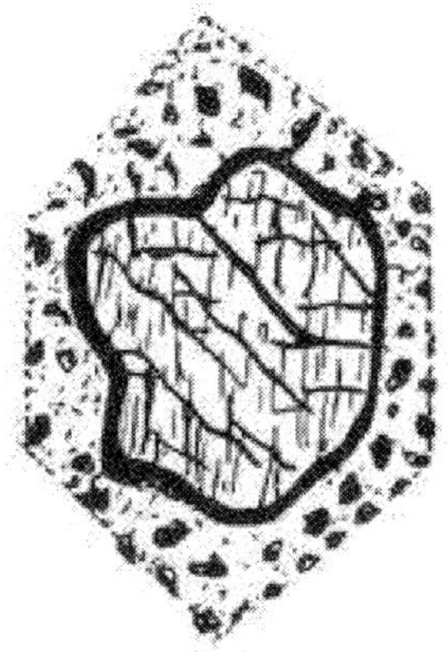

FIG. 14.—A crystal of augite within the mass of a rock altered in part to form a rim of the minerals hornblende and magnetite. Note the original outline of the augite crystal.

At other times a crystal which is imbedded in rock has been attacked upon its surface by the percolating solutions, and the dissolved materials have been deposited in place as a crown of new minerals which steadily widens its zone until the center is reached and the original crystal has been entirely transformed (Fig. 14). It is sometimes possible to say that the action by which these changes have been brought about has involved a nice adjustment of supply of the chemical constituents necessary to the formation of the new mineral or minerals. In rocks which are aggregates of several mineral species, a newly formed mineral may appear only at the common margin of certain of these species, thus showing that they supply those chemical elements which were necessary to the formation of the new substance (Fig. 15). Thus it is seen that below the earth's surface chemical reactions are constantly going on, and the earlier rocks are thus locally being transformed into others of a different mineral constitution.

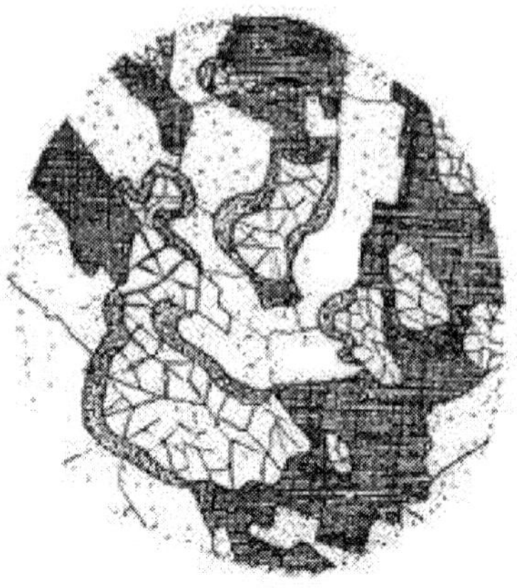

FIG. 15.—A new mineral (hornblende) forming as an intermediate "reaction rim" between the mineral having irregular fractures (olivine) and the dusty white mineral (lime-soda feldspar).

Near the earth's surface the carbon dioxide and the moisture which are present in the atmosphere are constantly changing the exposed portions of the lithosphere into carbonates, hydrates, and oxides. These compounds are more soluble than are the minerals out of which they were formed, and they are also more bulky and so tend to crack off from the parent mass on which they were formed. As we are to see, for both of these reasons the surface rocks of the lithosphere succumb to this attack from the atmosphere.

In connection with those wrinklings of the surface shell of the lithosphere from which mountains result, the underlying rocks are subjected to great strains, and even where no visible partings are produced, the rocks are deformed so that individual minerals may be bent into crescent-shaped or S-shaped forms, or they are parted into one or more fragments which remain imbedded within the rock.

READING REFERENCES FOR CHAPTER III

Theories of origin of the earth:—

THOMSON and TAIT. Natural Philosophy. 2d ed. Cambridge, 1883, pp. 422.

T. C. CHAMBERLIN. Chamberlin and Salisbury's Geology, vol. 2, pp. 1-81.

Rigidity of the earth:—

LORD KELVIN. The Internal Condition of the Earth as to Temperature, Fluidity, and Rigidity, Popular Lectures and Addresses, vol. 2, pp. 299-318; Review of evidence regarding the physical condition of the earth, *ibid.*, pp. 238-272.

HOBBS. Earthquakes (Appleton, New York, 1907), Chapters xvi and xvii.

Composition of the earth's core and shell:—

O. C. FARRINGTON. The Preterrestrial History of Meteorites, Jour. Geol., vol. 9, 1901, pp. 623-236.

E. S. DANA. Minerals and How to Study Them (a book for beginners in mineralogy). Wiley, New York, 1895.

On the nature of crystals:—

VICTOR GOLDSCHMIDT. Ueber das Wesen der Krystalle, Ostwalds Annalen der Naturphilosophie, vol. 9, 1909-1910, pp. 120-139, 368-419.

CHAPTER IV

THE ROCKS OF THE EARTH'S SURFACE SHELL

The processes by which rocks are formed.—Rocks may be formed in any one of several ways. When a portion of the molten lithosphere, so-called *magma*, cools and consolidates, the product is *igneous* rock. Either igneous or other rock may become disintegrated at the earth's surface, and after more or less extended travel, either in the air, in water, or in ice, be laid down as a sediment. Such sediments, whether cemented into a coherent mass or not, are described as *sedimentary* or *clastic* rocks. If the fluid from which they were deposited was the atmosphere, they are known as *subaërial* or *eolian* sediments; but if water, they are known as *subaqueous* deposits. Still another class are ice-deposited and are known as *glacial* deposits.

FIG. 16.—Laminated structure of sedimentary rock, Western Kansas (after a photograph by E. S. Tucker).

But, as we have learned, rocks may undergo transformations through mineral alteration, in which case they are known as *metamorphic* rocks. When these changes consist chiefly in the production of more soluble minerals at the surface, accompanied by thorough disintegration, due to the direct attack of the atmosphere, the resulting rocks are called *residual* rocks.

The marks of origin.—Each of the three great classes of rocks, the igneous, sedimentary, and metamorphic, is characterized by both coarser and finer structures, in the examination of which they may be identified. The igneous rocks having been produced from magmas, which are essentially homogeneous, are usually without definite directional structures due to an arrangement of their constituents, and are said to have a *massive* structure. Sedimentary rocks, upon the other hand, have been formed by an assorting process, the larger and heavier fragments having been laid down when there was high velocity of either wind or water current, and the smaller and lighter fragments during intermediate periods. They are therefore more or less banded, and are said to have a *bedded* or *laminated* structure (Fig. 16).

Again, igneous rocks, being due to a process of crystallization, are composed of mineral individuals which are bounded either by crystal planes or by irregular surfaces along which neighboring crystals have interfered with each other; but in either case the grains possess sharply angular boundaries. Quite different has been the result of the attrition between grains in the transportation and deposition of sediments, for it is characteristic of the sedimentary rocks that their constituent grains are well rounded. Eolian sediments have usually more perfectly rounded grains than subaqueous deposits.

Glacial deposits, if laid down directly by the ice, are unstratified, relatively coarse, and contain pebbles which are both faceted and striated. Such deposits are described as till or tillite. If glacier-derived material is taken up by the streams of thaw water and is by them redeposited, the sediments are assorted or stratified, and they are described as *fluvio-glacial* deposits.

The metamorphic rocks.—Both the coarser structures and the finer textures of the metamorphic rocks are intermediate between those of the igneous and the sedimentary classes. A metamorphosed sedimentary rock, in proportion to its alteration, loses the perfect lamination and the rounded grain which were its distinguishing characters; while an igneous rock takes on in the process an imperfect banding, and the sharp angles of its constituent grains become rounded off by a sort of peripheral crushing or granulation. Metamorphic rocks are therefore characterized by an imperfectly banded structure described as *schistosity* or *gneiss banding*, and the constituent grains may be either angular or rounded. If the metamorphism has not been too intense or too long continued, it is generally possible to determine, particularly with the aid of the polarizing microscope, whether the original rock from which it was derived was of igneous or of sedimentary origin. There are, however, many examples which have defied a reliable verdict concerning their origin.

Characteristic textures of the igneous rocks.—In addition to the massiveness of their general aspect and the angular boundaries of their constituents, there are many additional textures which are characteristic of the igneous rocks. While those that have consolidated below the earth's surface, the *intrusive* rocks, are notably compact, the magmas which arrive at the surface of the lithosphere before their consolidation reveal special structures dependent either upon the expansion of steam and other gases within them, or upon the conditions of flow over the earth's surface. Magmas which thus reach the surface of the earth are described as *lavas*, and the rocks produced by their consolidation are *extrusive* or *volcanic* rocks. The steam included in the lava expands into bubbles or vesicles which may be large or small, few or many. According to the number and the size of these cavities, the rock is said to have a *vesicular*, *scoriaceous*, or *pumiceous* texture.

Most lavas, when they arrive at the earth's surface, contain crystals which are more or less disseminated throughout the molten mass. The tourist who visits Mount Vesuvius at the time of a light eruption may thrust his staff into the stream of lava and extract a portion of the viscous substance in which are seen beautiful white crystals of the mineral leucite, each bounded by twenty-four crystal faces. It is clear that these crystals must have developed by a slow growth within the magma while it was still below the surface, and when the inclosing lava has consolidated, these earlier crystals lie scattered within a *groundmass* of glassy or minutely crystalline material. This scattering of crystals belonging to an earlier generation within a groundmass due to later consolidation is thus an indication of interruption in the process of crystallization, and the texture which results is described as *porphyritic* (Fig. 17 *b*). Should the lava arrive at the surface before any crystals have been generated and consolidate rapidly as a rock glass, its texture is described as *glassy* (Fig. 17 *c*).

When the crystals of the earlier generation are numerous and needle-like in form, as is very often the case, they arrange themselves "end on" during the rock flow, so that when consolidation has occurred, the rock has a kind of puckered lamination which is the characteristic of the *fluxion* or *flow* texture. This texture has sometimes been confused with the lamination of the sedimentary rocks, so that wrong conclusions have been reached regarding origin. At other times the same needle-like crystals within the lava have grouped themselves radially to form rounded nodules called spherulites. Such nodules give to the rock a *spherulitic* texture, which is nowhere better displayed than in the beautiful glassy lavas of Obsidian Cliff in the Yellowstone National Park.

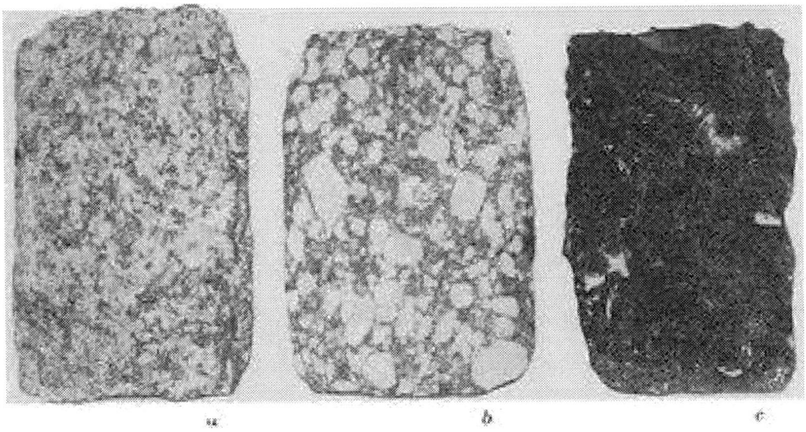

FIG. 17.—Characteristic textures of igneous rocks. *a*, granitic texture characteristic of the deep-seated intrusive rocks; *b*, porphyritic texture characteristic of the extrusive and of the near-surface intrusive rocks; *c*, glassy texture of an extrusive rock.

Those intrusive rocks which consolidate deep below the earth's surface, part with their heat but slowly, and so the process of crystallization is continued without interruption. Starting from many centers, the crystals continue to grow until they mutually intersect in an interlocking complex known as the *granitic* texture (Fig. 17 *a*).

Classification of rocks.—In tabular form rocks may thus be classified as follows:—

Igneous. Massive and with sharply angular grains.	*Intrusive.* Granitic or porphyritic texture.
	Extrusive. Glassy or porphyritic texture; often also with vesicular, scoriaceous, pumiceous, fluxion, or spherulitic textures.
Sedimentary. Laminate and with rounded grains.	*Subaërial.* Sands and loess.
	Subaqueous. (See below.)
	Glacial. Coarse, unstratified deposits with faceted pebbles. Till and tillite.
	Fluvio-glacial. Stratified sands and gravels with "worked over" glacial characters.
	Metamorphic proper. Due to below surface changes.

Metamorphic. Schistose and with grains either angular or rounded.	*Residual.* Disintegrated at or near surface.

Subdivisions of the sedimentary rocks.—While the eolian sediments are all the product of a purely mechanical process of lifting, transportation, and deposition of rock particles, this is not always the case with the subaqueous sediments, since water has the power of dissolving mineral substance, as it has also of furnishing a home for animal and vegetable life. Deposited materials which have been in solution in water are described as *chemical* deposits, and those which have played a part in the life process as *organic* deposits. The organic deposits from vegetable sources are peat and the coals, while limestones and marls are the chief depositories of the remains of the animal life of the water. The tabular classification of the sediments is as follows:—

Classification of Sediments.

Mechanical	*Subaqueous* Deposited by water.	Conglomerate, sandstone and shale.
	Subaërial or *Eolian* *Deposited by wind.*	Sandstone and loess.
	Glacial Deposited by ice.	Till and tillite.
	Fluvio-glacial Glacier-water deposits.	Sands and gravels.
Chemical	Calcareous tufa	Deposited in springs and rivers.
	Oölitic limestone	Deposited at the mouths of rivers between high and low tide.
Organic	Formed of plant remains.	Peats and coals.
	Formed of animal remains.	Limestones and marls.

Winds are under favorable conditions capable of transporting both dust and sand, but not the larger rock fragments. The dust deposits are found accumulating outside the borders of deserts as the so-called *loess* (Fig. 216), though the sand is never carried beyond the desert border, near which it collects in wide belts of ridges described as dunes. When this sand has been cemented into a coherent mass, it is known as eolian sandstone. A section of the appendix (B) is devoted to an outline description of some of the commoner rock types.

The different deposits of ocean, lake, and river.—Of the subaqueous sediments, there are three distinct types resulting: (1) from sedimentation in rivers, the *fluviatile* deposits; (2) from sedimentation in lakes, the *lacustrine* deposits; and (3) from sedimentation in the ocean, *marine* deposits. Again, the widest range of character is displayed by the deposits which are laid down in the different parts of the course of a stream. Near the source of a river, coarse river gravels may be found; in the middle course the finer silts; and in the mouth or delta region, where the deposits enter the sea or a lake, there is found an assortment of silts and clays. Except within the delta region, where the area of deposition begins to broaden, the deposits of rivers are stretched out in long and relatively narrow zones, and are so distinguished from the far more important lacustrine and marine deposits.

Lakes and oceans have this in common that both are bodies of standing as contrasted with flowing water; and both are subject to the periodical rhythmic motions and alongshore currents due to the waves raised by the wind. About their margins, the deposits of lake and ocean are thus in large part wrested by the waves from the neighboring land. Their distribution is always such that the coarsest materials are laid down nearest to the shore, and the deposits become ever finer in the direction of deeper water. Relatively far from shore may be found the finest sands and muds or calcareous deposits, while near the shore are sands, and, finally, along the beach, beds of beach pebbles or shingle. When cemented into coherent rocks, these deposits become shales or limestones, sandstones, and conglomerates, respectively.

As regards the limestones, their origin is involved in considerable uncertainty. Some, like the shell limestone or coquina of the Florida coast, are an aggregation of remains of mollusks which live near the border of the sea. Other limestones are deposited directly from carbonate of lime in solution in the water. A deposit of this nature is forming in southern Florida, both as a flocculent calcareous mud and as crystals of lime carbonate upon a limestone surface. Again, there is the reef limestone which is built up of the stony parts of the coral animal, and, lastly, the calcareous ooze of the deep-sea deposits.

The marine sediments which are derived from the continents, the so-called *terrigenous* deposits, are found only upon the continental shelf and upon the continental slope just outside it. Of these terrigenous deposits, it is customary to distinguish: (1) *littoral* or alongshore deposits, which are laid down between high and low tide levels; (2) *shoal water* deposits, which are found between low-water mark and the edge of the continental shelf; and (3) aktian or offshore deposits, which are found upon the continental slope. The littoral and shoal water deposits are mainly gravels and sands, while the offshore deposits are principally muds or lime deposits.

Special marks of littoral deposits.—The marks of ripples are often left in the sand of a beach, and may be preserved in the sandstone which results from the cementation of such deposits (pl. 11 A). Very similar markings are, however, quite characteristic of the surface of wind-blown sand. For the reason that deposits are subject to many vicissitudes in their subsequent history, so that they sometimes stand at steep angles or are even overturned, it is important to observe the curves of sand ripples so as to distinguish the upper from the lower surface.

In the finer sands and muds of sheltered tidal flats may be preserved the impressions from raindrops or of the feet of animals which have wandered over the flat during an ebb tide. When the tide is at flood, new material is laid down upon the surface and the impressions are filled, but though hardened into rock, these surfaces are those upon which the rock is easily parted, and so the impressions are preserved. In the sandstones of the Connecticut valley there has been preserved a quite remarkable record in the footprints of animals belonging to extinct species, which at the time these deposits were laid down must have been abundant upon the neighboring shores.

Between the tides muds may dry out and crack in intersecting lines like the walls of a honeycomb, and when the cracks have been filled at high tide, a structure is produced which may later be recognized and is usually referred to as "mud-crack" structure. This structure is of special service in distinguishing marine deposits from the subaërial or continental deposits.

A variation in the direction of winds of successive storms may be responsible for the piling up of the beach sand in a peculiar "plunge and flow" or "cross-bedded" structure, a structure which is extremely common in littoral deposits, though simulated in rocks of eolian origin.

The order of deposition during a transgression of the sea.—Many shore lines of the continents are almost constantly migrating either landward or seaward. When the shore line advances over the land, the coast is sinking, and marine deposits will be formed directly above what was recently the "dry land." Such an invasion of the land by the sea, due to a subsidence of the coast, is called a transgression of the sea, or simply a *transgression*. Though at any moment the littoral, shoal water, and offshore deposits are each being laid down in a particular zone, it is evident that each must advance in turn in the direction of the shore and so be deposited above the zones nearer shore.

Thus there comes to be a definite series of continuous beds, one above the other, provided only that the process is continued (Fig. 18). At the very bottom of this series there will usually be found a thin bed of pebbly beach materials, which later will harden into the so-called *basal conglomerate.* If the size of the pebbles is such as to make possible an identification, it will generally be found that these represent the ruins of the rock over which the sea has advanced upon the land.

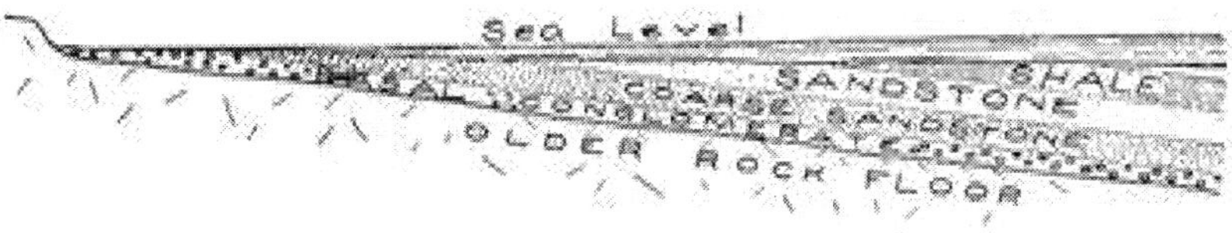

FIG. 18.—Diagram to show the order of the sediments laid down during a transgression of the sea.

Next in order above the basal conglomerate, will follow the coarser and then the finer sands, upon which in turn will be laid down the offshore sediments—the muds and the lime deposits. Later, when cemented together, these become in order, coarser and finer sandstones, shales, and limestones. The order of superposition, reading from the bottom to the top, thus gives the order of decreasing age of the formations.

A subsequent uplift of the coast will be accompanied by a recession of the sea, and when later dissected by nature for our inspection, the order of superposition and the individual character of each of the deposits may be studied at leisure. From such studies it has been found that along with the inorganic deposits there are often found the remains of life in the hard parts of such invertebrate animals as the mollusks and the crustacea. These so-called *fossils* represent animals which were gradually developed from simpler to more and more complex forms; and they thus serve the purpose of successive page numbers in arranging the order of disturbed strata, at the same time that they supply the most secure foundation upon which rests the great doctrine of evolution.

The basins of earlier ages.—It was the great Viennese geologist, Professor Suess, who first pointed out that in mountain regions there are found the thickest and the most complete series of the marine deposits; whereas outside these provinces the formations are separated by wide gaps representing periods when no deposits were laid down because the sea had retired from the region. The completeness of the series of deposits in the mountain districts can only be interpreted to mean that where these but lately formed mountains rise to-day, were for long preceding ages the basins for deposition of terrigenous sediments. It would seem that the lithosphere in its adjustment had selected these earlier sea basins with their heavy layers of sediment for zones of special uplift.

The deposits of the deep sea.—Outside the continental slope, whose base marks the limit of the terrigenous deposits, lies the deeper sea, for the most part a series of broad plains, but varied by more profound steep-walled basins, the so-called "deeps" of the ocean. As shown by the dredgings of the *Challenger* expedition and others of more recent date, the deposits upon the ocean floor are of a wholly different character from those which are derived from the continents. Except in the great deeps, or between depths of five hundred and fifteen hundred fathoms, these deposits are the so-called "ooze", composed of the calcareous or chitinous parts of algæ and of minute animal organisms. The pelagic or surface waters of the ocean are, as it were, a great meadow of these plant forms, upon which the minute crustacea, such as globigerina, foraminifera, and the pteropods, feed in countless myriads. The hard parts of both plant and animal organisms descend to the bottom and there form the ooze in which are sometimes found the ear bones of whales and the teeth of sharks.

In the deeps of the ocean, none of these vegetable or animal deposits are being laid down, but only the so-called "red clay", which is believed to represent decomposed volcanic material deposited by the winds as fine dust on the surface of the ocean, or the product of submarine volcanic eruption. From the absence of the ooze in these profound depths, the conclusion is forced upon us that the hard parts of the minute organisms are dissolved while falling through three or four miles of the ocean water.

Reading References for Chapter IV

J. S. Diller. The Educational Series of Rock Specimens collected and distributed by the United States Geological Survey, Bull. 150 U. S. Geol. Surv., 1898, pp. 1-400.

L. V. Pirsson. Rocks and Rock Minerals. Wiley, New York, 1908.

Sir John Murray. Deep-sea Deposits, Reports of the Challenger expedition, Chapter iii.

L. W. Collet. Les dépôts marins. Doin, Paris, 1907 (Encyclopédie Scientifique).

CHAPTER V

CONTORTIONS OF THE STRATA WITHIN THE ZONE OF FLOW

The zones of fracture and flow.—It is easy to think of the atmosphere and the hydrosphere as each sustaining at any point the load of the superincumbent material. At the sea level the weight of air upon each square inch of surface is about fifteen pounds, whereas upon the floor of the hydrosphere in the more profound deeps the load upon the square inch must be measured in tons. Near the lithosphere surface the rocks support by their strength the load of rock above them, but at greater depths they are unable to do this, for the load bears upon each portion of the rock with a pressure equivalent to the weight of a rock column which extends upward to the surface. The average specific gravity of rock is 2.7, and it is thus easy to calculate the length of the inch square column which has a weight equivalent to the crushing strength of any given rock. At the depth represented by the length of such a column, rocks cannot yield to pressure by fracture, for the opening of a crack implies that the rock upon either side is strong enough to prevent the walls from closing. At this depth, rock must therefore yield to pressure not by fracture, as it would at the surface, but by flow after the manner of a liquid; and so the zone below this critical level is referred to as the *zone of flow*.

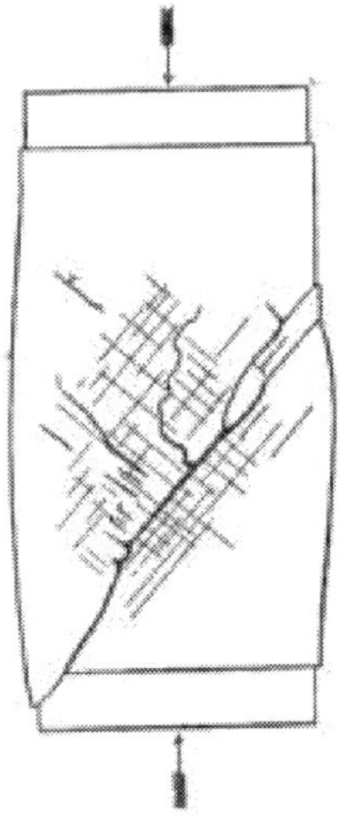

FIG. 19.—Two intersecting parallel series of fractures produced upon each free surface of a prismatic block of stiff molders' wax when broken by compression from the ends (after Daubrée and Tresca).

In contrast, the near-surface zone is called the *zone of fracture.* But different rocks possess different strengths, and these are subject to modifications from other conditions, such, for example, as the proximity of an uncooled magma. The zone of flow is therefore joined to the zone of fracture, not upon a definite surface, but in an intermediate zone described as the *zone of fracture and flow.*

Experiments which illustrate the fracture and flow of solid bodies.—A prismatic block prepared from stiff molders' wax, if crushed between the jaws of a testing machine, yields a system of intersecting fractures which are perpendicular to the free surfaces of the block and take two directions each inclined by half of a right angle to the direction of compression (Fig. 19). This experiment may illustrate the manner in which fractures are produced by the compression within the zone of fracture of the lithosphere, as its core continues to contract.

To reproduce the conditions within the zone of flow, it will be necessary to load the lateral surfaces of the block instead of leaving them unconstrained as in the above-described experiment. The experiment is best devised as in Fig. 20. Here a series of layers having varying degrees of rigidity is prepared from beeswax as a base, either stiffened by admixture of varying proportions of plaster of Paris, or weakened by the use of Venice turpentine. Such a series of layers may represent rocks of as widely different characters as limestone and shale. The load which is to represent superincumbent rock is supplied in the experiment by a deep layer of shot.

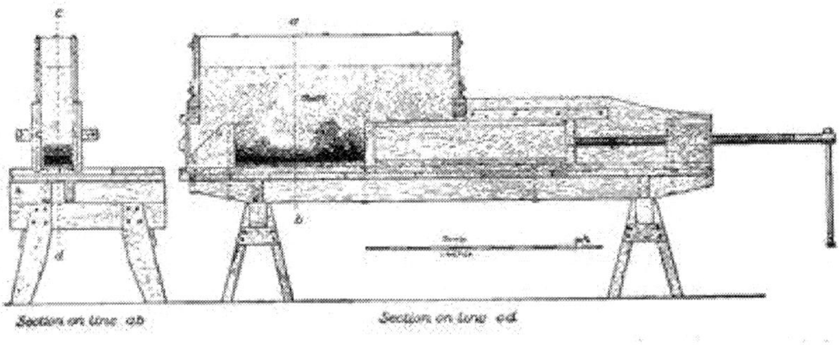

FIG. 20.—Apparatus to illustrate the folding of strata within the zone of flow (after Willis).

When compression is applied to the layers from the ends, these normally solid materials, instead of fracturing, are bent into a series of folds. The stiffer, or more competent, layers are found to be less contorted than are the weaker layers, particularly if the latter have been protected under an arch of the more competent layer (pl. 2 A).

The arches and troughs of the folded strata.—Every series of folds is made up of alternating arches and troughs. The arches of the strata the geologist calls *anticlines* or *anticlinal folds*, and the troughs he calls *synclines* or *synclinal folds* (Fig. 21). When a stratum is merely dropped in a bend to a lower level without producing a complete arch or a complete trough, this half fold is termed a *monocline.*

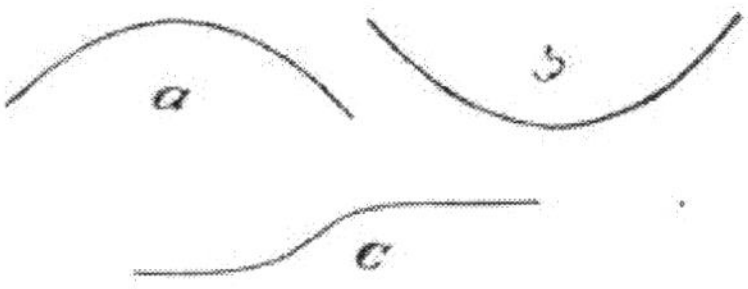

FIG. 21.—Diagrams representing *a*, an anticline; *b*, a syncline; and *c*, a monocline.

Any flexuring of the strata implies a reduction of their surface area, or, considering a single section, a shortening. If the arches and troughs are low and broad, the deformation of the strata is slight, the shortening is comparatively small, and the folds are described as *open* (Fig. 22 *b*). If they be relatively both high and narrow, the deformation is considerable, a larger amount of crustal shortening has gone on, and the folds are described as *close* (Fig. 22 *c*). This closing up of the folds may continue until their sides have practically the same slope, in which case they are said to be *isoclinal* (Fig. 22 *d*).

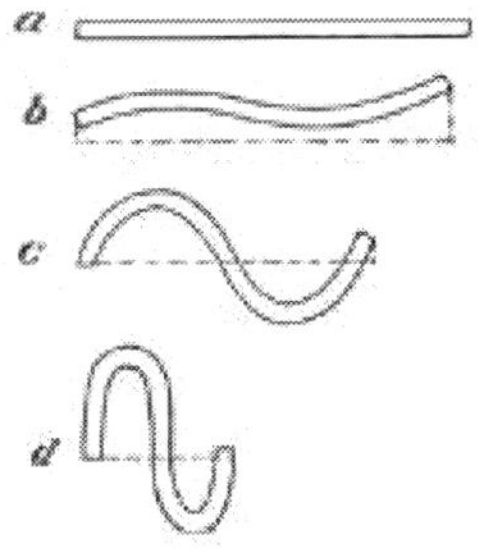

FIG. 22.—A comparison of folds to express increasing degrees of crustal shortening or progressive deformation within the zone of flow: *a*, stratum before folding; *b*, open folds; *c*, close folds; *d*, isoclinal folds.

The elements of folds.—Folds must always be thought of as having extension in each of the three dimensions of space (Fig. 23), and not as properly included within a single plane like the cross sections which we so often use in illustration. A fold may be conceived of as divided into equal parts by a plane which passes along the middle of either the arch or the trough, and is called the *axial plane*. The line in which this plane intersects the arch or the trough is the *axis*, which may be called the *crestline* in an anticline, and the *troughline* in a syncline.

In the case of many open folds the axis is practically horizontal, but in more complexly folded regions this is seldom true. The departure of the axis from the horizontal is called the *pitch*, and folds of this type are described as *pitching folds* or *plunging folds*. The axis is in reality in these cases thrown into a series of undulations or "longitudinal folds", and hence pitch will vary along the axis.

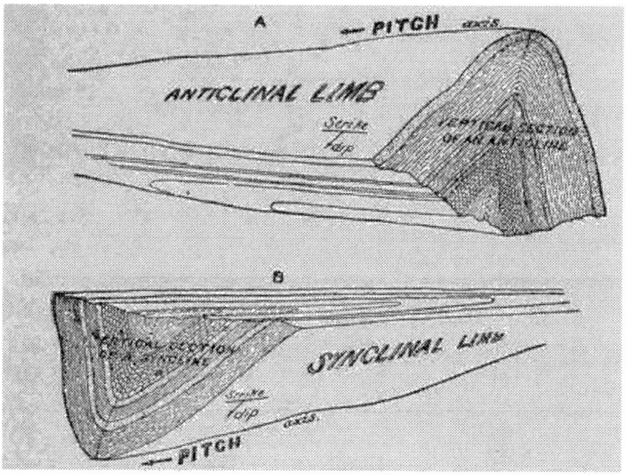

FIG. 23.—Anticlinal and synclinal folds in strata (after Willis).

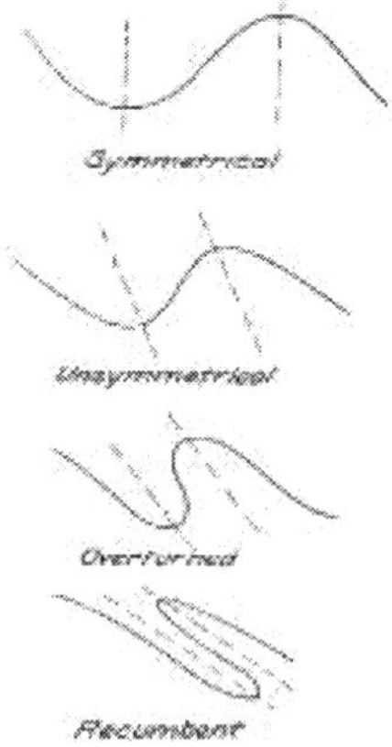

FIG. 24.—Diagrams to illustrate the different shapes of rock folds.

The shapes of rock folds.—By the axial plane each fold is divided into two parts which are called its *limbs*, which may have either the same or different average inclinations. To describe now the shapes of rock folds and not the degree of compression of the district, some additional terms are necessary. Anticlines or synclines whose limbs have about the same inclinations are known as *upright* or *symmetrical folds*. The axial plane of the symmetrical fold is vertical (Fig. 24). If this plane is inclined to the vertical, the folds are *unsymmetrical*. So soon as the steeper of the two limbs has passed the vertical position and inclines in the same direction as the flatter limb, the fold is said to be *overturned*. The departure from symmetry may go so far that the axial plane of the fold lies at a very flat angle, and the fold is then said to be *recumbent*. The observant traveler by train along any of the routes which enter the Alps may from his car window find illustrations of most of these types of rock folds, as he may also, though generally less easily, in passing through the Appalachian Mountains.

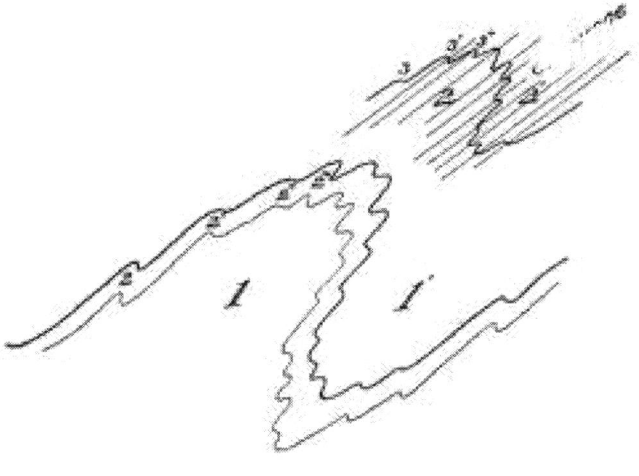

FIG. 25.—Secondary and tertiary flexures superimposed upon the primary ones.

In regions which have been closely folded the larger flexures of the strata may be found with folds of a smaller order of magnitude superimposed upon them, and these in turn may show crumplings of still lower orders. It has been found that the folds of the smaller orders of magnitude possess the shapes of the larger flexures, and much is therefore to be learned from their careful study (Fig. 25). It is also quite generally discovered that parallel planes of ready parting, which are described as *rock cleavage*, take their course parallel to the axial plane within each minor fold. As was long ago shown by the pioneer British geologists, these planes of cleavage are essentially parallel and follow the fold axes throughout large areas.

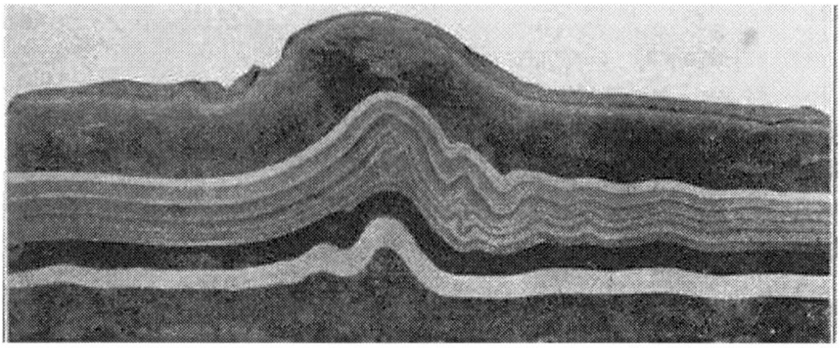

A. Layers compressed in experiments and showing the effect of a competent layer in the process of folding (after Willis).

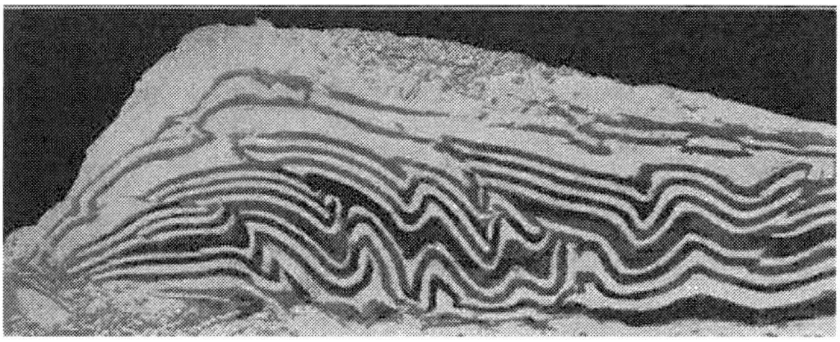

B. Experimental production of a series of parallel thrusts within closely folded strata (after Willis).

C. Apparatus to illustrate shearing action within the overturned limb of a fold.

The overthrust fold.—Whenever a stratum is bent, there is a tendency for its particles to be separated upon the convex side of the bend, at the same time that those upon the concave side are crowded closer together—there is tension in the former case and compression in the latter (Fig. 26). Within an unsymmetrical or an overturned fold, the peculiar distortions in the different sections of the stratum are less simple and are best illustrated by pl. 2 C. This apparatus shows two similar piles of paper sheets, upon the edges of each of which a series of circles has been drawn. When now one of the piles is bent into an unsymmetrical fold, it is seen that through an accommodation by the paper sheets sliding each over its neighbor large distortions of the circles have occurred. In that steeper limb which with closer folding will be overturned the circles have been drawn out into long and narrow ellipses, and this indicates that those rock particles which before the bending were included in the circle have been moved past each other in the manner of the blades of a pair of shears. Such extreme "shearing" action is thus localized in the underturned limb of the fold, and a time must come with continuation of the compression when the fold will rupture at this critical place along a plane parallel to the longest axis of the ellipses or nearly parallel to the axial plane of the anticline. Such structures probably occur in the zone of combined fracture and flow, up into which the beds are forced in cases of close compression. Relief thus being found upon this plane of fracture, the upper portion of the fold will now ride over the lower, and the displacement is described as a *thrust* or *overthrust*.

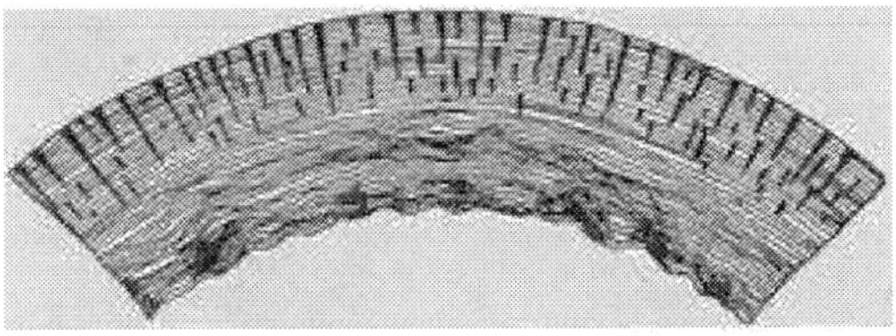

FIG. 26.—A bent stratum to illustrate tension upon the convex and compression upon the concave side (after Van Hise).

In the long series of experiments conducted by Mr. Bailey Willis of the United States Geological Survey, all the stages between the overturned fold and the overthrust fold were reproduced. Where a series of folds was closely compressed, a parallel series of thrusts developed (pl. 2 B), so that a series of slices cutting across neighboring strata was slid in succession, each over the other, like the scales upon a fish or the shingles upon a roof. Quite remarkable structures of this kind have been discovered in rocks of such closely folded districts as the Northwest Highlands of Scotland, where the overriding is measured in miles. Near the thrust planes the rocks show a crushing of the grains, and the planes themselves are sometimes corrugated and polished by the movement.

Restoration of mutilated folds.—Since flexuring of the rocks takes place within the zone of flow at a distance of several miles below the earth's surface, it is quite obvious that the results of the process can be studied only after some thousands of feet of superincumbent strata have been removed. We are a little later to see by what processes this lowering of the surface is accomplished, but for the present it may be sufficient to accept the fact, realizing that before foldings in the strata can reach the surface, they must have passed through the upper zone of fracture.

It might perhaps be supposed that the anticlines would appear as the mountains upon the surface, and occasionally this is true; as, for example, in the folded Jura Mountains of western Europe. More generally, the mountains have a synclinal structure and the valleys an anticlinal one; but as no general rule can be applied, it is necessary to make a restoration of the truncated folds in each district before their character can be known.

The geological map and section.—The earth's surface is in most regions in large part covered with soil or with other incoherent rock material, so that over considerable areas the hard rocks are hidden from view. Each locality at which the rock is found at the earth's surface "in place" is described as an *outcropping* or *exposure*. In a study of the region each such exposure must be examined to determine the nature of the rock, especially for the purpose of correlation with neighboring exposures, and, in addition, both the probable direction in which it is continued along the surface—the *strike*—and the inclination of its beds—the *dip*. If the outcroppings are sufficiently numerous, and rock type, strike and dip, may all be determined, the folds of the district may be restored with almost as much accuracy as though their curves were everywhere exposed to view. A cross section through the surface which represents the observed outcrops with their inclinations and the assumed intermediate strata in their probable attitudes is described as a *geological section* (Fig. 27). A map upon which the data have been entered in their correct locations, either with or without assumptions concerning the covered areas, is known as a *geological map*.

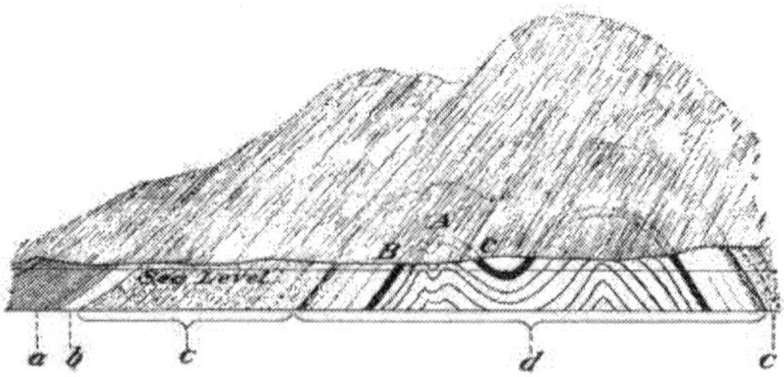

FIG. 27.—A geological section based upon observations at outcrops, but with the truncated arches restored.

If the axes of folds are absolutely horizontal, and the surface of the earth be represented as a plain, the lines of intersection of the truncated strata with the ground, or with any horizontal surface, will give the directions of continuation of the individual strata. This strike direction is usually determined at each exposure by use of a compass provided with a spirit level. When that edge of the leveled compass which is parallel to the north-south line upon the dial is held against the sloping rock stratum, the angle of strike is measured in degrees by the compass needle. If the cardinal directions have been placed in their correct positions upon the compass dial, the needle will point to the northwest when the strike is northeast, and *vice versa* (Fig. 28 *a*). Upon the geologist's compass it is therefore customary to reverse the initials which represent the east and west directions, in order that the correct strike may be read directly from the dial (Fig. 28 *b*).

FIG. 28.—Diagram to illustrate the manner of determining the strike of rock beds at an outcropping. *a*, a compass which has the cardinal directions in their natural positions; *b*, a compass with the east and west initials reversed upon the dial; *c*, home-made clinometer in position to determine the dip.

By the dip is meant the inclination of the stratum at any exposure, and this must obviously be measured in a vertical plane along the steepest line in the bedding plane. The dip angle is always referred to a horizontal plane, and hence vertical beds have a dip of 90°. The device for measuring this angle of dip, the *clinometer*, is merely a simple pendulum which serves as an indicator and is centered at the corner of a graduated quadrant. A home-made variety is easily constructed from a square piece of board and an attached paper quadrant (Fig. 28 *c*), but the geologist's compass is always provided with a clinometer attachment to the dial.

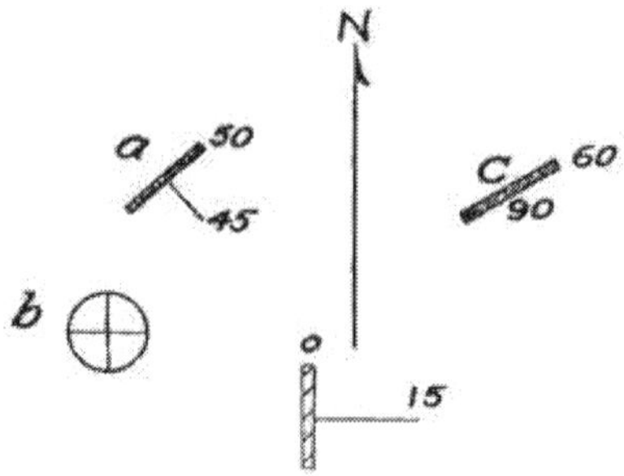

FIG. 29.—Diagram to show the use of T symbols to indicate the dip and strike of outcroppings.

Since the strike is the intersection of the bedding plane with a horizontal surface, and the dip is the intersection with that particular vertical plane which gives the steepest inclination, the strike and dip are perpendicular to each other. To represent them upon maps, it is more or less customary to use the so-called T symbols, the top of the T giving the direction of the strike and the shank that of the dip. If meridians are drawn upon the map, the direction or attitude of the T can be found by the use of a simple protractor; and when entered upon the map, the exact angle of the strike may be supplied by a figure near the top of the T, and the dip angle by a figure at the end of the shank. It is the custom, also, to make the length of the shank inversely proportional to the steepness of the dip, so that in a broad way the attitudes of the strata may be taken in at a glance (Fig. 29). It is further of advantage to make the top of the T a double line, so that some symbol or color may show the correlations of the different exposures. To illustrate, in Fig. 29, the symbol marked *a* represents an outcrop of limestone, the strike of which is 50° east of north (N. 50° E.), and the dip of which is 45° southeast. In the same figure *b* represents a shale outcrop in horizontal beds, which have in consequence a universal strike and a dip of 0°. An exposure of limestone in vertical beds which strike N. 60° E. is shown at *c*, etc.

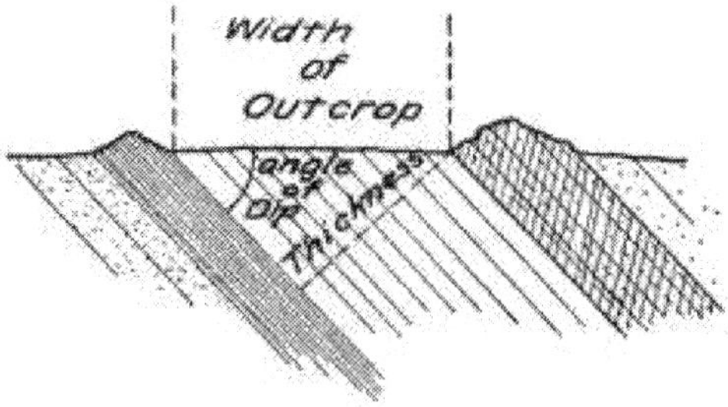

FIG. 30.—Diagram to show how the thickness of a formation may be obtained from the angle of the dip and the width of the exposures.

Measurement of the thickness of formations.—When formations still lie in horizontal beds, we may sometimes learn their thickness directly either from the depth of borings to the underlying rock, or by measurements upon steep cañon walls. If the beds stand vertically, the matter is exceedingly simple, for in this case the thickness is the width of the outcrops of the formation between the beds which bound it upon either side. In the general case, in which the beds are neither horizontal nor vertical, the thickness must be obtained indirectly from the width of the exposures and the angle of the dip. The factor by which the exposure width must be multiplied is known as the sine of the dip angle (Fig. 30), which is given with sufficient accuracy for most purposes in the following table. It is obvious that in order to obtain the full thickness of a formation it is necessary to measure from the contact with the adjacent formation upon the one side to a similar contact with the nearest formation upon the other.

Natural Sines

0°	.00	35°	.57	70°	.94
5°	.09	40°	.64	75°	.97
10°	.17	45°	.71	80°	.98
15°	.26	50°	.77	85°	1.00
20°	.34	55°	.82	90°	1.00
25°	.42	60°	.87		
30°	.50	65°	.91		

FIG. 31.—Combined surface and sectional views of a plunging anticline (after Willis).

FIG. 32.—Combined surface and sectional views of a plunging syncline (after Willis).

The detection of plunging folds.—When the axis of a fold is horizontal, its outcrops upon a plain will continue to have the same strike until the formation comes to an end. Upon a generally level surface, therefore, any regular progressive variation in the strike direction is an indication that the folds have a plunging or pitching character. Many serious mistakes of interpretation have been made because of a failure to recognize this evidence of plunging folds. The way in which the strikes are progressively modified will be made clear by the diagrams of Figs. 31 and 32, the first representing a pitching anticline and the second a pitching syncline. In both these reciprocal cases the strikes of the beds undergo the same changes, and the dip directions serve to distinguish which of the two structures is present in a given case. There is, however, one further difference in that the hard layers of the plunging anticline, where they disappear below the surface in the axis, will present a domed surface sloping forward like the back of a whale as it rises above the surface of the sea. Plunging folds in series will thus appear in the topography as a series of sharply zigzagging ranges at those localities where the harder layers intersect the surface. Such features are encountered in eastern Pennsylvania, where the hard formations of the Appalachian Mountain system plunge northeastward under the later formations. The pitch of the larger fold is often disclosed by that of the minor puckerings superimposed upon it.

FIG. 33.—Unconformity between a lower and an upper series of beds upon the coast of California. Note how the hard layer stands in relief upon the connecting surface (after Fairbanks).

The meaning of an unconformity.—The rock beds, which are deposited one above the other during a transgression of the sea, are usually parallel and thus represent a continuous process of deposition. Such beds are said to be *conformable*. Where, upon the other hand, two series of deposits which are not parallel to each other are separated by a break, they are said to form *unconformable* series, and the break or surface of junction is an *unconformity* (Fig. 33).

Here it is evident that the sediments which compose the lower series of beds have been folded in the zone of flow, though the upper series has evidently escaped this vicissitude. Furthermore, the surface which delimits the lower series from the upper is somewhat irregular and shows a hard layer standing in relief, as it would if it had opposed greater resistance to the attacks of the atmosphere upon it.

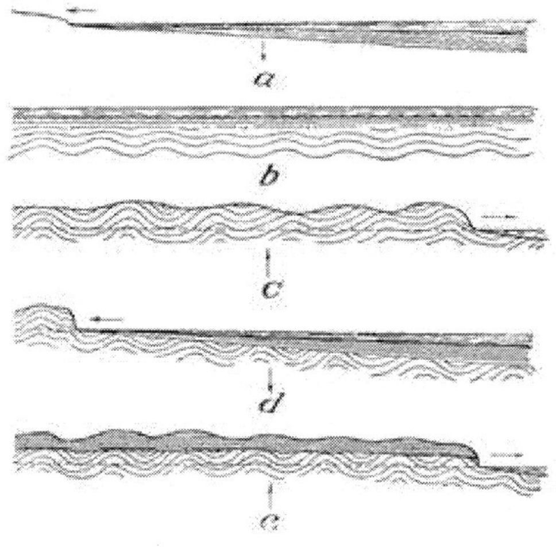

FIG. 34.—Series of diagrams to illustrate in succession the episodes involved in the historical development of an angular unconformity. The vertical arrows indicate direction of movement of the land, and the horizontal arrows the direction of shore migration.

In reality, an unconformity between formations must be interpreted to mean that the lower series is not only older than the upper, as shown by the order of superposition, but that the time of its deposition was separated from that of the upper by a hiatus in which important changes took place in the lower series. The stages or episodes in the history of the beds represented in Fig. 33 may be read as follows (see Fig. 34 *a-e*):—

(*a*) Deposition of the lower series during a transgression of the sea.

(*b*) Continued subsidence and burial of the lower series beneath overlying sediments, and flexuring in the zone of flow.

(*c*) Elevation of the combined deposits to and far above sea level and removal by erosion of vast thicknesses of the upper sediments.

(*d*) A new subsidence of the truncated lower series and deposition of the upper series across its eroded surface.

(*e*) A new elevation of the double series to its present position above sea level.

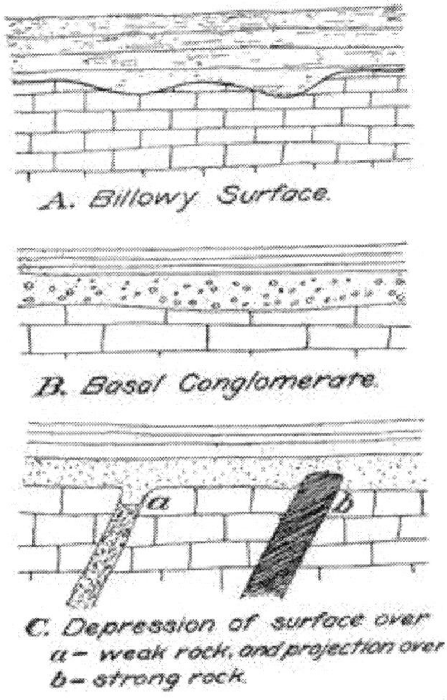

FIG. 35.—Types of deceptive or erosional unconformities.

From this succession of episodes it is seen that a break of this kind between two series of deposits involves a double oscillation of subsidence followed by elevation—a large depression followed by a large elevation, a smaller subsidence followed by elevation. The time interval which must have been represented by these repeated operations is so vast as at first to stagger the mind in contemplating it. When, as in this instance, the dips of the lower series of beds differ from those of the upper, we have to do with an *angular unconformity*. It may be, however, that the lower series was not so far depressed as to enter the zone of flow, and that its beds meet those of the upper series with apparent conformity. Such an unconformity is often extremely difficult to recognize, and it is described as a *deceptive* or *erosional unconformity.*

With a deceptive unconformity the clew to its real nature is usually some fact which indicates that the lower series of sediments had been raised above the level of the sea before the upper series was deposited upon it. This may be apparent either in the irregularity of the surface on which the two series are joined, in some evidence of the action of waves such as would be furnished by a basal conglomerate in the upper series, or some indication of different resistance of different rocks of the lower series to attacks of the atmosphere upon them (Figs. 33 and 35 *a-c*).

In most cases, at least, the lowest member of the upper series will be a different type of rock from the uppermost member of the lower series, hence the frequent occurrence of the discordant cross bedding in sandstone should not deceive even the novice into the assumption of an unconformity.

READING REFERENCES TO CHAPTER V

The zones of fracture and flow:—

C. R. VAN HISE. Principles of North American Precambrian Geology, 16th Ann. Rept. U.S. Geol. Surv., 1895, Pt. I, pp. 581-603.

BAILEY WILLIS. Mechanics of Appalachian Structure, 13th Ann. Rept. U.S. Geol. Surv., 1893, Pt. II, pp. 217-253.

A. DAUBRÉE. Études Synthétiques de Géologie Expérimentale. Paris, 1879, pp. 306-328, pl. II.

W. PRINZ. Quelques remarques générales à propos de l'essai de carte tectonique de la belgique, etc., Bull. Soc. Belge Geol., vol. 18, 1904, p. 143, pl. V.

Analysis of folds:—

VAN HISE and WILLIS as above; DE MARGERIE et HEIM; Les dislocations de l'écorce terrestre (in French and German languages). Zurich, 1888.

Geological maps:—

WM. H. HOBBS. The Mapping of the Crystalline Schists, Jour. Geol., vol. 10, 1902, pp. 780-792, 858-890.

CHAPTER VI

THE ARCHITECTURE OF THE FRACTURED SUPERSTRUCTURE

FIG. 36.—A set of master joints developed in shale upon the shores of Cayuga Lake near Ithaca, New York (after U. S. G. S.).

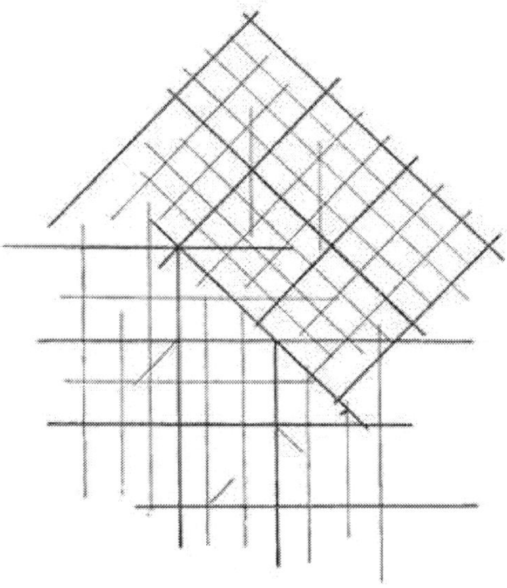

FIG. 37.—Diagram to show how sets of master joints differing in direction by half a right angle may abruptly replace each other.

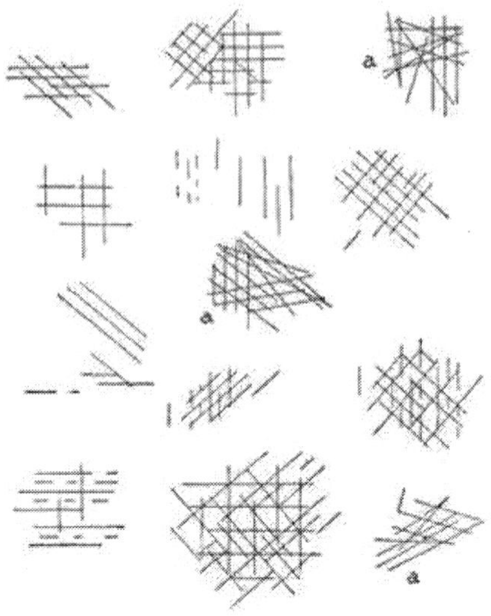

FIG. 38.—Diagram to show the different combinations of the series composing two double sets of master joints, and in *a*, *a*, *a* additional disorderly fractures.

The system of the fractures.—In referring to experiments made upon the fracture of solid blocks under compression (p. 41), it was shown that two series of parallel fractures develop perpendicular to each free surface of the block, and that these series are each of them inclined by half of a right angle to the direction of compression, and thus perpendicular to each other. The fragments into which a block with one free surface would thus tend to be divided should be square prisms perpendicular to the free surface. It would be interesting, if it were practicable, to learn from experiment how these prisms would be further fractured by a continuation of the compression. From mechanical considerations involving the resolution of forces with reference to the ready-formed fractures, it seems probable that the next series of fractures to form would bisect the angles of the first double series or set. Wherever rocks are found exposed in their original attitudes, they are, in fact, seen to be intersected by two parallel series of fractures which are perpendicular to the earth's surface and to each other and are described as *joints*. In many cases more than two series of such fractures are found, yet even in these cases two more perfectly developed series are prominent and almost exactly perpendicular to each other as well as to the earth's surface. This omnipresent double series or *set* of joints is the well-known set of *master joints*, and very often it is found developed practically alone (Fig. 36). Over large areas, the direction of the set of master joints may remain practically constant, or this set may quite suddenly give place to a similar set which is, however, turned through half a right angle from the first (Fig. 37). Not infrequently two such sets of master joints are found together bisecting each other's angles within the same rocks, and to them are sometimes added additional though less perfect series of joint planes.

Studied throughout a considerable district, the various series which make up these two sets of master joints may be seen locally developed in different combinations as well as in association with additional fissure planes which are not easily reduced to any simple law of arrangement (Fig. 38 *a, a, a*). Only rarely are regular joint series observed which do not stand perpendicular to the original attitude of the rock beds. In a few localities, however, rectangular joint sets have been discovered which divide the rock into prisms parallel to the earth's surface and with the joint series inclined to it each by half a right angle. Where the rock beds have been much disturbed, the complex of joints may be such as to defy all attempts at orderly arrangement.

FIG. 39.—View on the shore at Holstensborg, West Greenland, to show the subequal spacing of the joints (after Kornerup).

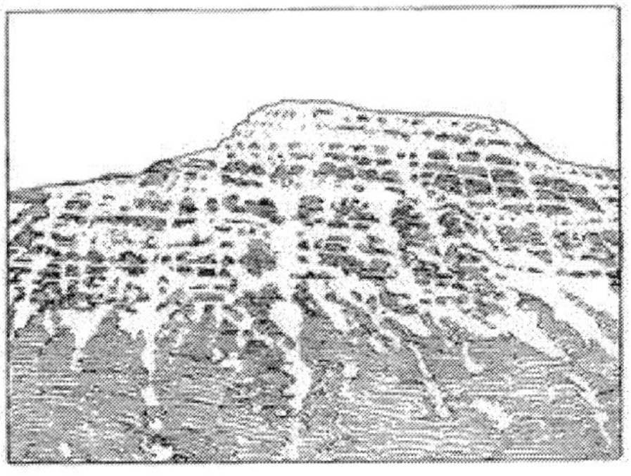

FIG. 40.—View of an exposed hillside in Iceland upon which the snow collected in crannies along the joints brings out to advantage both

the larger and the smaller intervals of the joint system (after Thoroddsen).

The space intervals of joints.—The same kind of subequal spacing which characterizes the fractures near the surface of the block in Daubrée's experiment (Fig. 19, p. 41) is found simulated by the rock joints (Fig. 39). Such unit intervals between fractures may be grouped together into larger units which are separated by fractures of unusual perfection. We may think of such larger space units as having the smaller ones superimposed upon them (Fig. 40).

The displacements upon joints—faults.—In the vast majority of cases, the joint fractures when carefully examined betray no evidence of any appreciable movement of the two walls upon each other. Generally the rock layers are seen to cross the joints without apparent displacement. Joints are therefore planes of disjunction only, and not planes of displacement.

FIG. 41.—Faulted blocks of basalt divided by joints near Woodbury, Connecticut. To show the structure of the rock, some of the foliage has been removed in preparing the sketch from a photograph.

Within many districts, however, a displacement may be seen to have occurred upon certain of the joint planes, and these are then described as *faults*. Such displacements of necessity imply a differential movement of sections or blocks of the earth's crust, the so-called *orographic blocks*, which are bounded by the joint planes and play individual rôles in the movement. A simple case of such displacements in rocks intersected by a single set of master joints is represented in the model of plate 4 C. The most prominent fault represented by this model runs lengthwise through the middle, and the displacement which is measured upon it not only varies between wide limits, but is marked by abrupt changes at the margins of the larger blocks. This vertical displacement upon the fault is called its *throw*. Though not illustrated by the model, horizontal displacements may likewise occur, and these will be more fully discussed when the subject of earthquakes is considered in the following chapter. An actual example of blocks displaced by vertical adjustment is represented in Fig. 41, a simple type of faulting which has taken place in rocks but slightly disturbed from their original attitude, but intersected by a relatively simple system of master joints. In those regions where the beds have been folded and perhaps overthrust before their elevation into the zone of fracture, and which are further intersected by disorderly fissure planes, the results are far more complex. In such cases the planes of individual displacement may not be vertical, though they are generally steeper than 45°. For their description it is necessary to make use of additional technical terms (Fig. 42). The inclination of a sloping fault plane measured against the vertical is called the *hade* of the fault. The *total displacement* is measured along the plane of the fault from a point upon one limb to the point from which it was separated in the other. The additional terms are made sufficiently clear by the diagram.

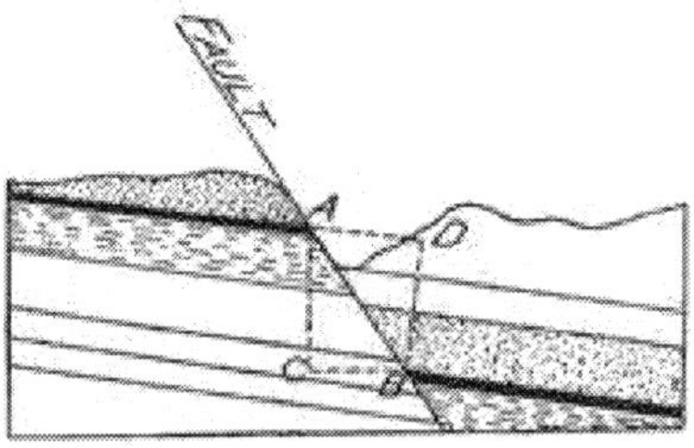

FIG. 42.—A fault in previously disturbed strata. *AB*, displacement; *AC*, throw; *BD*, stratigraphic throw; *BC*, heave; angle *CAB*, hade.

Methods of detecting faults.—The first effect of a fault is usually to produce a crack at the surface of the earth; and, provided there is a vertical displacement or throw, an escarpment which rises upon the upthrown side of the fault. In general it may be said that escarpments which appear at the earth's surface as plane surfaces probably represent planes of fracture, though not necessarily planes of faulting. In many cases the actual displacements lie buried under loose rock débris near to and paralleling the escarpment, and in some cases as a result of the erosional processes working upon alternately hard and soft layers of rock, the escarpment may later appear upon the downthrown side or limb of the fault (Fig. 43). As an illustration of a fault escarpment, the façade of El Capitan and many other rock faces of the Yosemite valley may be instanced.

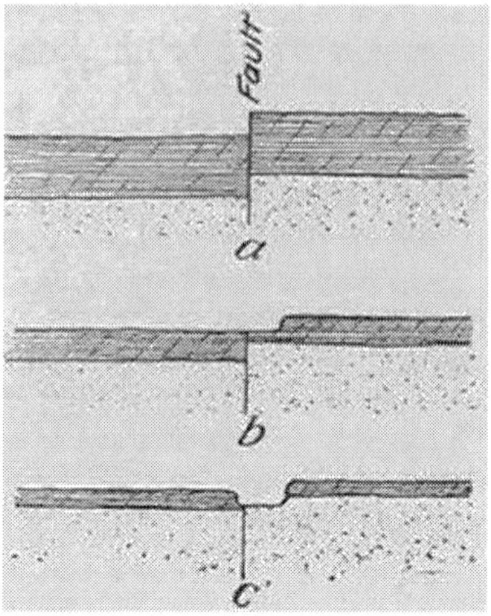

FIG. 43.—Diagrams to show how an escarpment originally on the upthrown side of the fault may, through erosion, appear upon the downthrown side.

FIG. 44.—A fault plane exhibiting "drag." The opening is artificial (after Scott).

When we have further studied the erosional processes at the earth's surface, it will be appreciated that faults tend to quickly bury themselves from sight, whereas fold structures will long remain in evidence. Many faults will thus be overlooked, and too great weight is likely to be ascribed to the folds in accounting for the existing attitudes and positions of the rock masses.

Faults must therefore be sought out if mistakes of interpretation are to be avoided.

The most satisfactory evidence of a fault is the discovery of a rock bed which may be easily identified, and which is actually seen displaced on a plane of fracture which intersects it (Fig. 42, p. 59). When such an easily recognizable layer is not to be found, the plane of displacement may perhaps be discovered as a narrow zone composed of angular fragments of the rock cemented together by minerals which form out of solution in water. Such a fractured rock zone which follows a plane of faulting is a *fault breccia.* If the fault breccia, or vein rock, is much stronger than the rock on either side, it may eventually stand in relief at the surface like a dike or wall. At other times the displacement produces little fracture of the walls, but they slide over each other in such a manner as to yield either a smoothly corrugated or an evenly polished surface which is described as "slickensides." It may be, however, that during the movement either one or both of the walls have "dragged", and so are curled back in the immediate neighborhood of the fault plane (Fig. 44).

When, as is quite generally the case, the actual plane of displacement of a fault is not open to inspection, the movement may be proven by the observation of abrupt, as contrasted with gradual, changes in the strikes and dips of neighboring exposures (Fig. 45); or by noting that some easily recognized formation has been sharply offset in its outcrops (Fig. 46).

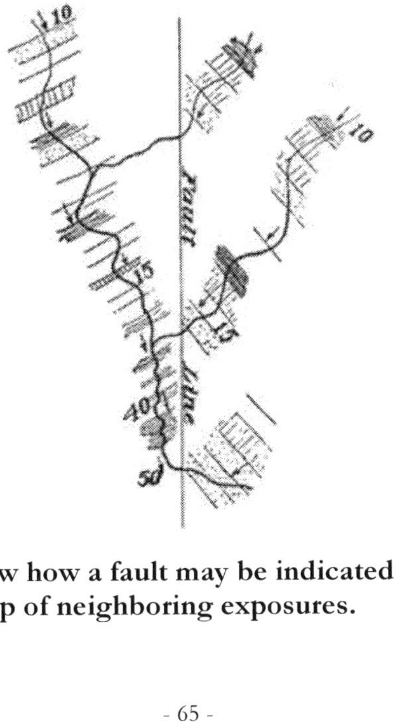

FIG. 45.—Map to show how a fault may be indicated in abrupt changes of the strike and dip of neighboring exposures.

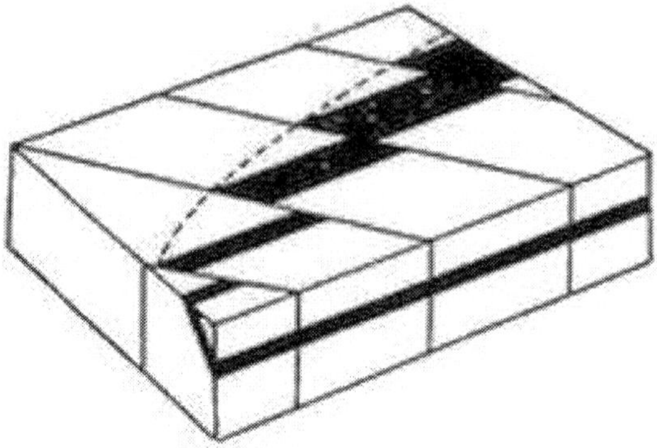

FIG. 46.—A series of parallel faults indicated by successive offsets in the course of an easily recognizable rock formation.

There are in addition many indications rather than proofs of the presence of faults, which must be taken account of in every general study of the geology of a district. Thus the outcrops of all neighboring formations may terminate abruptly upon a straight line which intersects all alike. Deep-seated fissure springs may be aligned in a striking manner, and so indicate the course of a prominent fracture, though not necessarily of a fault. Much the same may be said of the dikes of cooled magma which have been injected along preëxisting fractures.

The base of the geological map.—Modern topographic maps form an important part of the library of the serious student of physiography; they are the gazetteer of this branch of science. Every civilized nation has to-day either completed a topographic atlas of its territory, or it is vigorously prosecuting a survey to furnish maps which represent the relief with some detail, and publishing the results in the form of an atlas of quadrangles. Thus a relief map will erelong be obtainable of any part of the civilized world, and may be purchased in separate sections. Nowhere is this work being taken up with greater vigor than in the United States, where a vast domain representing every type of topographic peculiarity is being attacked from many centers. Here and elsewhere the relief of the land is being expressed by so-called contours or lines of equal altitude upon the earth's surface. It is as though a series of horizontal planes, separated by uniform intervals of 20 or 40 or 100 feet, had been made to intersect the surface, and the intersection curves, after consecutive numeration, had been dropped into a single plane for printing.

Where the slopes are steep, the contour lines in the topographic map will appear crowded together and so produce a deep shade upon the map; whereas with relatively flat surfaces white patches will stand out prominently upon the map. More and more the topographic map is coming into use, and for the student of nature in particular it is important to acquire facility in interpreting the relief from the topographic map. To further this end, a special model has been devised, and its use is described in appendix C. Usually before any satisfactory geological map can be prepared, a contoured topographic map of the district to be studied must be available.

The field map and the areal geological map.—As the atlas of topographic maps is the physiographic gazetteer, so geological maps together constitute the reference dictionary of descriptive geology. Not only are topographic maps of many districts now generally available, but more and more it has become the policy of governments to supply geological maps in the same quadrangle form which is the unit of the topographic map. The geological map is, however, a complex of so many conventional symbols, that without some practical experience in the actual preparation of one, it is exceedingly difficult for the student to comprehend its significance. A modern geological map is usually a rectangular sheet printed in color, upon which are many irregular areas of individual hue joined to each other like the parts of a child's picture puzzle.

The colored areas upon the geological map are each supposed to indicate where a certain rock type or formation lies immediately below the surface, and this distribution represents the best judgment of the geologist who, after a study of the district, has prepared the map. Unfortunately the conventions in use are such that his observation and his theory have been hopelessly intermingled in the finished product. Armed with the geological map, the student who visits the district finds spread out before him, it may be, a landscape of hill and valley, of green forest and brown farming land, which is as different as may be from the colored puzzle which he holds in his hand. Hidden under the farm vegetation or masked by the woods are scattered outcroppings of rock which have been the basis of the geologist's judgment in preparing the map. Experience shows that in order to bridge the wide gap between the geology in the landscape and the patches of color upon the map something more than mere examination of the colored sheet is necessary. We shall therefore describe, with the aid of laboratory models, the various stages necessary to the preparation of a geological map, and every student should be advised to follow this by practical study of some small area where rocks are found in outcrop.

Though the published *areal geological map* represents both fact and theory, the map maker retains an unpublished *field map* or map of observations, upon which the final map has been based. This field map shows the location of each outcrop that has been studied, with a record of the kind of rock and of such observations as strike, dip, and pitch. Our task will therefore be to prepare: (1) a field map; (2) an areal geological map; and (3) some typical geological sections.

Laboratory models for the study of geological maps.—In order to represent in the laboratory the disposition of rock outcrops in the field, special laboratory tables are prepared with removable covers and with fixed tops, which are divided into squares numbered like the township sections of the national domain (Fig. 47). To represent the rock outcrops, blocks are prepared which may be fixed in any desired position by fitting a pin into a small augur hole bored through the table. The outcrop blocks for the sedimentary rock types are so constructed as to show the strike and dip of the beds. (See Appendix D.)

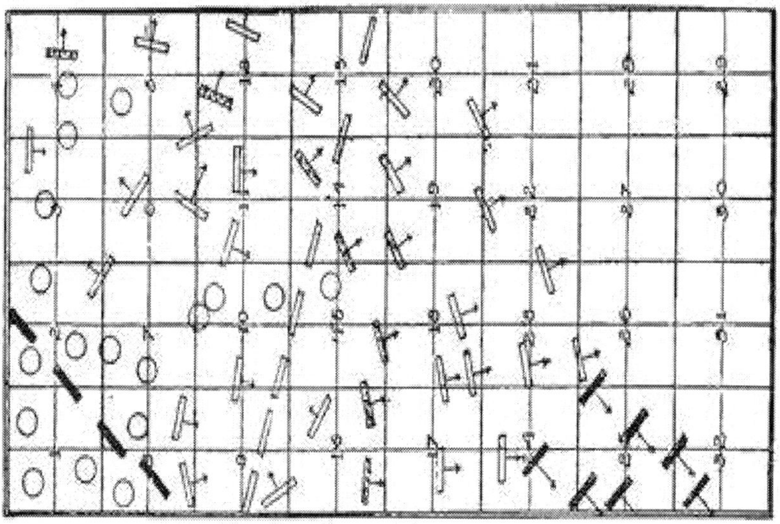

FIG. 47.—Field map prepared from a laboratory table.

The method of preparing the map.—To prepare the map, use is made of a geological compass with clinometer attachment, a protractor, and a map base divided into sections like the top of the table, and on the scale of one inch to the foot. Each exposure represented upon the table is "visited" and then located upon the base map in its proper position and attitude. The result is the field map (Fig. 47), which thus represents the facts only, unless there have been uncertainties in the correlation of exposures or in determining the position of the bedding plane.

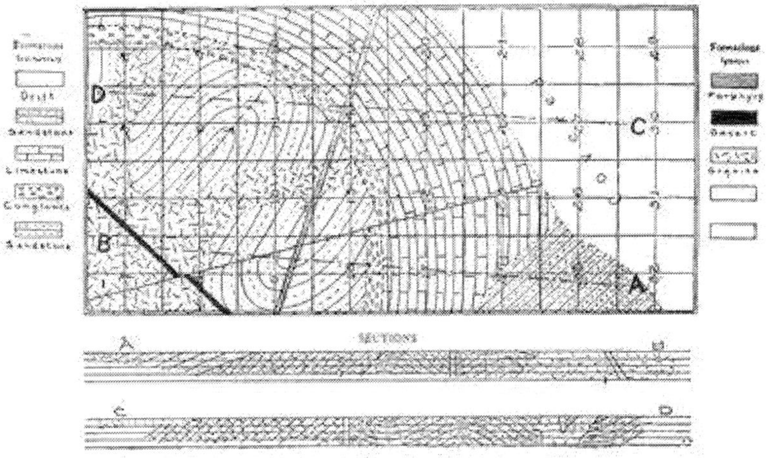

FIG. 48.—Areal geological map constructed from the field map of Fig. 47, with two selected geological sections.

To prepare the areal geological map from the field map, it is first necessary to fix the *boundaries* which separate formations at the surface; and now perhaps for the first time it is realized how large an element of uncertainty may enter if the exposures were widely separated. It is clear that no two persons will draw these lines in the same positions throughout, though certain portions of them—where the facts are more nearly adequate—may correspond. In Fig. 48 is represented the areal geological map constructed from the field map, with the doubtful area at one side left blank.

Some conclusions from this map may now be profitably considered. The complexly folded sandstone formation at the left of the map appears as the oldest member represented, since its area has been cut through by the intrusive granite which does not intrude other formations, and is unconformably overlaid by the limestone and its basal layer of conglomerate. The limestone in turn is unconformably overlaid by the merely tilted sandstone beds at the right of the map. These three sedimentary formations clearly represent decreasing amounts of close folding, from which it is clear that each earlier formation has passed through an episode not shared by that of next younger age. Of the other intrusive rocks, the dike of porphyry is younger than all the other formations, with the possible exception of the upper sandstone. Offsetting of the formations has disclosed the course of a fault, and from its relations to the dikes we may learn that of these the porphyry is younger and the basalt older than the date of the faulting.

The dashed lines upon the map (*AB* and *CD*) have been selected as appropriate lines along which to construct geological sections (Fig. 48, below map), and from these sections the *exposed* thicknesses of the different formations may be calculated. In one instance only, that of the conglomerate, can we be sure that this exposed thickness measures the entire formation.

Fold *versus* fault topography.—The more resistant or "stronger" rock beds, as regards attacks of the atmosphere, in the course of time come to stand in relief, separated by depressions which overlie the "weaker" formations. Simple open folds which are not plunging exercise an influence upon topography by producing generally long and straight ridges. More complex flexures, since they generally plunge, make themselves apparent by features which in the map are represented by curves. Fracture structures, and especially block displacements, are differentiated from these curving features by the dominance of straight or nearly rectilinear lines upon the map. The effect of erosion is to reduce the asperity of features and to mold them with flowing curves. The fracture structures are for this reason much more likely to be overlooked, and if they are not to elude the observer, they must be sought out with care. Fold and fracture structures may both be revealed upon the same map.

READING REFERENCES TO CHAPTER VI

Joint systems:—

JOHN PHILLIPS. Observations made in the Neighborhood of Ferrybridge in the Years 1826-1828, Phil. Mag., 2d ser., vol. 4, 1828, pp. 401-409; Illustrations of the geology of Yorkshire, Pt. II, The Limestone District. London, 1836, pp. 90-98.

SAMUEL HAUGHTON. On the Physical Structure of the Old Red Sandstone of the County of Waterford, considered with reference to cleavage, joint surfaces, and faults, Trans. Roy. Soc. London, vol. 148, 1858, pp. 333-348.

W. C. BRÖGGER. Spaltenverwerfungen in der Gegend Langesund-Skien, Nyt Magazin for Naturvidernskaberne, vol. 28, 1884, pp. 253-419.

WM. H. HOBBS. The Newark System of the Pomperaug Valley, Connecticut, 21st Ann. Rept. U. S. Geol. Surv., Pt. III, 1901, pp. 85-143.

Geological map:—

WM. H. HOBBS. The Interpretation of Geological Maps, School Science and Mathematics, vol. 9, 1909, pp. 644-653.

CHAPTER VII

THE INTERRUPTED CHARACTER OF EARTH MOVEMENTS: EARTHQUAKES AND SEAQUAKES

Nature of earthquake shocks.—Man's belief in the stability of Mother Earth—the *terra firma*—is so inbred in his nature that even a light shock of earthquake brings a rude awakening. The terror which it inspires is no doubt largely to be explained by this disillusionment from the most fundamental of his beliefs. Were he better advised, the long periods of quiet which separate earthquakes, and not the lighter shocks which follow all grander disturbances, would occasion him concern.

FIG. 49.—View of a portion of the ruins of Messina after the earthquake of December 28, 1908.

Earthquakes are the sensible manifestations of changes in level or of lateral adjustments of portions of the continents, and the seismic disturbances upon the sea—seaquakes and seismic sea waves—relate to similar changes upon the floor of the ocean.

During the grander or catastrophic earthquakes, the changes are indeed terrifying, and have usually been accompanied by losses to life and property, which are only to be compared with those of great conflagrations or of inundations on thickly populated plains. The conflagration has all too frequently been an aftermath of the great historic earthquakes. The earthquake of December 28, 1908, in southern Italy, destroyed almost the entire population of a great city, and left of its massive buildings only a confused heap of rubble (Fig. 49). Two years later a heavy earthquake resulted in great damage to cities in Costa Rica (Fig. 50), while two years earlier our own country was first really awakened to the danger in which it stands from these convulsive earth throes; though, as we shall see, these dangers can be largely met through proper methods of construction.

FIG. 50.—Ruins of the Carnegie Palace of Peace at Cartago, Costa Rica, destroyed when almost completed by the great earthquake of May 4, 1910 (after a photograph by Rear-Admiral Singer, U.S.N.).

Earthquakes are usually preceded for a brief instant by subterranean rumblings whose intensity appears to bear no relation to the shocks which follow. The ground then rocks in wavelike motions, which, if of large amplitude, may induce nausea, prevent animals from keeping upon their feet, and wreck all structures not specially adapted to withstand them. Heavy bodies are sometimes thrown up from the ground (Fig. 51), and at other times similar heavy masses are, apparently because of their inertia, more deeply imbedded in the earth. Thus gravestones and heavy stone posts are often sunk more deeply in the ground and are surrounded by a hollow and perhaps by small open cracks in the surface (Fig. 52). When bodies are thrown upward, it would imply that a quick upward movement of the ground had been suddenly arrested, while the burial of heavy bodies in the earth is probably due to a movement which begins suddenly and is less abruptly terminated.

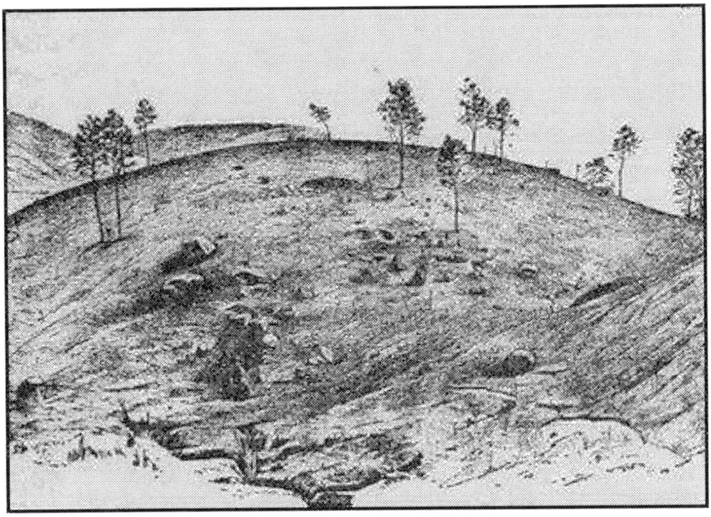

FIG. 51.—Bowlders thrown into the air and overturned during the Assam earthquake of 1897 (after R. D. Oldham).

FIG. 52.—Heavy post sunk deeper into the ground during the Charleston earthquake of August 31, 1886 (after Dutton).

Seaquakes and seismic sea waves.—Upon the ocean the quakes which emanate from the sea floor are felt on shipboard as sudden joltings which produce the impression that the ship has struck upon a shoal, though in most instances there is no visible commotion in the water. The distribution of these shocks, as indicated either by the experiences of neighboring ships at the time of a particular shock, or by the records of vessels which at different times have sailed over an area of frequent seismic disturbance, appears to be limited to narrow zones or lines (Fig. 53). The same tendency of under-sea disturbances to be localized upon definite straight lines has been often illustrated by the behavior of deep-sea cables which are laid in proximity to one another and which have been known to part simultaneously at points ranged upon a straight line.

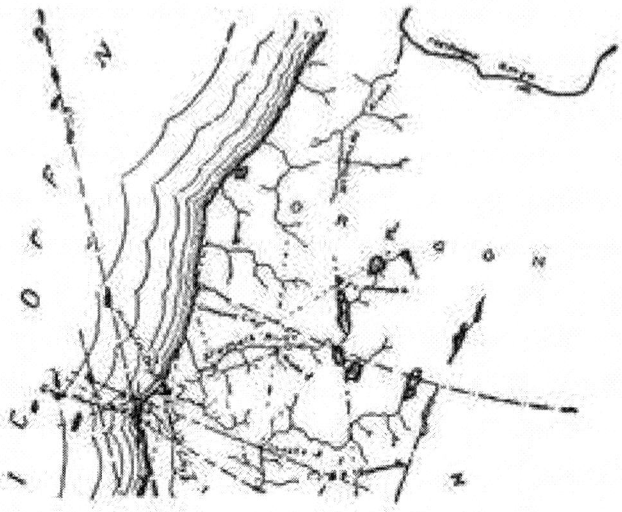

FIG. 53.—Map showing the localities at which shocks have been reported at sea off Cape Mendocino, California.

Far grander disturbances upon the floor of the ocean have been revealed by the great sea waves—the so-called "tidal waves", properly referred to as *tsunamis*—which recur in those sea districts which adjoin the special earthquake zones upon the continents (p. 86). The forerunner of such a sea wave approaching the shore is usually a sudden withdrawal of the water so as to lay bare a portion of the bottom, but this is well-recognized to be the premonition of a gigantic oncoming wave which sweeps all before it and is only halted when it has rolled over all the low-lying country and encountered a mountain wall. Such seismic waves have been especially common upon the Pacific shore of South America and upon the Japanese littoral (Fig. 54). These waves proceed from above the great deeps upon the ocean bottom, and clearly result from the grander earth movements to which these depressions owe their exceptional depth. The withdrawal of the water from neighboring shores may be presumed to be connected with a descent of the floor of the depression and the consequent drawing-in of the ocean surface above. The later high wave would thus represent the dispersion of the mountain of water which is raised by the meeting of the waters from the different sides of the depression.

FIG. 54.—Effect of a seismic water wave at Kamaishi, Japan, in 1896 (after E. R. Scidmore).

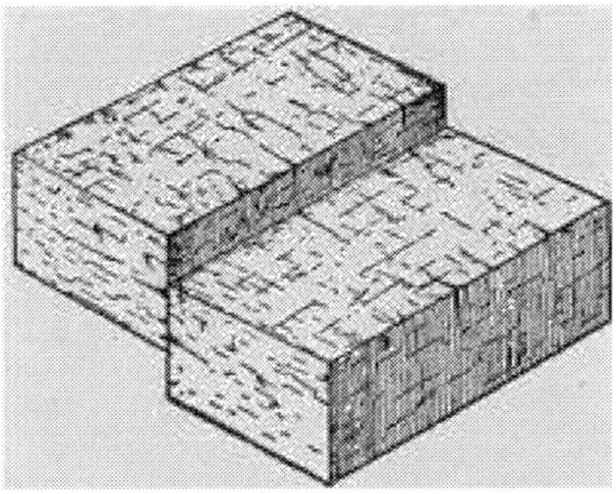

FIG. 55.—A fault of vertical displacement.

The grander and the lesser earth movements.—Upon the land the grander and so-called catastrophic earthquakes are usually the accompaniment of important changes in the surface of the ground that will be discussed in later sections. Those shocks which do little damage to structures produce no visible changes in the earth's surface, except, it may be, to shake down some water-soaked masses of earth upon the steeper slopes. Still other movements, and these too slight to be felt even in the night when the animal world is at rest, may yet be distinguished by their sounds, the unmistakable rumblings which are characteristic alike of the heaviest and the lightest of earthquake shocks.

FIG. 56.—Escarpment produced by an earthquake fault of vertical displacement which cut across the Chedrang River and thus produced a waterfall, Assam earthquake of 1897 (after R. D. Oldham).

Changes in the earth's surface during earthquakes—faults and fissures.—Each of the grander among historic earthquakes has been accompanied by noteworthy changes in the configuration of the earth's surface within the district where the shocks were most intense. A section of the ground is usually found to have moved with reference to another upon the other side of a vertical plane which is usually to be seen; we have here to do with the actual making of a fault or displacement such as we find the fossil examples of within the rocks. The displacement, or throw, upon the fault plane may be either upward or downward or laterally in one direction or the other, or these movements may be combined. A movement of adjacent sections of the ground upward or downward with reference to each other (Fig. 55) has been often observed, notably at Midori after the great Japanese earthquake of 1891, and in the Chedrang valley of Assam after the earthquake of 1897 (Fig. 56).

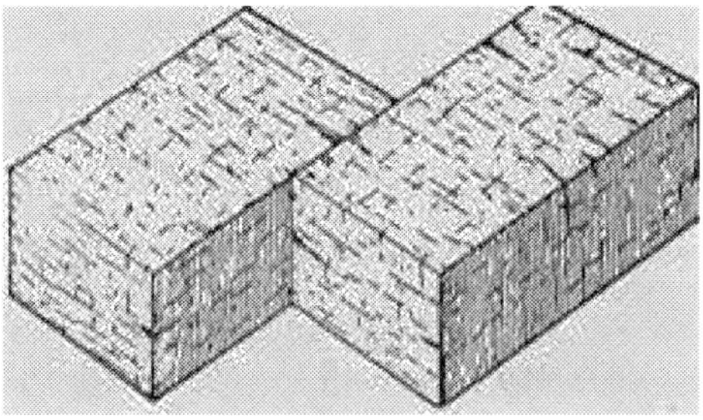

FIG. 57.—A fault of lateral displacement.

FIG. 58.—Fence parted and displaced fifteen feet by a transverse fault formed during the California earthquake of 1906 (after W. B. Scott).

FIG. 59.—Fault with vertical and lateral displacements combined.

A lateral throw, unaccompanied by appreciable vertical displacement (Fig. 57), is especially well illustrated by the fault in California which was formed during the earthquake of 1906 (Fig. 58). A combination of the two types of displacement in one (Fig. 59) is exemplified by the Baishiko fault of Formosa at the place shown in plate 3 A.

The measure of displacement.—To afford some measure of the displacements which have been observed upon earthquake faults, it may be stated that the maximum vertical throw measured upon the fault in the Neo valley of Japan (1891) was 18 feet, in the Chedrang valley of Assam (1897) 35 feet, and of the Alaskan coast (1899) 47 feet. Large sections of land were bodily uplifted in these cases within the space of a few seconds, or at most a few minutes, by the amounts given. The largest recorded lateral displacement measured upon an earthquake fault is about 21 feet upon the California rift after the earthquake of 1906; though an amount only slightly less than this is indicated in the shifting of roads and arroyas dating from the earthquake of 1872 in the Owens valley, California. Fault lines once established are planes of special weakness and become later the seat of repeated movements of the same kind.

PLATE 3.

A. **An earthquake fault opened in Formosa in 1906, with vertical and lateral displacements combined (after Omori).**

B. **Earthquake faults opened in Alaska in 1889, on which vertical slices of the earth's shell have undergone individual adjustments (after Tarr and Martin).**

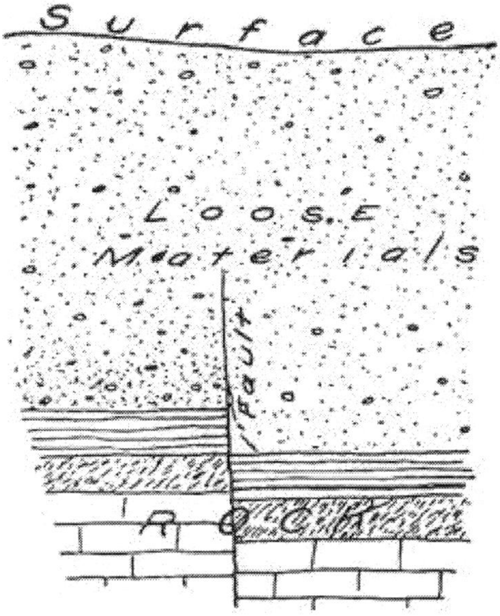

FIG. 60.—Diagram to show how small faults in the rock basement may be masked at the surface through adjustments within the loose rock mantle.

The greater number of earthquake faults are found in the loose rock cover which so generally mantles the firmer rock basement, and it is almost certain that the throws within the solid rock are considerably larger than those which are here measured at the surface, owing to the adjustments which so readily take place in the looser materials. Those lighter shocks of earthquake which are accompanied by no visible displacements at the surface do, however, in some instances affect in a measure the flow of water upon the surface, and thus indicate that small changes of surface level have occurred without breaks sufficiently sharp to be perceived (Fig. 60). Intermediate between the steep escarpment and the masked displacement just described is the so-called "mole-hill" effect,—a rounded and variously cracked slope or ridge above the position of a buried fault (Fig. 61).

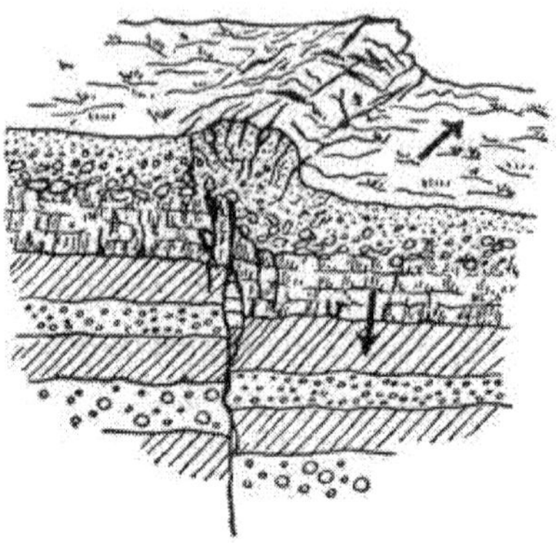

FIG. 61.—Diagram to show the appearance of a "mole hill" above a buried earthquake fault (after Kotô).

The escarpments due to earthquake faults in loose materials at the earth's surface can obviously retain their steepness for a few years or decades at the most; for because of their verticality they must gradually disappear in rounded slopes under the action of the elements. Smaller displacements within a rock which rapidly disintegrates under the action of frost and sun will likewise before long be effaced. In those exceptional instances where a resistant rock type has had all altered upper layers planed away until a fresh and hard surface is exposed, and has further been protected from the frost and sun beneath a thin layer of soil, its original surface may be retained unaltered for many centuries. Upon such a surface the lightest of sensible shocks, or even the smaller earth movements which are not perceived at the time, may leave an almost indelible record. Such records particularly show that the movements which they register occur upon the planes of jointing within the rock, and that these ready formed cracks have probably been the seats of repeated and cumulative adjustments (Fig. 62).

FIG. 62.—Post-glacial earthquake faults of small but cumulative displacement, eastern New York (after Woodworth).

FIG. 63.—Earthquake cracks in Colorado desert (after a photograph by Sauerven).

Contraction of the earth's surface during earthquakes.—The wide variations in the amount of the lateral displacement upon earthquake faults, like those opened in California in 1906, show that at the time of a heavy earthquake there must be large local changes in the density of the surface materials. Literally, thousands of fissures may appear in the lowlands, many of them no doubt a secondary effect of the shaking, but others, like the *quebradas* of the southern Andes or the "earthquake cracks" in the Colorado desert (Fig. 63), may have a deeper-seated origin. Many facts go to show, however, that though local expansion does occur in some localities, a surface contraction is a far more general consequence of earth movement. In civilized countries of high industrial development, where lines of metal of one kind or another run for long distances beneath or upon the surface of the ground, such general contraction of the surface may be easily proven. Comparatively seldom are lines of metal pulled apart in such a way as to show an expansion of the surface; whereas bucklings and kinkings of the lines appear in many places to prove that the area within which they are found has, as a whole, been reduced.

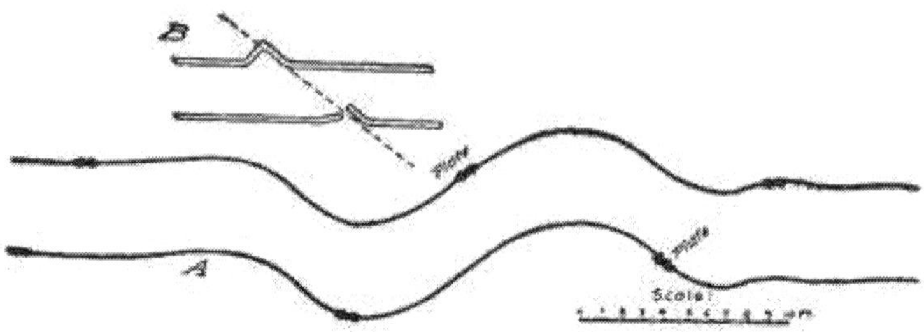

FIG. 64.—Diagrams to show how railway tracks are either broken or buckled locally within the district visited by an earthquake.

FIG. 65.—The Biwajima railroad bridge in Japan after the earthquake of 1891 (after Milne and Burton).

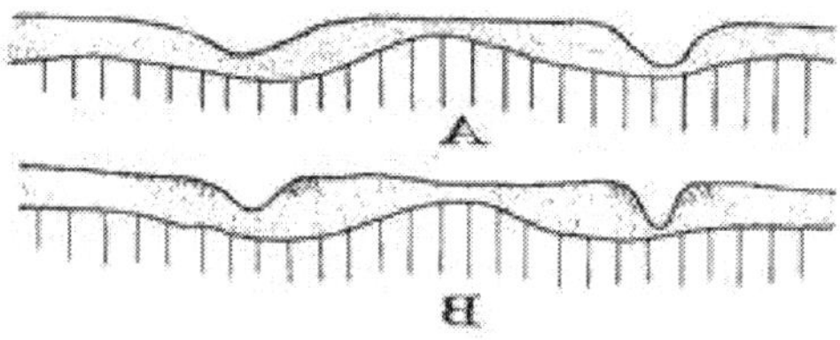

FIG. 66.—Diagrams to show how the compression of a district and its consequent contraction during an earthquake may close up the joint spaces within the rock basement and concentrate the contraction of the overlying mantle where this is partially cut through and so weakened in the valley sections.

Water pipes laid in the ground at a depth of some feet may be bowed up into an arch which appears above the surface; lines of curbing are raised into broken arches, and the tracks of railways are thrown into local loops and kinks which imply a very considerable local contraction of the surface (Fig. 64). With unvarying regularity railway or other bridges which cross rivers or ravines, if the structures are seriously damaged, indicate that the river banks have drawn nearer together at the time of the disturbance. In such cases, whenever the bridge girder has remained in place upon its abutments, these have either been broken or back-tilted as a whole in such a manner as to indicate an approach of the foundations which was prevented at the top by the stiffness of the girder (Fig. 65).

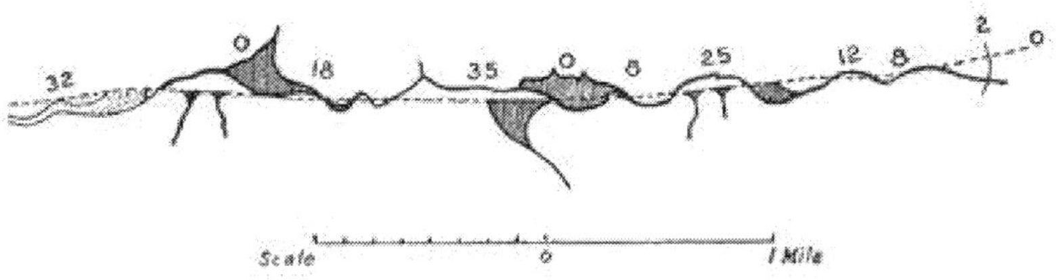

FIG. 67.—Map of the Chedrang fault which made its appearance during the Assam earthquake of 1897. The figures give the amounts of the local vertical displacement measured in feet (after R. D. Oldham).

The simplest explanation of such an approach of the banks at the sides of the valleys cut in loose surface material is to be found in a general closing up of the joint spaces within the underlying rock, and an adjustment of the mantle upon the floor mainly in the valley sections (Fig. 66).

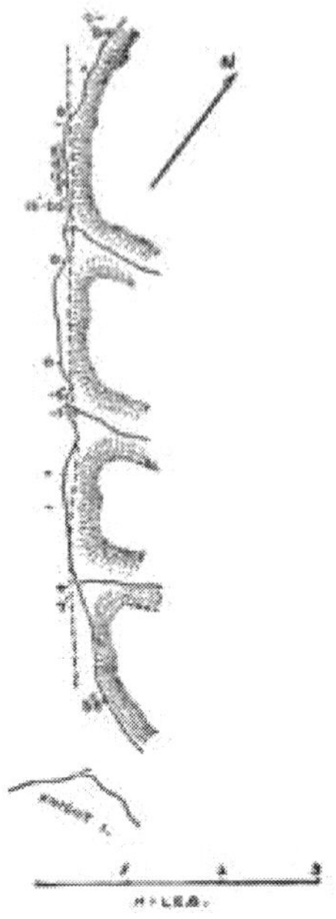

FIG. 68.—Map giving the displacements in feet measured along an earthquake fault formed in Alaska in 1899 (after Tarr and Martin).

The plan of an earthquake fault.—In our consideration of earthquake faults we have thus far given our attention to the displacement as viewed at a single locality only. Such displacements are, however, continued for many miles, and sometimes for hundreds of miles; and when now we examine a map or plan of such a line of faulting, new facts of large significance make their appearance. This may be well illustrated by a study of the plan of the Chedrang fault which appeared at the time of the Assam earthquake of 1897 (Fig. 67). From this map it will be noticed that the upward or downward displacement upon the perpendicular plane of the fault is not uniform, but is subject to large and *sudden* changes. Thus in order the measurements in feet are 32, 0, 18, 35, 0, 8, 25, 12, 8, 2, 0. The fault formed in 1899 upon the shores of Russell Fjord in Alaska (Fig. 68) reveals similar sudden changes of throw, only that here the direction of the movement is often reversed; or, otherwise expressed, the upthrow is suddenly transferred from one side of the fault to the other. Such abrupt changes in the direction of the displacement have been observed upon many earthquake faults, and a particularly striking one is represented in Fig. 69.

FIG. 69.—Abrupt change in the direction of throw upon an earthquake fault which was formed in the Owens valley, California, in 1872. The observer looks directly along the course of the fault from the left foreground to the cliff beyond and to the left of the impounded water (after a photograph by W. D. Johnson).

The block movements of the disturbed district.—The displacements upon earthquake faults are thus seen to be subdivided into sections, each of which differs from its neighbors upon either side and is sharply separated from them, at least in many instances. These points of abrupt change of displacement are, in many cases at least, the intersection points with transverse faults (Fig. 69). Such points of abrupt change in the degree or in the direction of the displacement may be, when looked at from above, abrupt turning points in the direction of extension of the fault, whose course upon the map appears as a zigzag line made up of straight sections connected by sharp elbows (Fig. 70).

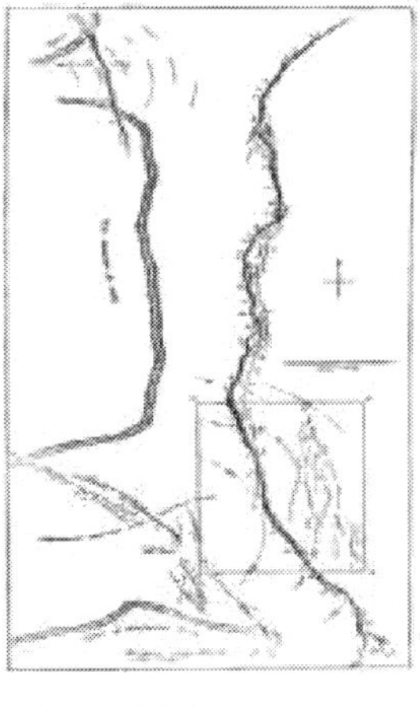

FIG. 70.—Map of the faults within an area of the Owens valley, California, formed in part during the earthquake of 1872, and in part due to early disturbances, In the western portions the displacements cut across firm rock and alluvial deposits alike without deviation of direction (after a map by W. D. Johnson).

Such a grouping of surface faults as are represented upon the map is evidence that the area of the earth's shell, which is included, has at the time of the earthquake been subject to adjustments as a series of separate units or blocks, certain of the boundaries of which are the fault lines represented. The changes in displacement measured upon the larger faults make it clear that the observed faults can represent but a fraction of the total number of lines of displacement, the others being masked by variations in the compactness of the loose mantling deposits. Could we but have this mantle removed, we should doubtless find a rock floor separated into parts like an ancient Pompeiian pavement, the individual blocks in which have been thrown, some upward and some downward, by varying amounts. Less than a hundred miles away to the eastward from the Owens Valley, a portion of this pavement has been uncovered in the extensive operations of the Tonapah Mining District, so that there we may study in all its detail the elaborate pattern of earth marquetry (Fig. 71) which for the floor of the Owens valley is as yet denied us.

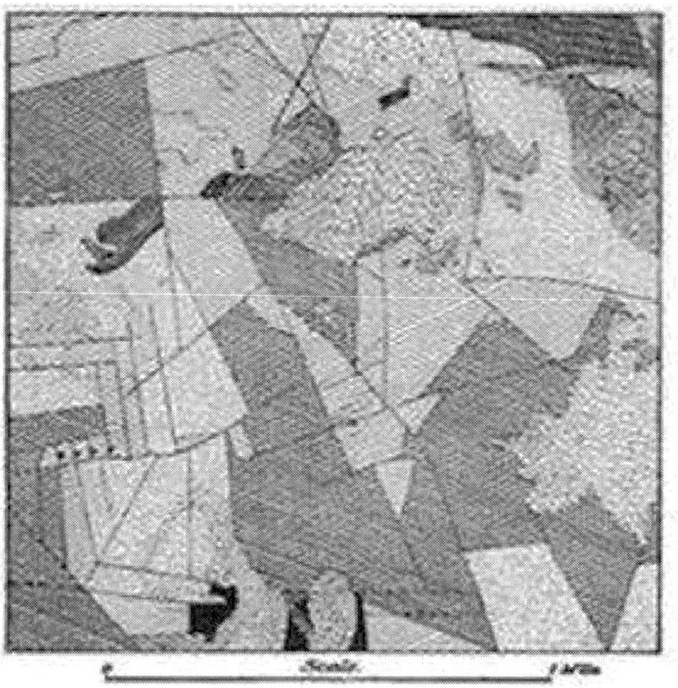

FIG. 71.—Marquetry of the rock floor of the Tonapah Mining District, Nevada (after Spurr).

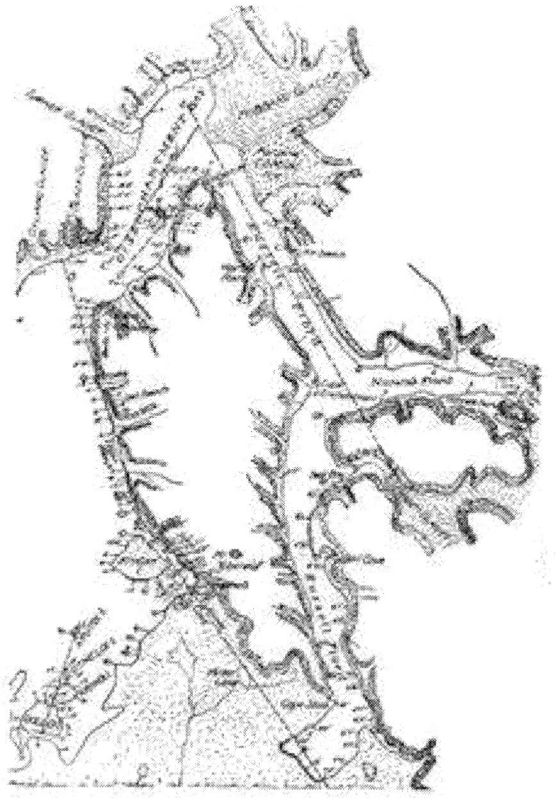

FIG. 72.—Map of a portion of the Alaskan coast to show the adjustments in level during the earthquake of 1899 (after Tarr and Martin).

The earth blocks adjusted during the Alaskan earthquake of 1899.—For a study of the adjustments which take place between neighboring earth blocks during a great earthquake, the recent Alaskan disturbance has offered the advantage that the most affected district was upon the seacoast, where changes of level could be referred to the datum of the sea's surface. Here a great island and large sections of the neighboring shore underwent movements both as a whole in large blocks and in adjustments of their subordinate parts among themselves (Fig. 72). Some sections of the coast were here elevated by as much as 47 feet, while neighboring sections were uplifted by smaller amounts (Fig. 73), and certain smaller sections were even dropped below the level of the sea.

FIG. 73.—View on Haencke Island, Disenchantment Bay, Alaska, revealing the shore that rose seventeen feet above the sea during the earthquake of 1899, and was found with barnacles still clinging to the rock (after Tarr and Martin).

The amount of such subsidence is, however, difficult to ascertain, for the reason that the former shore features are now covered with water and thus removed from observation. In favorable localities the minimum amount of submergence may sometimes be measured upon forest trees which are now flooded with sea water. In Fig. 74 a portion of the coast is represented where the beach sand is now extended back into the spruce forest, a distance of a hundred feet or more, and where sedgy beach grass is growing among trees whose roots are now laved in salt water. At the front of this forest the great storm waves overturn the trees and pile the wreckage in front of those that still remain standing.

FIG. 74.—Partially submerged forest upon the shore of Knight Island, Alaska, due to the sinking of a section of the coast during the earthquake of 1899 (after Tarr and Martin).

FIG. 75.—Settlement of a section of the shore at Port Royal, Jamaica, during the earthquake of January 14, 1907, adjacent to a similar but larger settlement of the near shore during the earthquake of 1692 (after a photograph by Brown).

Upon the glaciated rock surfaces of the Alaskan coast, exceptionally favorable opportunities are found for study of the intricate pattern of the earth mosaic which is under adjustment at the time of an earthquake. Upon Gannett Nunatak the surface was found divided by parallel faults into distinct slices which individually underwent small changes of level (plate 3 B).

CHAPTER VIII

THE INTERRUPTED CHARACTER OF EARTH MOVEMENTS: EARTHQUAKES AND SEAQUAKES (Concluded)

Experimental demonstration of earth movements.—The study of the Alaskan earthquake of 1899 showed that during this adjustment within the earth's shell some of the local blocks moved upward and by larger amounts than their neighbors, and that still others were actually depressed so that the sea flowed over them. It must be evident that such differential vertical movements of neighboring blocks at the earth's surface can only take place if lateral transfers of material are made beneath it. From under those strips of coast land which were depressed, material must have been moved so as to fill the void which would otherwise have formed beneath the sections that were uplifted. If we take into consideration much larger fractions upon the surface of our planet, we are taught by the great seaquakes which are now registered upon earthquake instruments at distant stations that large *downward* movements are to-day in progress beneath the sea much more than sufficient to compensate all extensions of the earth's surface within those districts where the land is rising in mountains. From under the offshore deeps of the ocean to beneath the growing mountains upon the shore, a transfer of earth material must be assumed to take place when disturbances are registered.

Within the time interval that separates the sudden adjustments of the surface which are manifested in earthquakes, the condition of strain which brings them about is steadily accumulating, due, as we generally assume, to earth contraction through loss of its heat. It seems probable that the resistance to an immediate adjustment is found in the rigidity of the shell because of the compression to which it is subjected. To illustrate: a row of blocks well fitted to each other may be held firmly as a bridge between the jaws of a vice, because so soon as each block starts to fall a large resistance from friction upon its surface is called into existence, a force which increases with the degree of compression.

It is thus possible upon this assumption crudely to demonstrate the adjustment of earth blocks by the simple device represented in plate 4 A. The construction of this experimental tank is so simple that little explanation is necessary. Wooden blocks of different heights are supported in water within a tank having a glass front, and are kept in a strained condition at other than their natural positions of flotation by the compression of a simple vice at the top. Held firmly in this position, they may thus represent the neighboring blocks within the earth's outer shell which are supported upon relatively yielding materials beneath, and prevented from at once adjusting themselves to their natural positions through the compression to which they are subjected. Held as they now are, the water near the ends of the tank is forced up beneath the blocks to higher than its natural level, and thus tends to flow from both ends toward the center. Such a movement would permit the end blocks to drop and force the middle ones to rise. The end blocks are, let us say, the sections of Alaskan coast line which sunk during the earthquake, as the center blocks are the sections which rose the full measure of 47 feet. Upon a larger scale the end blocks may equally well be considered as the floor of the great deeps off the Alaskan coast, whose sinking at the time of the earthquake was the cause of the great sea wave. Upon this assumption the center blocks would represent the Alaskan coast regarded as a whole, which underwent a general uplift.

Though we may not, in our experiment, vary the tendency to adjustment by any contractional changes in either the water or the blocks, we may reduce the compression of the vice, which leads to the same general result. As the compression of the vice is slowly relaxed, a point is at last reached at which friction upon the block surfaces is no longer sufficient to prevent an adjustment taking place, and this now suddenly occurs with the result shown in plate 4 B. In the case of the earth blocks, this sudden adjustment is accompanied by mass movements of the ground separated by faults, and these movements produce successional vibrations that are particularly large near the block margins, and other frictional vibrations of such small measure as to be generally appreciated by sounds only. The jolt of the adjustments has thrown some blocks beyond their natural position of rest, and these sink and rise subsequently in order to readjust themselves with lighter vibrations, which may be repeated and continued for some time. In the case of the earth these later adjustments are the so-called *aftershocks* which usually continue throughout a considerable period following every great earthquake. Gradually they fall off in intensity and frequency until they can no longer be felt, and are thereafter continued for a time as rumblings only.

PLATE 4.

A Experimental tank to illustrate the earth movements which are manifested in earthquakes. The sections of the earth's shell are here represented before adjustment has taken place.

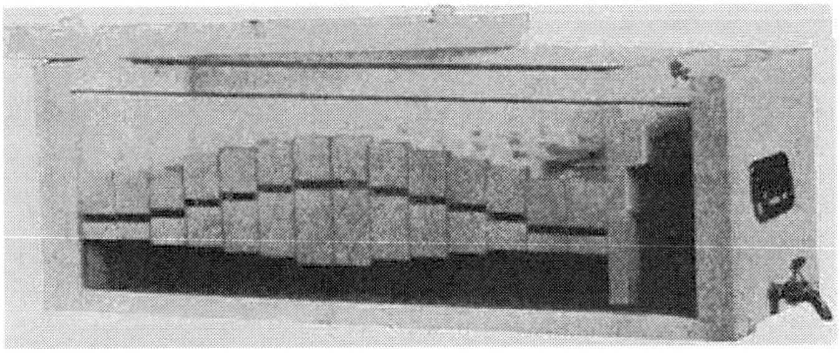

B. The same apparatus after a sudden adjustment.

C. Model to illustrate a block displacement in rocks which are intersected by master joints.

Derangement of water flow by earth movement.—The water which supported the blocks in our experiment has represented the more mobile portion of the earth's substance beneath its outer zone of fracture. The surface water layers in the tank may, however, be considered in a different way, since their behavior is remarkably like that of the water within and upon the earth's surface during an earth adjustment. At the instant when adjustment takes place in the tank, water frequently spurts upward from the cracks between the sinking end blocks; and if in place of one of the higher center blocks we insert one whose top is below the level of the water in the tank, a "lake" will be formed above it. When the adjustment occurs, this lake is immediately drained by outflow of the water at its bottom along one of the cracks between the blocks (Fig. 76).

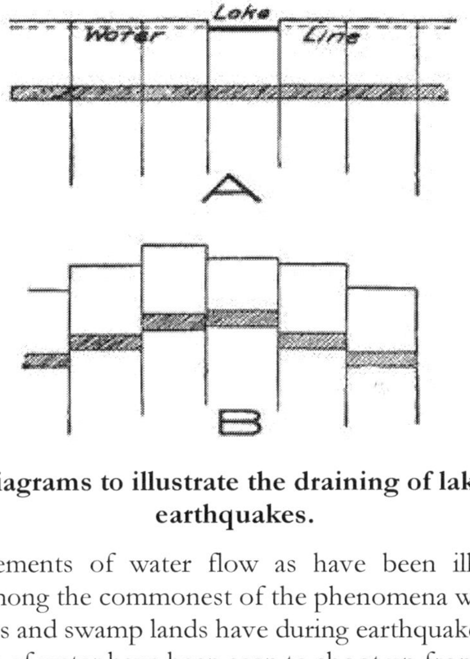

FIG. 76.—Diagrams to illustrate the draining of lakes during earthquakes.

Such derangements of water flow as have been illustrated by the experiment are among the commonest of the phenomena which accompany earthquakes. Lakes and swamp lands have during earthquakes been suddenly drained, fountains of water have been seen to shoot up from the surface and have played for some minutes or hours before their sudden disappearance in a sucking down of the water with later readjustment. During the great earthquake of the lower Mississippi valley in 1811, known as the New Madrid earthquake, the earlier Lake Eulalie was completely drained, and upon the now exposed bed there appeared parallel fissures on which were ranged funnel-like openings down which the water had been sucked. In other sections of the affected region the water shot up in sheets along fissures to the tops of high trees. Areas where such spurting up of the water has been observed have in most cases been shown to correspond to areas of depression, and such areas have sometimes been left flooded with water. During the Indian earthquake of 1819 an area of some 200 square miles suddenly sank and was transformed into a lake.

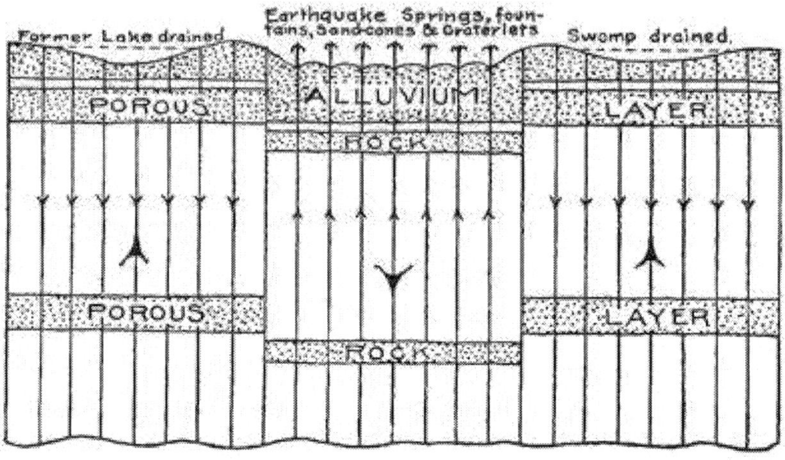

FIG. 77.—Diagram to illustrate-the derangements of flow of water at the time of an earthquake; water issuing at the surface over downthrown rocks, and being sucked down in upthrown blocks.

FIG. 78.—Mud cones aligned upon a fissure opened at Moraza, Servia, during the earthquake of April 4, 1904 (after Michailovitch).

Sand or mud cones and craterlets.—From a very moderate depth below the surface to that of several miles, all pore spaces and all larger openings within the rock are completely filled with water, the "trunk lines" of whose circulation is by way of the joints or along the bedding planes of the rocks. The principal reservoirs, so to speak, of this water inclosed within the rock are the porous sand formations. When, now, during an earthquake a block of the earth's shell is suddenly sunk and as suddenly arrested in its downward movement, the effect is to compress the porous layers and so force the contained water upward along the joints to the surface, carrying with it large quantities of the sand (Fig. 77).

FIG. 79.—One of the many craterlets formed near Charleston, South Carolina, during the earthquake of August 31, 1886. The opening is twenty feet across, and the leaves about it are encased in sand as were those upon the branches of the overhanging trees to a height of some twenty feet (after Dutton).

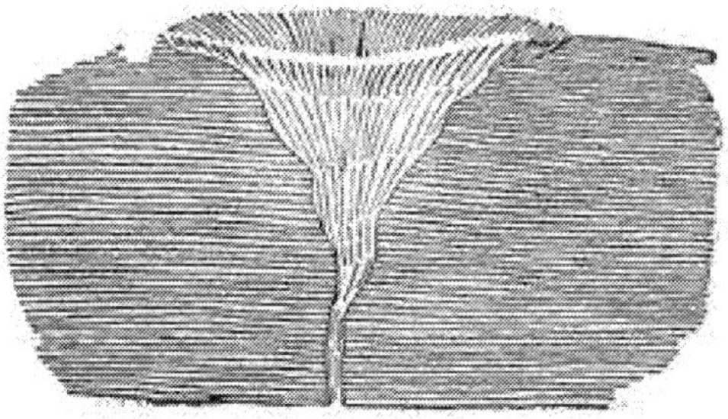

FIG. 80.—Cross section of a craterlet to show the trumpet-like form of the sand column.

Ejected at the surface this water appears in fountains usually arranged in line over joints, or even in continuous sheets, and the sand collecting about the jets builds up lines of *sand* or *mud cones* sometimes described as "mud volcanoes" (Fig. 78). The amount of sand thus poured out is sometimes so great that blankets of quicksand are spread over large sections of the country. Most frequently, however, the sand is not built above the general level of the surface, but forms a series of *craterlets* which are largely shaped as the water is sucked down at the time of the readjustment with which the play of such earthquake fountains is terminated (Fig. 79). Subsequent excavations made about such craterlets have shown them to have the form of a trumpet, and that in the sand which so largely fills them there are generally found scales of mica and such light bodies as would be picked out from the heterogeneous materials of the sand layers and carried upward in the rush of water to the surface (Fig. 80).

The earth's zones of heavy earthquake.—Since earthquakes give notice of a change of level of the ground, the special danger zones from this source are the growing mountain systems which are usually found near the borders of the sea. Such lines of mountains are to-day rising where for long periods in the past were the basins of deposition of former seas. They thus represent the zones upon the earth's surface which are the most unstable—which in the recent period have undergone the greatest changes of level.

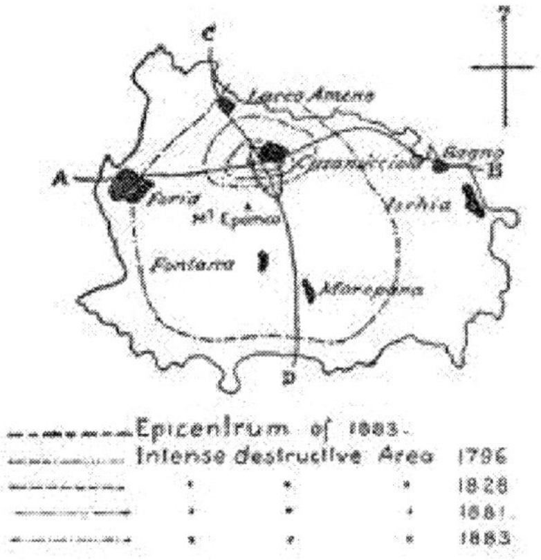

FIG. 81.—Map of the island of Ischia to show how the shocks of recent earthquakes have been concentrated at the crossing point of two fractures (after Mercalli and Johnston-Lavis).

By far the most unstable belt upon the earth's surface is the rim surrounding the Pacific Ocean, within which margin it has been estimated that about 54 per cent of the recorded shocks of earthquake have occurred. Next in importance for seismic instability is the zone which borders both the Mediterranean Sea and the Caribbean—the American Mediterranean—and is extended across central Asia through the Himalayas into Malaysia. Both zones approximate to great circles upon the earth's surface and intersect each other at an angle of about 67°. It has been estimated that about 95 per cent of the recorded continental earthquakes have emanated from these belts.

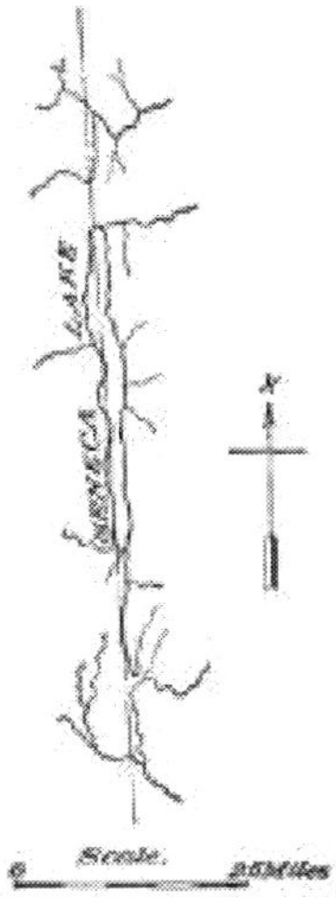

FIG. 82.—A line of earth fracture indicated in the plan of the relief, which may at any time become the seat of movement and resultant shock.

The special lines of heavy shock.—Within any earthquake district the shocks are not felt with equal severity at all places, but there are, on the contrary, definite lines which the disturbance seems to search out for special damage. From their relations to the relief of the land these lines would appear to be lines of fracture upon the boundaries of those sections of the crust that play individual rôles in the block adjustment which takes place. More or less masked as these lines are beneath the rounded curves of the landscape, they are given an altogether unenviable prominence with each succeeding earthquake. At such times we may think of the earth's surface as specially sensitized for laying bare its hidden structure, as is the sensitized plate under the magical influence of the X rays.

When, at the time of an earthquake, blocks are adjusted with reference to their neighbors, the movements of oscillation are greatest in those marginal portions of direct contact. Corners of blocks—the intersecting points of the important faults—should for the same reason be shaken with a double violence, and this assumption appears to be confirmed by observation. Upon the island of Ischia, off the Bay of Naples, the shocks from recent earthquakes have been strangely concentrated near the town of Casamicciola, which was last destroyed in 1883. This unfortunate city lies at the crossing point of important fractures whose course upon the island is marked by numerous springs and *suffioni* (Fig. 81).

Seismotectonic lines.—The lines of important earth fractures, as will be more clearly shown in the sequel (p. 227), are often indicated with some clearness by straight lines in the plan of the surface relief (Fig. 82). Lines of this nature are easily made out upon the map of the West Indies, and if we represent upon it by circles of different diameters the combined intensities of the recorded earthquakes in the various cities, it appears that the heavily shaken localities are ranged upon lines stamped out in the relief, with the most severely damaged places at their intersections (Fig. 83). These lines of exceptional instability are known as *seismotectonic lines*—earthquake structure lines.

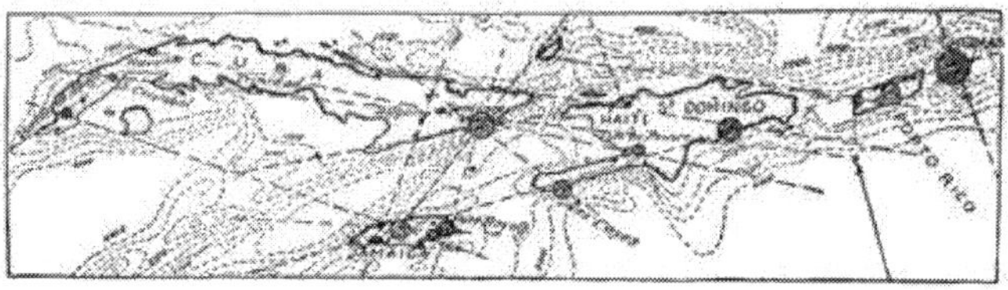

FIG. 83.—Seismotectonic lines of the West Indies.

The heavy shocks above loose foundations.—It is characteristic of faults that they soon bury themselves from sight under loose materials, and are thus made difficult of inspection. The escarpment which is the direct consequence of a vertical displacement upon a fault tends to migrate from the place of its formation, rounding the surface as it does so and burying the fault line beneath its deposits (Fig. 43, p. 60).

This is not, however, the sole reason why loose foundations should be places of special danger at the time of earth shocks, for the reason that earthquake waves are sent out in all directions from the surfaces of displacement through the medium of the underlying rock. These waves travel within the firm rock for considerable distances with only a gradual dissipation of their energy, but with their entry into the loose surface deposits their energy is quickly used up in local vibrations of large amplitude, and hence destructive to buildings.

FIG. 84.—Device to illustrate the different effects upon the transmission and the character of shocks which are produced by firm rock and by loose materials.

The essential difference between firm rock and such loose materials as are found upon a river bottom or in the "made land" about our cities may be illustrated by the simple device which is represented in Fig. 84. Two similar metal pans are suspended from a firm support by bands of steel and "elastic" braid of similar size and shape, and carry each a small block of wood standing upon its end. Similar light blows are now administered directly to the pans with the effect of upsetting that block which is supported by the loose braid because of the large range or amplitude of movement that is imparted to the pan. The "elastic" braid, because of these large vibrations of which it is susceptible, may represent the loose materials when an earthquake wave passes into them. In the case of the steel support, the energy of the blow, instead of being dissipated in local swingings of the pan, is to a large extent transmitted through the elastic metal to materials beyond. The steel thus resembles in its high elasticity the firmer rock basement, which receives and transmits the earthquake shocks, but except when ruptured in a fault is subject to vibrations of small amplitude only.

Construction in earthquake regions.—Wherever earthquakes have been felt, they are certain to occur again; and wherever mountains are growing or changes of level are in progress, there no record of past earthquakes is required in order to forecast the future seismic history. Although the future earthquakes may be predicted, the time of their coming is, fortunately or unfortunately, still hidden from us. If one's lot is to be cast in an earthquake country, the only sane course to pursue is to build with due regard to future contingencies.

The danger, from destructive fires may to-day be largely met by methods of construction which levy an additional burden of cost. Though the danger from seismic disturbances can hardly be met as fully as that from fire, yet it is true that buildings may be so constructed as to withstand all save those heaviest shocks in the immediate vicinity of the lines of large displacement. Here, also, a considerable additional expense is involved in the method of construction, in the case of residences particularly.

From what has been said, it is obvious that much of the danger from earthquakes can be met by a choice of site away from lines of important fracture and from areas of relatively loose foundation. The choice of building materials is next of importance. Those buildings which succumb to earthquakes are in most cases racked or shaken apart, and thus they become a prey to their own inherent properties of inertia. Each part of a structure may be regarded as a weight which is balanced upon a stiff rod and pivoted upon the ground. When shocks arrive, each part tends to be thrown into vibration after the manner of an inverted pendulum. In proportion, therefore, as the weights are large and rest upon long supports, the danger of overthrow and of tearing apart is increased. In general, structures are best constructed of light materials whose weight is concentrated near the ground. Masonry structures, and especially high ones, are, therefore, the least suited for resisting earthquakes, of which the late complete destruction of the city of Messina is a grewsome reminder. Despite repeated warnings in the past, the buildings of that stricken city were generally constructed of heavy rubble, which in addition had been poorly cemented (Fig. 49, p. 67). Such structures are usually first ruptured at the edges and corners, since here the vibrations which tend to tear the building asunder are resisted by no supports and are reënforced from neighboring walls.

FIG. 85.—House wrecked in San Francisco earthquake of 1906 because the floors and partitions were not securely fastened to the walls (after R. L. Humphrey).

An advantage of the first importance is evidently secured if the rods of the pendulum, of which the building is conceived to be composed, have sufficient elasticity to be considerably distorted without rupture and to again recover their original position. This is the supreme advantage of structural steel for all large buildings, which is coupled, however, with the disadvantage that the riveted fastenings are apt to be quickly sheered off under the vibrations. Large and high buildings, when sufficiently elastic, have fortunately the property of destroying the earth waves by interference before they have traveled above the lower stories.

For large structures in which wood cannot be used, strongly reënforced concrete is well adapted, for it has in general the same advantages as steel with somewhat reduced elasticity, but with a more effective binding together of the parts. This requirement of thorough bracing and tying together of the several parts of a building causes it to vibrate, not as many pendulums, but as one body. If met, it removes largely the danger from racking strains, and for small structures particularly it is the requirement which is most easily complied with. For such buildings it is therefore necessary that the framework should be built in a close network with every joint firmly braced and with all parts securely tied together. Especial attention should be given to the fastenings of floor and partition ends. The house shown in Fig. 85 could not have been subjected to heavy shocks, for though the walls are thrown down, the floors and partitions have been left near their original positions.

FIG. 86.—Building wrecked at San Mateo, California, during the late earthquake. The heavy roof and upper floor, acting as a unit, have battered down the upper walls (after J. C. Branner).

This tendency of the walls, floors, partitions, and roof to act as individual units in the vibration, is one that must be reckoned with and be met by specially effective bracing and tying at the junctions. Otherwise these larger parts of the structure may act like battering rams to throw over the walls or portions of them (Fig. 86).

READING REFERENCES FOR CHAPTERS VII AND VIII

General works:—

JOHN MILNE. Seismology. London, 1898, pp. 320.

C. E. DUTTON. Earthquakes in the Light of the New Seismology. Putnam, New York, 1904, pp. 314.

A. SIEBERG. Handbuch der Erdbebenkunde. Braunschweig, 1904, pp. 362.

COUNT F. DE MONTESSUS DE BALLORE. Les Tremblements de Terre, Géographie Séismologique. Paris, 1906, pp. 475; La Science Séismologique. Paris, 1907, pp. 579.

WILLIAM HERBERT HOBBS. Earthquakes, an Introduction to Seismic Geology. Appleton, New York, 1907, pp. 336.

C. G. KNOTT. The Physics of Earthquake Phenomena. Clarendon Press, Oxford, 1908, pp. 283.

E. RUDOLPH. Ueber Submarine Erdbeben und Eruptionen, Beiträge zur Geophysik, vol. 1, 1887, pp. 133-365; vol. 2, 1895, pp. 537-666; vol. 3, 1898, pp. 273-536.

Descriptive reports of some important earthquakes:—

C. E. DUTTON. The Charleston Earthquake of August 31, 1886, 9th Ann. Rept. U. S. Geol. Surv., 1889, pp. 203-528.

B. KOTÔ. On the Cause of the Great Earthquake in Central Japan, 1891, Jour. Coll. Sci. Imp. Univ., Tokyo, Japan, vol. 5, 1893, pp. 295-353, pls. 28-35.

JOHN MILNE and W. K. BURTON. The Great Earthquake of Central Japan. 1891, pp. 10, pls. 30.

R. D. OLDHAM. Report on the Great Earthquake of 12th June, 1897, Mem. Geol. Surv. India. Vol. 29, 1899, pp. 379, pls. 42.

A. C. LAWSON, and others. The California Earthquake of April 18, 1906, Report of the State Earthquake Investigation Commission, three quarto vols. (Carnegie Institution of Washington); many plates and figures.

Italian Photographic Society, Messina and Reggio before and after the Earthquake of December 28, 1908 (an interesting collection of pictures). Florence, 1909.

R. S. TARR and L. MARTIN. Recent Changes of Level in the Yakutat Bay Region, Alaska, Bull. Geol. Soc. Am., vol. 17, 1906, pp. 29-64, pls. 12-23.

WILLIAM HERBERT HOBBS. The Earthquake of 1872 in the Owens Valley, California, Beiträge zur Geophysik, vol. 10, 1910, pp. 352-385, pls, 10-23.

Faults in connection with earthquakes:—

WILLIAM H. HOBBS. On Some Principles of Seismic Geology, Beiträge zur Geophysik, vol. 8, 1907, Chapters iv-v.

Expansion or contraction of the earth's surface during earthquakes:—

WILLIAM H. HOBBS. A Study of the Damage to Bridges during Earthquakes, Jour. Geol., vol. 16, 1908, pp. 636-653; The Evolution and the Outlook of Seismic Geology, Proc. Am. Phil. Soc., vol. 48, 1909, pp. 27-29.

Earthquake construction:—

JOHN MILNE. Construction in Earthquake Countries, Trans. Seis. Soc., Japan, vol. 14, 1889-1890, pp. 1-246.

F. DE MONTESSUS DE BALLORE. L'art de bâtir dans les pays à tremblements de terre (34th Congress of French Architects), L'Architecture, 193 Année, 1906, pp. 1-31.

GILBERT, HUMPHREY, SEWELL, and SOULÉ. The San Francisco Earthquake and Fire of April 18, 1906, and their Effects on Structures and Structural Materials, Bull. 324, U. S. Geol. Surv., 1907, pp. 1-170, pls. 1-57.

WILLIAM H. HOBBS. Construction in Earthquake Countries, The Engineering Magazine, vol. 37, 1909, pp. 1-19.

LEWIS ALDEN ESTES. Earthquake-proof Construction, a discussion of the effects of earthquakes on building construction with special reference to structures of reënforced concrete, published by Trussed Concrete Steel Company. Detroit, 1911, pp. 46.

CHAPTER IX

THE RISE OF MOLTEN ROCK TO THE EARTH'S SURFACE

VOLCANIC MOUNTAINS OF EXUDATION

Prevalent misconceptions about volcanoes.—The more or less common impression that a volcano is a "burning mountain" or a "smoking mountain" has been much fostered by the school texts in physical geography in use during an earlier period. The best introduction to a discussion of volcanoes is, therefore, a disillusionment from this notion. Far from being burning or smoking, there is normally no combustion whatever in connection with a volcanic eruption. The unsophisticated tourist who, looking out from Naples, sees the steam cap which overhangs the Vesuvian crater tinged with brown, easily receives the impression that the material of the cloud is smoke. Even more at night, when a bright glow is reflected to his eye and soon fades away, only to again glow brightly after a few moments have passed, is it difficult to remove the impression that one is watching an intermittent combustion within the crater. The cloud which floats away from the crest of the mountain is in reality composed of steam with which is admixed a larger or smaller proportion of fine rock powder which gives to the cloud its brownish tone. The glow observed at night is only a reflection from molten lava within the crater, and the variation of its brightness is explained by the alternating rise and fall of the lava surface by a process presently to be explained.

Not only is there no combustion in connection with volcanic eruptions, but so far as the volcano is a mountain it is a product of its own action. The grandest of volcanic eruptions have produced no mountains whatever, but only vast plains or plateaus of consolidated molten rock, and every volcanic mountain at some time in its history has risen out of a relatively level surface.

When the traditional notions about volcanoes grew up, it was supposed that the solid earth was merely a "crust" enveloping still molten material. As has already been pointed out in an earlier chapter, this view is no longer tenable, for we now know that the condition of matter within the earth's interior, while perhaps not directly comparable to any that is known, yet has properties most resembling known matter in a solid state; it is much more rigid than the best tool steel. While there must be reservoirs of molten rock beneath active volcanoes, it is none the less clear that they are small, local, and temporary. This is shown by the comparative study of volcanic outlets within any circumscribed district.

It is perhaps not easy to frame a definition of a volcano, but its essential part, instead of being a mountain, is rather a vent or channel which opens up connection between a subsurface reservoir of molten rock and the surface of the earth. An eruption occurs whenever there is a rise of this material, together with more or less steam and admixed gases, to the surface. Such molten rock arriving at the surface is designated *lava.* The changes in pressure upon this material during its elevation induce secondary phenomena as the surface is approached, and these manifestations are often most awe inspiring. While often locally destructive, the geological importance of such phenomena is by reason of their terrifying aspect likely to be greatly exaggerated.

Early views concerning volcanic mountains.—As already pointed out, a volcano at its birth is not a mountain at all, but only, so to speak, a shaft or channel of communication between the surface and a subterranean reservoir of molten rock. By bringing this melted rock to the surface there is built up a local elevation which may be designated a mountain, except where the volume of the material is so large and is spread to such distances as to produce a plain (see fissure eruptions below).

In the early history of geology it was the view of the great German geologist von Buch and his friend and colleague von Humboldt, that a volcanic mountain was produced in much the same manner as is a blister upon the body. The fluids which push up the cuticle in the blister were here replaced by fluid rock which elevated the sedimentary rock layers at the surface into a dome or mound which was open at the top—the so-called *crater.* This "elevation-crater" theory of volcanoes long held the stage in geological science, although it ignored the very patent fact that the layers on the flanks of volcanic cones are not of sedimentary rock at all, but, on the contrary, of the volcanic materials which are brought up to the surface during the eruption. The observational phase of science was, however, dawning, and the English geologists Scrope and Lyell were able to show by study of volcanic mountains that the mound about the volcanic vent was due to the accumulation of once molten rock which had been either exuded or ejected. Making use of data derived from New Zealand, Scrope showed that, instead of being elevated during the formation of a volcanic mountain, the sedimentary strata of the vicinity may be depressed near the volcanic vent (Fig. 87).

FIG. 87.—Breached volcanic cone near Auckland, New Zealand, showing the bending down of the sedimentary strata in the neighborhood of the vent (after Heaphy and Scrope).

The birth of volcanoes.—To confirm the impression that the formation of the volcanic mountain is in reality a secondary phenomenon connected with eruptions, we may cite the observed birth of a number of volcanoes. On the 20th of September, 1538, a new volcano, since known as Monte Nuovo (new mountain), rose on the border of the ancient Lake Lucrinus to the westward of Naples. This small mountain attained a height of 440 feet, and is still to be seen on the shore of the bay of Naples. From Mexico have been recorded the births of several new volcanoes: Jorullo in 1759, Pochutla in 1870, and in 1881 a new volcano in the Ajusco Mountains about midway between the Gulf of Mexico and the Pacific Ocean. The latest of new volcanoes is that raised in Japan on November 9, 1910, in connection with the eruption of Usu-san. This "New Mountain" reached an elevation of 690 feet.

FIG. 88.—View of the new Camiguin volcano from the sea. It was formed in 1871 over a nearly level plain. The town of Catarman appears at the right near the shore (after an unpublished photograph by Professor Dean C. Worcester).

As described by von Humboldt, Jorullo rose in the night of the 28th of September, 1759, from a fissure which opened in a broad plain at a point 35 miles distant from any then existing volcano. The most remarkable of new volcanoes rose in 1871 on the island of Camiguin northward from Mindanao in the Philippine archipelago. This mountain was visited by the *Challenger* expedition in 1875, and was first ascended and studied thirty years later by a party under the leadership of Professor Dean C. Worcester, the Secretary of the Interior of the Philippine Islands, to whom the writer is indebted for this description and the accompanying illustration of this largest and most interesting of new-born volcanoes. As in the case of Jorullo, the eruption began with the formation of a fissure in a level plain, some 400 yards distant from the town of Catarman (Fig. 88). The eruption continued for four years, at the end of which time the height of the summit was estimated by the *Challenger* expedition to be 1900 feet. At the time of the first ascent in 1905, the height was determined by aneroid as 1750 feet, with sharp rock pinnacles projecting some 50 or 75 feet higher.

Active and extinct volcanoes.—The terms "active" and "extinct" have come into more or less common use to describe respectively those volcanoes which show signs of eruptive activity, and those which are not at the time active. The term "dormant" is applied to volcanoes recently active and supposed to be in a doubtfully extinct condition. From a well-known volcano in the vicinity of Naples, volcanoes which no longer erupt lava or cinder, but show gaseous emanations (*fumeroles*) are said to be in the *solfatara* condition, or to show *solfataric* activity.

Experience shows that the term "extinct", while useful, must always be interpreted to mean apparently extinct. This may be illustrated by the history of Mount Vesuvius, which before the Christian era was forested in the crater and showed no signs of activity; and in fact it is known that for several centuries no eruption of the volcano had taken place. Following a premonitory earthquake felt in the year 63, the mountain burst out in grand explosive eruption in 79 A.D. This eruption profoundly altered the aspect of the mountain and buried the cities of Pompeii, Stabeii, and Herculaneum from sight. Once more, this time during the middle ages, for nearly five centuries (1139 to 1631) there was complete inactivity, if we except a light ash eruption in the year 1500. During this period of rest the crater was again forested, but the repose was suddenly terminated by one of the grandest eruptions in the mountain's history.

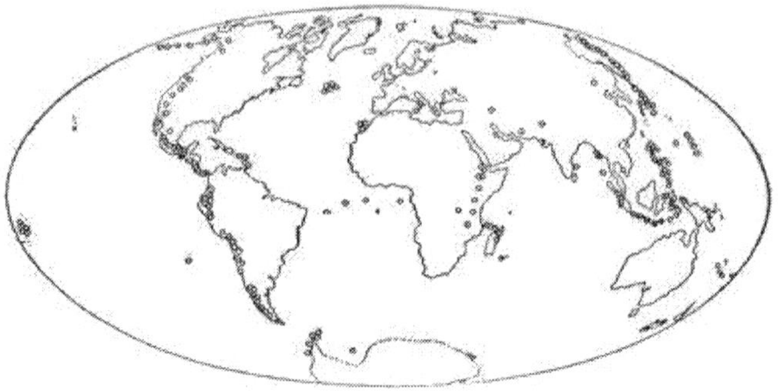

FIG. 89.—Map showing the location of the belts of active volcanoes.

The earth's volcano belts.—The distribution of volcanoes is not uniform, but, on the contrary, volcanic vents appear in definite zones or belts, either upon the margins of the continents or included within the oceanic areas (Fig. 89). The most important of these belts girdles the Pacific Ocean, and is represented either by chains or by more widely spaced volcanic mountains throughout the Cordilleran Mountain system of South and Central America and Mexico, by the volcanoes of the Coast and Cascade ranges of North America, the festooned volcanic chain of the Aleutian Islands, and the similar island arcs off the eastern coast of the Eurasian continent. The belt is further continued through the islands of Malaysia to New Zealand, and on the Pacific's southern margin are found the volcanoes of Victoria Land, King Edward Land, and West Antarctica.

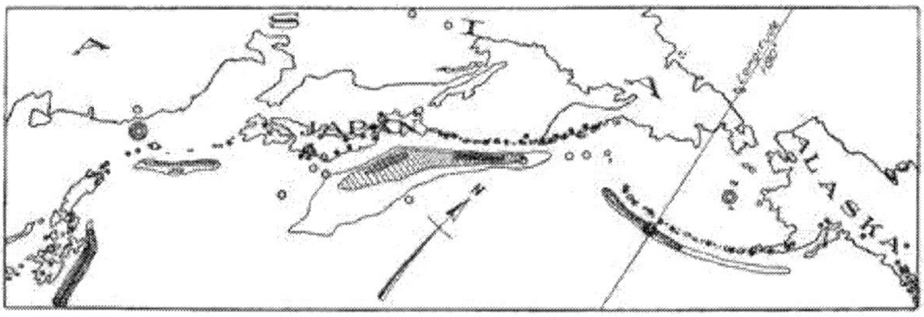

FIG. 90.—A portion of the "fire girdle" of the Pacific, showing the relation of the chains of volcanic mountains to the deeps of the neighboring ocean floor.

This volcano girdle is by no means a perfect one, for in addition to the principal festoons of the western border there are many secondary ones, and still other arcs are found well toward the center of the oceanic area. Another broad belt of volcanoes borders the Mediterranean Sea, and is extended westward into the Atlantic Ocean. Narrower belts are found in both the northern and southern portions of the Atlantic Ocean, on the margins of the Caribbean Sea, etc. The fact of greatest significance in the distribution seems to be that bands of active volcanoes are to be found wherever mountain ranges are paralleled by deeps on the neighboring ocean floor (Fig. 90). As has been already pointed out in the chapter upon earthquakes, it is just such places as these which are the seat of earthquakes; these are zones of the earth's crust which are undergoing the most rapid changes of level at the present time. Thus the rise of the land in mountains is proceeding simultaneously with the sinking of the sea floor to form the neighboring deeps.

FIG. 91.—Volcanic cones formed in 1783 above the Skaptár fissure in Iceland (after Helland).

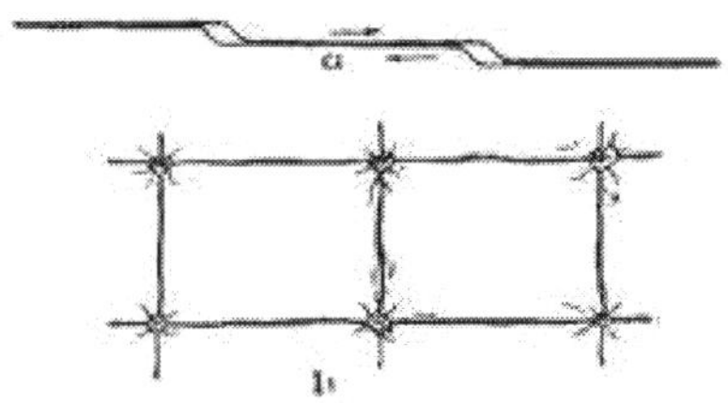

FIG. 92.—Diagrams to illustrate the location of volcanic vents upon fissure lines, *a*, openings caused by lateral movement of fissure walls; *b*, openings formed at fissure intersections.

Arrangement of volcanic vents along fissures and especially at their intersections.—Within those districts in which volcanoes are widely separated from their neighbors, the law of their arrangement is difficult to decipher, but the view that volcanic vents are aligned over fissures is now supported by so much evidence that illustrations may be supplied from many regions. An exceptionally perfect line of small cones is found along the Skaptár cleft in Iceland, upon which stands the large volcano of Laki. This fissure reopened in 1783, and great volumes of lava were exuded. Over the cleft there was left a long line of volcanic cones (Fig. 91). There are in Iceland two dominating series of parallel fissures of the same character which take their directions respectively northeast-southwest and north-south. Many such fissures are traceable at the surface as deep and nearly straight clefts or *gjás*, usually a few yards in width, but extending for many miles. The Eldgjá has a length of more than 18 English miles and a depth varying from 400 to

600 feet. On some of these fissures no lava has risen to the surface, whereas others have at numerous points exuded molten rock. Sometimes one end only of a fissure, the more widely gaping portion, has supplied the conduits for the molten lava. This is well illustrated by the cratered monticules raised by the common ant over the cracks which separate the blocks of cement sidewalk, the hillocks being located where the most favorable channel was found for the elevation of the materials.

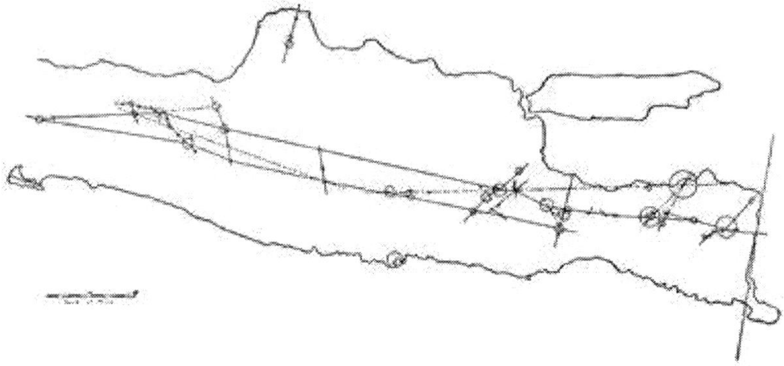

FIG. 93.—Outline map of the eastern portion of the island of Java, displaying the arrangement of volcanic vents in alignment upon fissures with the larger mountains at fissure intersections (after Verbeek).

Those places upon fissures which become lava conduits appear to be the ones where the cleft gapes widest so as to furnish the widest channel. Wherever a differential lateral movement of the walls has occurred, openings will be found in the neighborhood of each minor variation from a straight line (Fig. 92 *a*). Wherever there are two or more series of fissures, and this would appear to be the normal condition, places favorable for lava conduits occur at fissure intersections. Within such veritable volcano gardens as are to be found in Malaysia, the law of volcano distribution became apparent so soon as accurate maps had been prepared. Thus the outline map of a portion of the island of Java (Fig. 93) shows us that while the volcanoes of the island present at first sight a more or less irregular band or zone, there are a number of fissures intersecting in a network, and that the volcanoes are aligned upon the fissures with the larger cones located at the intersections. So also in Iceland, the great eruption of Askja in 1875 occurred at the intersection of two lines of fissure.

Outside these closely packed volcanic regions, similar though less marked networks are indicated; as, for example, in and near the Gulf of

Guinea. If now, instead of reducing the scale of our volcano maps, we increase it, the same law of distribution is no less clearly brought out. The monticules or small volcanic cones which form upon the flanks of larger volcanic mountains are likewise built up over fissures which on numerous occasions have been observed to open and the cones to form upon them.

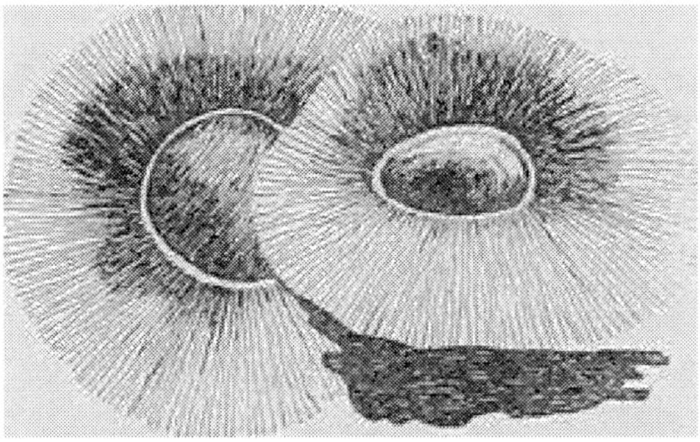

FIG. 94.—Map of the Puy Pariou in the Auvergne of central France. The seat of eruption has migrated along the fissure upon which the earlier cone had been built up (after Scrope).

Still further reducing now the area of our studies and considering for the moment the "frozen" surface of the boiling lava within the caldron of Kilauea, this when observed at night reveals in great perfection the sudden formation of fissures in the crust with the appearance of miniature volcanoes rising successively at more or less regular intervals along them.

It not infrequently happens that after a volcanic vent has become established above some conduit in a fissure, the conduit migrates along the fissure, thus establishing a new cone with more or less complete destruction of the old one (Fig. 94).

The so-called fissure eruptions.—The grandest of all volcanic eruptions have been those in which the entire length and breadth of the fissures have been the passageway for the upwelling lava. Such grander eruptions have been for the most part prehistoric, and in later geologic history have occurred chiefly in India, in Abyssinia, in northwestern Europe, and in the northwestern United States. In western India the singularly horizontal plateaus of basaltic lava, the Dekkan traps, cover some 200,000 square miles and are more than a mile in depth. The underlying basement where it appears about the margins of the basalt is in many places intersected by dikes or fissure fillings of the same material. No cones or definite vents have been found.

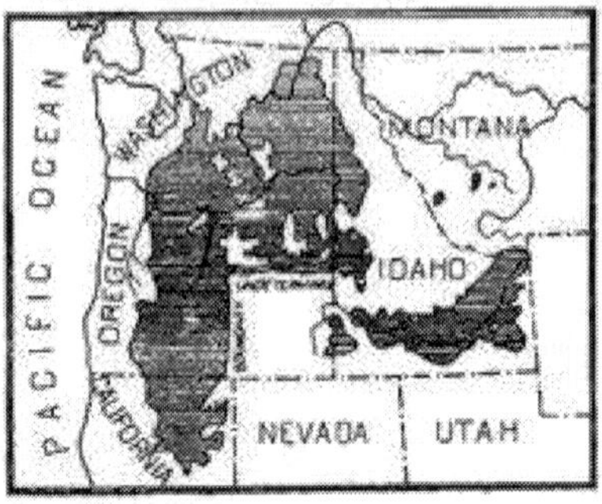

FIG. 95.—Basaltic plateau of the northwestern United States due to fissure eruptions of lava.

The larger portion of the northwestern British Isles would appear to have been at one time similarly blanketed by nearly horizontal beds of basaltic lava, which beds extended northwestward across the sea through the Orkney and Faroe islands to Iceland. Remnants of this vast plateau are to-day found in all the island groups as well as in large areas of northeastern Ireland, and fissure fillings of the same material occur throughout large areas of the British Isles. In many cases these dikes represent once molten rock which may never have communicated with the surface at the time of the lava outpouring, yet they well illustrate what we might expect to find if the basalt sheets of Iceland or Ireland were to be removed.

The floods of basaltic lava which in the northwestern United States have yielded the barren plateau of the Cascade Mountains (Fig. 95) would appear to offer another example of fissure eruption, though cones appear upon the surface and perhaps indicate the position of lava outlets during the later phases of the eruptive period. The barrenness and desolation of these lava plains is suggested by Fig. 96.

FIG. 96.—Lava plains about the Snake River in Idaho.

Though the greater effusions of lava have occurred in prehistoric times, and the manner of extrusion has necessarily been largely inferred from the immense volume of the exuded materials and the existence of basaltic dikes in neighboring regions, yet in Iceland we are able to observe the connection between the dikes and the lava outflows. Professor Thoroddsen has stated that in the great basaltic plateau of Iceland, lava has welled out quietly from the whole length of fissures and often on both sides without giving rise to the formation of cones. At three wider portions of the great Eld cleft, lava welled out quietly without the formation of cones, though here in the southern prolongation of the fissure, where it was narrower, a row of low slag cones appeared. Where the lava outwellings occurred, an area of 270 square miles was flooded.

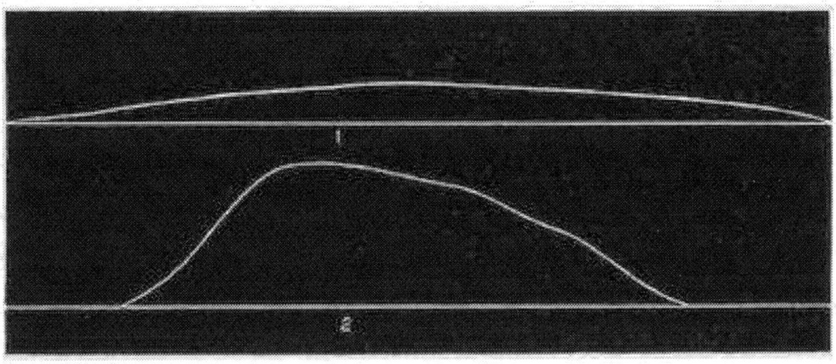

FIG. 97.—Characteristic profiles of lava volcanoes. 1, basaltic lava mountain; 2, mountain of siliceous lava (after Judd).

The composition and the properties of lava.—In our study of igneous rocks (Chapter IV) it was learned that they are composed for the most part of silicate minerals, and that in their chemical composition they represent various proportions of silica, alumina, iron, magnesia, lime, potash, and soda. The more abundant of these constituents is silica, which varies from 35 to 70 per cent of the whole. Whenever the content of silica is relatively low,—basic or basaltic lava,—the cooled rock is dark in color and relatively heavy. It melts at a relatively low temperature, and is in consequence relatively fluid at the temperatures which lavas usually have on reaching the earth's surface. Furthermore, from being more fluid, the water which is nearly always present in large quantity within the lava more readily makes its escape upon reaching the surface. Eruptions of such lava are for this reason without the violent aspects which belong to extrusions of more siliceous (more "acidic") lavas. For the same reason, also, basaltic lava flows more freely and can spread much farther before it has cooled sufficiently to consolidate. This is equivalent to saying that its surface will assume a flatter angle of slope, which in the case of basaltic lava seldom exceeds ten degrees and may be less than one degree (Fig. 97).

FIG. 98.—A driblet cone (after J. D. Dana).

Siliceous lavas, on the other hand, are, when consolidated, relatively light both in color and weight and melt at relatively high temperatures. They are, therefore, usually but partly fused and of a viscous consistency when they arrive at the earth's surface. Because of this viscosity they offer much resistance to the liberation of the contained water, which therefore is released only to the accompaniment of more or less violent explosions. The lava is blown into the air and usually falls as consolidated fragments of various degrees of coarseness.

FIG. 99.—View of Leffingwell crater, a cinder cone in the Owens valley, California (after an unpublished photograph by W. D. Johnson).

It must not, however, be assumed that the temperature of lava is always the same when it arrives at the surface, and hence it may happen that a siliceous lava is exuded at so high a temperature that it behaves like a normal basaltic lava. On the other hand, basaltic lavas may be extruded at unusually low temperatures, in which case their behavior may resemble that of the normal siliceous lavas. If, however, as is generally the case, the energy of explosion of a basaltic lava is relatively small, any ejected portions of the liquid lava travel to a moderate height only in the air, so that on falling they are still sufficiently pasty to adhere to rock surfaces and thus build up the remarkably steep cones and spines known as "spatter cones" or "driblet cones" (Fig. 98). When, on the other hand, the energy of explosion is great, as is normally the case with siliceous lavas, the portions of ejected lava have been fully consolidated before their fall to the surface, so that they build up the same type of accumulation as would sand falling in the same manner. The structures which they form are known as tuff, cinder, or ash cones (Fig. 99).

Whenever the contained water passes off from siliceous lavas without violent explosions, the lava may flow from the vent, but in contrast to basaltic lavas it travels a short distance only before consolidating. The resulting mountain is in consequence proportionately high and steep (Fig. 97). Eruptions characterized by violent explosions accompanied by a fall of cinder are described as *explosive* eruptions. Those which are relatively quiet, and in which the chief product is in the form of streams of flowing lava, are spoken of as *convulsive* eruptions.

The three main types of volcanic mountain.—If the eruptions at a volcanic vent are exclusively of the explosive type, the material of the mountain which results is throughout tuff or cinder, and the volcano is described as a *cinder cone.* If, on the other hand, the vent at every eruption exudes lava, a mountain of solid rock results which is a *lava dome.* It is, however, the exception for a volcano which has a long history to manifest but a single kind of eruption. At one time exuding lava comparatively quietly, at another the violence with which the steam is liberated yields only cinder, and the mountain is a composite of the two materials and is known as a *composite volcanic cone.*

The lava dome.—When successive lava flows come from a crater, the structure which results has the form of a more or less perfect dome. If the lava be of the basaltic or fluid type, the slopes are flat, seldom making an angle of as much as ten degrees with the horizon and flatter toward the summit (Fig. 101, p. 106). If of siliceous or viscous lava, on the other hand, the slopes are correspondingly steep and in some cases precipitous. To this latter class belong some of the *Kuppen* of Germany, the *puys* of central France, and the *mamelons* of the Island of Bourbon.

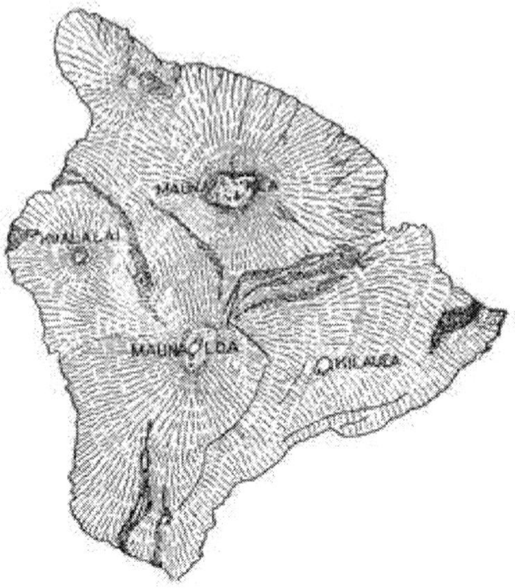

FIG. 100.—Map of Hawaii and the lava volcanoes of Mokuaweoweo (Mauna Loa) and Kilauea (after the government map by Alexander).

The basaltic lava domes of Hawaii.—At the "crossroads of the Pacific" rises a double line of lava volcanoes which reach from 20,000 to 30,000 feet above the floor of the ocean, some of them among the grandest volcanic mountains that are known. More than half the height and a much larger proportion of the bulk of the largest of these are hidden beneath the ocean's surface. The two great active vents are Mokuaweoweo (on Mauna Loa) and Kilauea, distinct volcanoes notwithstanding the fact that their lava extravasations have been merged in a single mass. The rim of the crater of Mauna Loa is at an elevation of 13,675 feet above the sea, whereas that of Kilauea is less than 4000 feet and appears to rest upon the flank of the larger mountain (Figs. 100 and 101). Although one crater is but 20 miles distant from the other and nearly 10,000 feet lower, their eruptions have apparently been unsympathetic. Nowhere have still active lava mountains been subjected to such frequent observations extending throughout a long period, and the dynamics of their eruptions are fairly well understood. To put this before the reader, it will be best to consider both mountains, for though they have much in common, the observations from one are strangely complementary to those of the other. The lower crater being easily accessible, Kilauea has been often visited, and there exists a long series of more or less consecutive observations upon it, which have been assembled and studied by Dana and Hitchcock. The place of outflow of the Kilauea lavas has not generally been visible, whereas Mokuaweoweo has slopes rising nearly 14,000 feet above the sea and displays the records of outflow of many eruptions, some of which were accompanied by the grandest of volcanic phenomena.

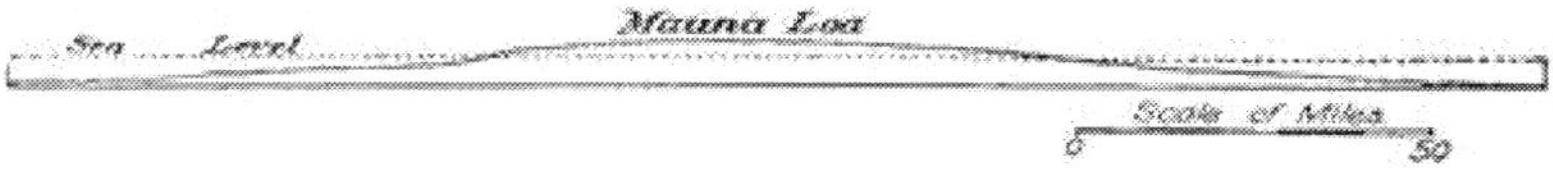

FIG. 101.—Section through Mauna Loa and Kilauea.

Lava movements within the caldron of Kilauea.—The craters of these mountains are the largest of active ones, each being in excess of seven miles in circumference. In shape they are irregularly elliptical and consist of a series of steps or terraces descending to a pit at the bottom, in which are open lakes of boiling lava. Enough is known of the history of Kilauea to state that the steep cliffs bounding the terraces are fault walls produced by inbreak of a frozen lava surface. The cliff below the so-called "black ledge" was produced by the falling in of the frozen lava surface at the time of the outflow of 1840, the lava issuing upon the eastern flank of the mountain and pouring into the sea near Nanawale. Since that date the floor of the pit below the level of this ledge has been essentially a movable platform of frozen lava of unknown and doubtless variable thickness which has risen and descended like the floor of an elevator car between its guiding ways (Fig. 102). The floor has, however, never been complete, for one or more open lakes are always to be seen, that of Halemaumau located near the southwestern margin having been much the most persistent. Within the open lakes the boiling lava is apparently white hot at the depth of but a few inches below the surface, and in the overturnings of the mass these hotter portions are brought to the surface and appear as white streaks marking the redder surface portions. From time to time the surface freezes over, then cracks open and erupt at favored points along the fissures, sending up jets and fountains of lava, the material of which falls in pasty fragments that build up driblet cones. Small fluid clots are shot out, carrying a threadlike line of lava glass behind them, the well-known "Pelé's hair." Sometimes the open lakes build up congealed walls, rising above the general level of the pit, and from their rim the lava spills over in cascades to spread out upon the frozen floor, thus increasing its thickness from above (Fig. 103). At other times a great dome of lava has been pushed up from the pit of Halemaumau under a frozen shell, the molten lava shining red through cracks in its surface and exuding so as to heal each widely opened fissure as it forms.

FIG. 102.—Schematic diagram to illustrate the moving platform of frozen lava which rises and falls in the crater of Kilauea.

At intervals of from a few years to nine or ten years the crater has been periodically drained, at which times the moving platform of frozen lava has sunk more or less rapidly to levels far below the black ledge and from 900 to 1700 feet below the crater rim. Following this descent a slow progressive rise is inaugurated, which has sometimes gone on at a rate of more than a hundred feet per year, though it is usually much slower than this. When the platform has reached a height varying from 700 to 350 feet below the crater rim, another sudden settlement occurs which again carries the pit floor downward a distance of from 300 to 700 feet.

FIG. 103.—View of the open lava lake of Halemaumau within the crater of Kilauea, the molten lava shown cascading over the raised lava walls on to the floor of the pit (after Pavlow).

The draining of the lava caldrons.—The changes which go on within the crater of Mokuaweoweo, though less studied than those of Kilauea, appear to be in some respects different. Here every eruption seems to be preceded by a more or less rapid influx of melted lava to the pit of the crater, this phenomenon being observed from a distance as a brilliant light above the crater—the reflection of the glow from overhanging vapor clouds. The uprising of the lava has often been accompanied by the formation of high lava fountains upon the surface, and the molten lava sometimes appears in fissures near the crater rim at levels well above the lava surface within the pit.

Although in many cases the lava which has thus flooded the crater has suddenly drained away without again becoming visible, it is probable that in such cases an outlet has been found to some submarine exit, since under-ocean discharge effects have been observed in connection with eruptions of each of the volcanoes.

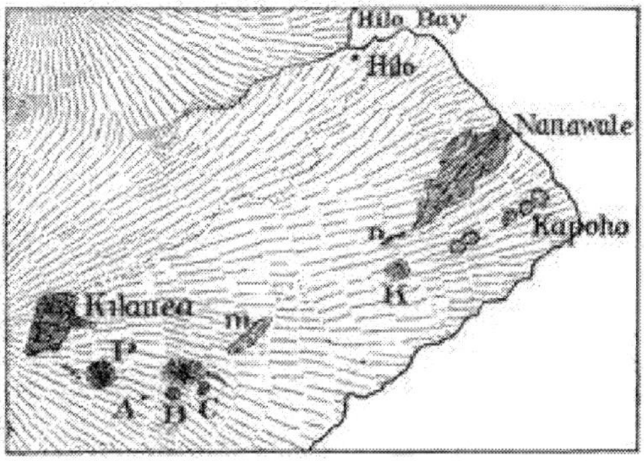

FIG. 104.—Map showing the manner of outflow of lava from Kilauea during the eruption of 1840. The outflowing lava made its appearance successively at the points *A*, *B*, *C*, *m*, *n*, and finally at a point below *n*, from whence it issued in volume and flowed down to the sea at Nanawale (after J. D. Dana).

Inasmuch as no earthquakes are felt in connection with such outflows as have been described, it is probable that the hot lava fuses a passageway for itself into some open channel underneath the flanks of the mountain. Such a course is well illustrated by the outflow of Kilauea in 1840, when, it will be remembered, occurred the great down-plunge of the crater that yielded the pit below the black ledge. At this time the lava first made its appearance upon the flanks of the mountain at the bottom of a small pit or inbreak crater which opened five miles southeast of the main crater of Kilauea (Fig. 104). Within this new crater the lava rose, and small ejections soon followed from fissures formed in its neighborhood. Some time after, the lava sank in the first new crater, only to reappear successively at other small openings (Fig. 104, *B*, *C*, *m*, *n*) and finally to issue in volume at a point eleven miles from the shore and flow thereafter *upon the surface* of the mountain until it had reached the sea. Only the slightest earth tremors were felt, and as no rumblings were heard, it is evident that the lava fused its way along a buried channel largely open at the time (see below, p. 112).

In a majority of the eruptions of Mokuaweoweo, when the outflowing lavas have become visible, the molten rock has apparently fused its way out to the surface of the mountain at points from 1000 to 3000 feet below the bottom of the crater, and this discharge has corresponded in time to the lowering of the lava surface within the crater. There are, however, three instances upon record in which the lava issued from definite rents which were formed upon the mountain flanks at comparatively low levels. In contrast to the formation of fused outlets, these ruptures of a portion of the mountain's flank were always accompanied by vigorous local earthquakes of short duration. In one instance (the eruption of 1851) such a rent appeared under the same conditions but at an elevation of 12,500 feet, or near the level of the lava in the crater.

The outflow of the lava floods.—In order to properly comprehend these and many otherwise puzzling phenomena connected with volcanoes, it is necessary to keep ever in mind the quite remarkable heat-insulating property of congealed lava. So soon as a thin crust has formed upon the surface of molten rock, the heat of the underlying fluid mass is given off with extreme slowness, so that lava streams no longer connected with their internal lava reservoirs may remain molten for decades.

FIG. 105.—Lava of Matavanu upon the Island of Savaii flowing down to the sea during the eruption of 1906. The course may be followed by the jets of steam escaping from the surface down to the great steam cloud which rises where the fluid lava discharges into the sea (after H. I. Jensen).

We have seen that for Mokuaweoweo and Kilauea, lava either quietly melts its way to the surface at the time of outflow, or else produces a rent for its egress to the accompaniment of vigorous local earthquakes. In either case if the lava issues at a point far below the crater, gigantic lava fountains arise at the point of outflow, the fluid rock shooting up to heights which range from 250 to 600 or more feet above the surface. A certain proportion of this fluid lava is sufficiently cooled to consolidate while traveling in the air, and falling, it builds up a cinder cone which is left as a location monument for the place of discharge. From this outlet the molten lava begins its journey down the slope of the mountain, and quickly freezes over to produce a tunnel, beneath the roof of which the fluid lava flows with comparatively slow further loss of heat. Save for occasional steam jets issuing from its surface, it may give little indication of its presence until it has reached the sea (Fig. 105).

FIG. 106.—Lava stream discharging into the sea from beneath the frozen roof of a lava tunnel. Eruption of Matavanu on Savaii in 1906 (after Sapper).

If sufficient in volume and the shore be not too distant, the stream of lava arrives at the sea, where, discharging from the mouth of its tunnel, it throws up vast volumes of steam and induces ebullition of the water over a wide area (Fig. 106). Professor Dana, who visited Hawaii a few months only after the great outflow of 1840, states that the lava, upon reaching the ocean, was shivered like melted glass and thrown up in millions of particles which darkened the sky and fell like hail over the surrounding country. The light was so bright that at a distance of forty miles fine print could be read at midnight.

FIG. 107.—Diagrammatic representation of the structure of the flanks of lava volcanoes as a result of the draining of frozen lava streams.

Protected from any extensive consolidation by its congealed cover, the lava within a stream may all drain away, leaving behind an empty lava tunnel, which in the case of the Hawaiian volcanoes sometimes has its roof hung with beautiful lava stalactites and its floor studded with thin lava spines. Later lava outflows over the same or neighboring courses bury such tunnels beneath others of similar nature, giving to the mountain flanks an elongated cellular structure illustrated schematically in Fig. 107. These buried channels may in the future be again utilized for outflows similar in character to that of Kilauea in 1840.

FIG. 108.—Diagram to show the manner of formation of mesas or table mountains by the outflow of lava in valleys and the subsequent more rapid erosion of the intervening ridges. *R*, earlier river valley; *R'R'*, later valleys.

While the formation of lava stalactites of such perfection and beauty is peculiar to the Hawaiian lava tunnels, the formation of the tunnel in connection with lava outflow is the rule wherever a dissipation at the end has permitted of drainage. A few hours only after the flow has begun, the frozen surface has usually a thickness of a few inches, and this cover may be walked over with the lava still molten below. At first in part supported by the molten lava, the tunnel roof sometimes caves in so soon as drainage has occurred.

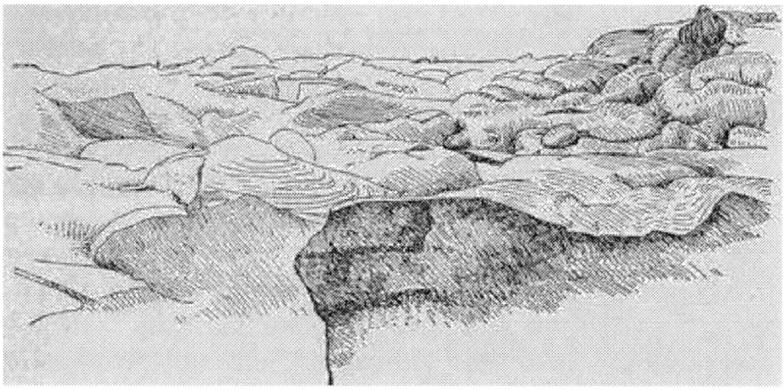

FIG. 109.—Surface of lava of the Pahoehoe type.

Wherever basaltic lava has spread out in valleys on the surface of more easily eroded material, either cinder or sedimentary formations, the softer intervening ridges are first carried away by the eroding agencies, leaving the lava as cappings upon residual elevations. Thus are derived a type of table mountain or *mesa* of the sort well illustrated upon the western slopes of the Sierra Nevadas in California (Fig. 108).

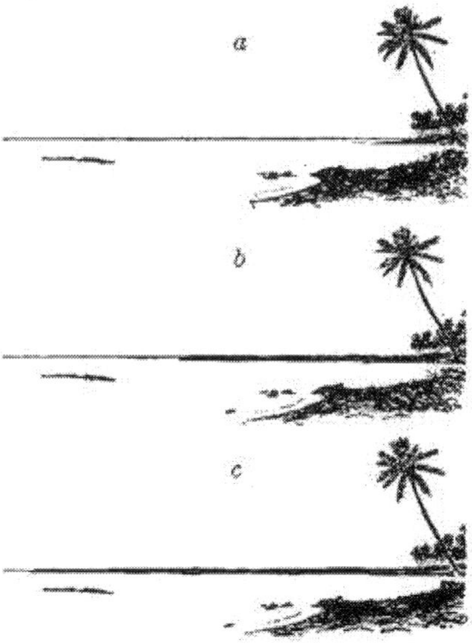

FIG. 110.—Three successive views to illustrate the growth of the Island of Savaii from the outflow of lava at Matavanu in the year 1906. *a*, near the beginning of the outflow; *b*, some weeks later than *a*; *c*, some weeks later than *b* (after H. I. Jensen).

The surface which flowing lava assumes, while subject to considerable variation, may yet be classified into two rather distinct types. On the one hand there is the billowy surface in which ellipsoidal or kidney-shaped masses, each with dimensions of from one to several feet, lie merged in one another, not unlike an irregular collection of sofa pillows. This type of lava has become known as the *Pahoehoe*, from the Hawaiian occurrence (Fig. 109). A variation from this type is the "corded" or "ropy" lava, the surface of which much resembles rope as it is coiled along the deck of a vessel, the coils being here the lines of scum or scoriæ arranged in this manner by the currents at the surface of the stream (Fig. 123, p. 124). A quite different type is the block lava (*Aa* type) which usually has a ragged scoriaceous surface and consists of more or less separate fragments of cooled lava (Fig. 131, p. 130).

Wherever lava flows into the sea in quantity, it extends the margin of the shore, often by considerable areas. The outflow of Kilauea in 1840 extended the shore of Hawaii outward for the distance of a quarter of a mile, and a more recent illustration of such extension of land masses is furnished by Fig. 110.

CHAPTER X

THE RISE OF MOLTEN ROCK TO THE EARTH'S SURFACE

VOLCANIC MOUNTAINS OF EJECTED MATERIALS

The mechanics of crater explosions.—If we now turn from the lava volcano to the active cinder cone, we encounter an entire change of scene. In place of the quiet flow and convulsive movements of the molten lava, we here meet with repeated explosions of greater or less violence. If we are to profitably study the manner of the explosions, considering the volcanic vent as a great experimental apparatus, it would be well to select for our purpose a volcano which is in a not too violent mood. The well-known cinder cone of Stromboli in the Eolian group of islands north of Sicily has, with short and unimportant interruptions, remained in a state of light explosive activity since the beginning of the Christian era. Rising as it does some three thousand feet directly out of the Mediterranean, and displaying by day a white steam cap and an intermittent glow by night, its summit can be seen for a distance of a hundred miles at sea and it has justly been called the "Lighthouse of the Mediterranean." The "flash" interval of this beacon may vary from one to twenty minutes, and it may show, furthermore, considerable variation of intensity.

For the reason that the crater of the mountain is located at one side and at a considerable distance below the actual summit, the opportunity here afforded of looking into the crater is most favorable whenever the direction of the wind is such as to push aside the overhanging steam cloud (Fig. 111). Long ago the Italian vulcanologist Spallanzani undertook to make observations from above the crater, and many others since his day have profited by his example.

Within the crater of the volcano there is seen a lava surface lightly frozen over and traversed by many cracks from which vapor jets are issuing. Here, as in the Kilauea crater, there are open pools of boiling lava. From some of these, lava is seen welling out to overflow the frozen surface; from others, steam is ejected in puffs as though from the stack of a locomotive. Within others lava is seen heaving up and down in violent ebullition, and at intervals a great bubble of steam is ejected with explosive violence, carrying up with it a considerable quantity of the still molten lava, together with its scumlike surface, to fall outside the crater and rattle down the mountain's slope into the sea. Following this explosion the lava surface in the pool is lowered and the agitation is renewed, to culminate after the further lapse of a few minutes in a second explosion of the same nature. The rise of the lava which precedes the ejection appears at night as a brighter reflection or glow from the overhanging steam cloud—the flash seen by the mariner from his vessel.

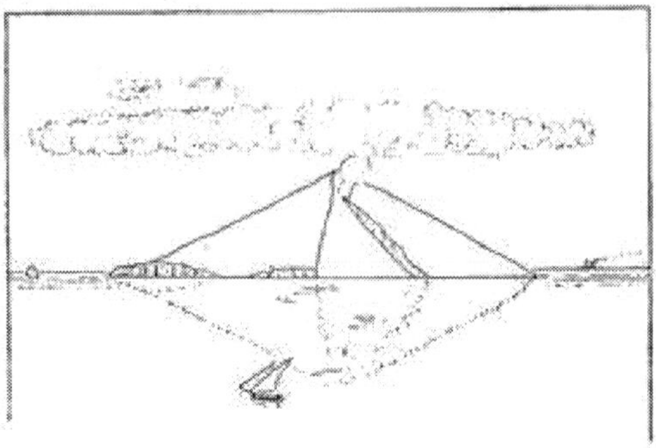

FIG. 111.—The volcano of Stromboli, showing the excentric position of the crater (after a sketch by Judd).

What is going on within the crater of Stromboli we may perhaps best illustrate by the boiling of a stiff porridge over a hot fire. Any one who has made corn mush over a hot camp fire is fully aware that in proportion as the mush becomes thicker by the addition of the meal, it is necessary to stir the mass with redoubled vigor if anything is to be retained within the kettle. The thickening of the mush increases its viscosity to such an extent that the steam which is generated within it is unable to make its escape unless aided by openings continually made for it by the stirring spoon. If the stirring motion be stopped for a moment, the steam expands to form great bubbles which soon eject the pasty mass from the kettle.

For the crater of Stromboli this process is illustrated by the series of diagrams in Fig. 112. As the lava rises toward the surface, presumably as a result of convectional currents within the chimney of the volcano, the contained steam is relieved from pressure, so that at some depth below the surface it begins to separate out in minute vesicles or bubbles, which, expanding as they rise, acquire a rapidly accelerating velocity. Soon they flow together with a quite sudden increase of their expansive energy, and now shooting upward with further accelerated velocity, a layer of liquid lava with its cover of scum is raised on the surface of a gigantic bubble and thrown high into the air. Cooled during their flight, the quickly congealed lava masses become the tuff or volcanic ash which is the material of the cinder cone.

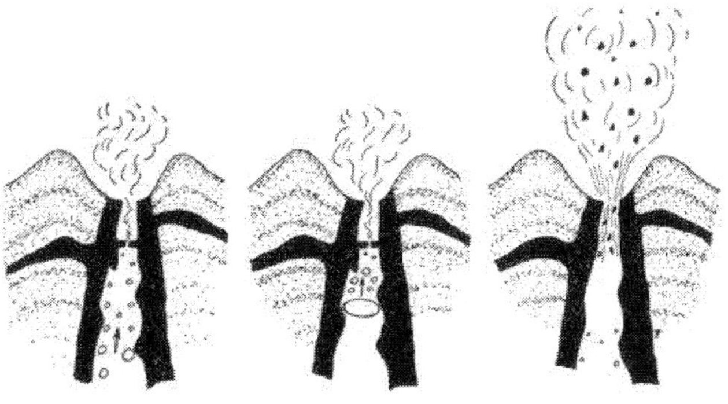

FIG. 112.—Diagrams to illustrate the nature of eruptions within the crater of Stromboli.

Grander volcanic eruptions of cinder cones.—Most cinder and composite cones, in the intervals between their grander eruptions, if not entirely quiescent, lapse into a period, of light activity during which their crater eruptions appear to be in all essential respects like the habitual explosions within the Strombolian crater. This phase of activity is, therefore, described as *Strombolian*. By contrast, the occasional grander eruptions which have punctuated the history of all larger volcanoes are described in the language of Mercalli as *Vulcanian* eruptions, from the best studied example.

Just what it is that at intervals brings on the grander Vulcanian outburst within a volcano is not known with certainty; but it is important to note that there is an approach to periodicity in the grander eruptions. It is generally possible to distinguish eruptions of at least two orders of intensity greater than the Strombolian phase; a grander one, the examples of which may be separated by centuries, and one or more orders of relatively moderate intensity which recur at intervals perhaps of decades, their time intervals subdividing the larger periods marked off by the eruptions of the first order.

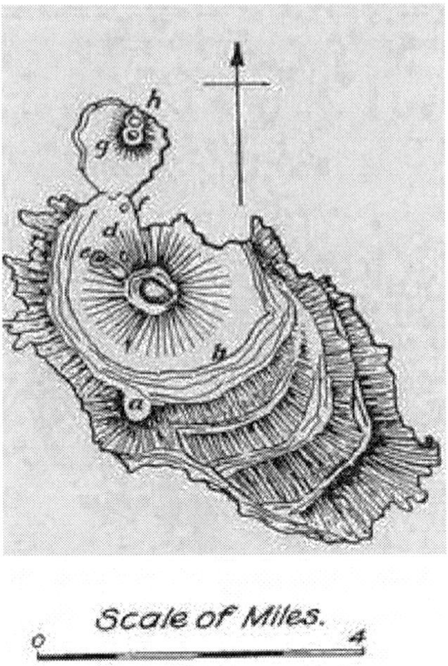

FIG. 113.—Map of Volcano in the Eolian group of islands. The smaller craters partially dissected by the waves belong to Vulcanello (after Judd).

The eruption of Volcano in 1888.—In the Eolian Islands to the north of Sicily was located the mythical forge of Vulcan. From this locality has come our word "volcano", and both the island and the mountain bear no other name to-day (Fig. 113). There is in the structure of the island the record of a somewhat complex volcanic history, but the form of the large central cinder cone was, according to Scrope, acquired during the eruption of 1786, at which time the crater is reported to have vomited ash for a period of fifteen days. Passing after this eruption into the solfatara condition, with the exception of a light eruption in 1873, the volcano remained quiet until 1886. So active had been the fumeroles within the crater during the latter part of this period that an extensive plant had been established there for the collection especially of boracic acid. In 1886 occurred a slight eruption, sufficient to clear out the bottom of the crater, though not seriously to disturb the English planter whose vineyards and fig orchards were in the valley or *atrio* near the point *d* upon the map (Fig. 113), nearly a mile from the crater rim. On the 3d of August, 1888, came the opening discharge of an eruption, which, while not of the first order of magnitude, was yet the greatest in more than a century of the mountain's history, and may serve us to illustrate the Vulcanian phase of activity within a cinder cone. During the day, to the accompaniment of explosions of considerable violence, projectiles fell outside the crater rim and rolled down the steep slopes toward the *atrio*. These explosions were repeated at intervals of from twenty to thirty minutes, each beginning in a great upward rush of steam and ash, accompanied by a low rumbling sound. During the following night the eruptions increased in violence, and the anxious planter remained on watch in his villa a mile from the crater. Falling asleep toward morning, he was rudely awakened by a rain of projectiles falling upon his roof. Hastily snatching up his two children he ran toward the door just as a red hot projectile, some two feet in diameter, descended through the roof, ceiling, and floor of the drawing room, setting fire to the building. A second projectile similar to the first was smashed into fragments at his feet as he was emerging from the house, burning one of the children. Making his escape to Vulcanello at the extremity of the island, the remainder of the night and the following day, until rescue came from Lipari, were spent just beyond the range of the falling masses.

FIG. 114.—"Bread-crust" lava projectile from the eruption of Volcano in 1888 (after Mercalli).

When the writer visited the island some months later, the eruption was still so vigorous that the crater could not be reached. The ruined villa, smashed and charred, stood with its walls half buried in ash and lapilli, among which were partly smashed pumiceous lava projectiles. The entire *atrio* about the mountain lay buried in cinder to the depth of several feet and was strewn with projectiles which varied in size from a man's fist to several feet in diameter (Fig. 114). The larger of these exhibited the peculiar "bread-crust" surface and had generally been smashed by the force of their fall after the manner of a pumpkin which has been thrown hard against the ground. One of these projectiles fully three feet in diameter was found at the distance of a mile and a half from the crater. Though diminished considerably in intensity, the rhythmic explosions within the crater still recurred at intervals varying from four minutes to half an hour, and were accompanied by a dull roar easily heard at Lipari on a neighboring island six miles away. Simultaneously, a dark cloud of "smoke", the peculiar "cauliflower cloud" or *pino* mounted for a couple of miles above the crater (Fig. 115), and the rise was succeeded by a rain of small lava fragments or *lapilli* outside the crater rim.

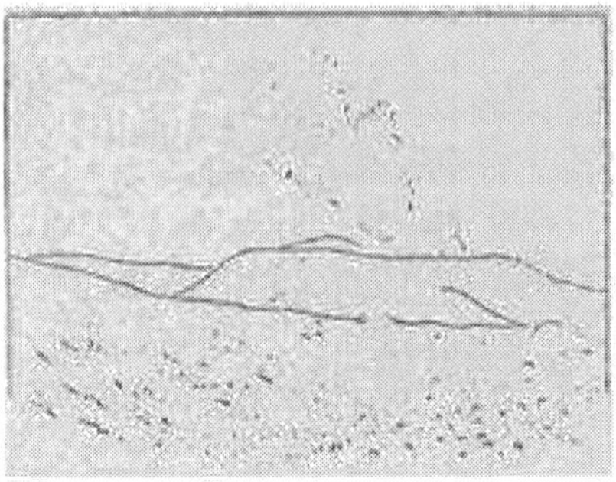

FIG. 115.—Peculiar "cauliflower cloud" or *pino* composed of steam and ash, rising above the cinder cone of Volcano during the waning phases of the explosive eruption of 1888 (after a photograph by B. Hobson).

There seems to be no good reason to doubt that Vulcanian cinder eruptions of this type differ chiefly in magnitude from the rhythmic explosion within the crater of Stromboli, if we except the elevation of a considerable quantity of accessory and older tuff which is derived from the inner walls of the crater and carried upward into the air together with the pasty cakes of fresh lava derived from the chimney. It is this accessory material which gives to the *pino* its dark or even black appearance.

FIG. 116.—Double explosive eruption of Taal volcano on the morning of January 30, 1911.

The eruption of Taal volcano on January 30, 1911.—The recent eruption of the cinder cone known as Taal volcano is of interest, not only because so fresh in mind, but because two neighboring vents erupted simultaneously with explosions of nearly equal violence (Fig. 116). This Philippine volcano lies near the center of a lake some fifteen miles in diameter and about fifty miles south of the city of Manila. After a period of rest extending over one hundred and fifty years, the symptoms of the coming eruption developed rapidly, and on the morning of January 30 grand explosions of steam and ash occurred simultaneously in the neighboring craters, and the condensed moisture brought down the ash in an avalanche of scalding mud which buried the entire island. Almost the entire population of the island, numbering several hundreds, was literally buried in the blistering mud (Fig. 117); and the gases from the explosions carried to the distant shores of the lake added to this number many hundred victims.

FIG. 117.—The thick mud veneer upon the island of Taal (after a photograph by Deniston).

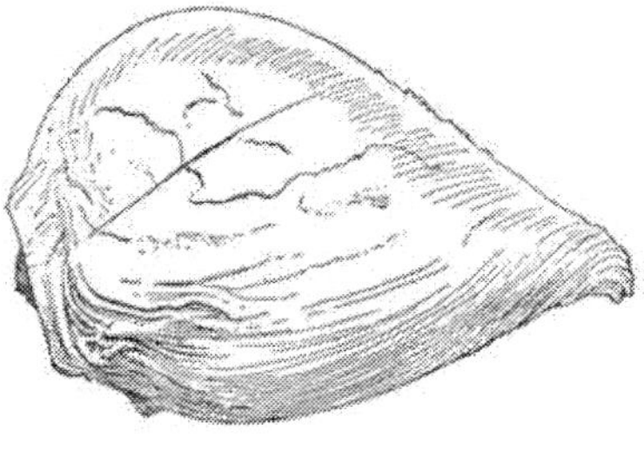

FIG. 118.—A pear-shaped lava projectile.

The shocks which accompanied the explosions raised a great wave upon the surface of the lake, which, advancing upon the shores, washed away structures for a distance of nearly a half mile.

The materials and the structure of cinder cones.—Obviously the materials which compose cinder cones are the cooled lava fragments of various degrees of coarseness which have been ejected from the crater. If larger than a finger joint, such fragments are referred to as *volcanic projectiles*, or, incorrectly, as "volcanic bombs." Of the larger masses it is often true that the force of expulsion has not been applied opposite the center of mass of the body. Thus it follows that they undergo complex whirling motions during their flight, and being still semiliquid, they develop curious pear-shaped or less regular forms (Fig. 118). When crystals have already separated out in the lava before its rise in the chimney of the volcano, the surrounding fluid lava may be blown to finely divided volcanic dust which floats away upon the wind, thus leaving the crystals intact to descend as a crystal rain about the crater. Such a shower occurred in connection with the eruption of Etna in 1669, and the black augite crystals may to-day be gathered by the handful from the slopes of the Monti Rossi (Fig. 125, p. 125).

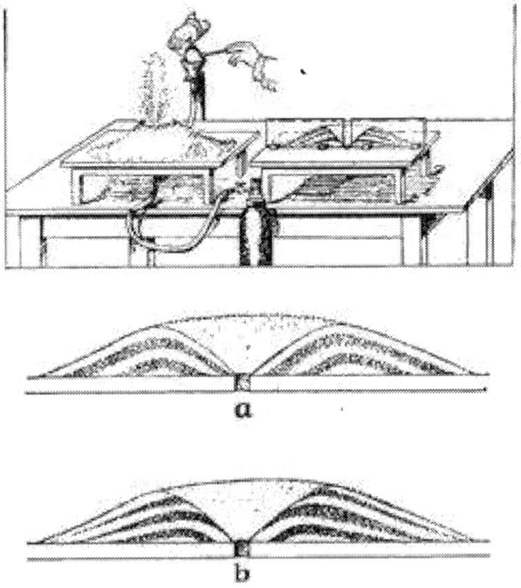

FIG. 119.—Artificial production of the structure of a cinder cone with use of colored sands carried up in alternation by a current of air (after G. Linck).

The term *lapilli*, or sometimes *rapilli*, is applied to the ejected lava fragments when of the average size of a finger joint. This is the material which still partially covers the unexhumed portions of the city of Pompeii.

Volcanic *sand*, *ash*, and *dust* are terms applied in order to increasingly fine particles of the ejected lava. The finest material, the volcanic dust, is often carried for hundreds and sometimes even for thousands of miles from the crater in the high-level currents of the atmosphere. Inasmuch as this material is deposited far from the crater and in layers more or less horizontal, such material plays a small rôle in the formation of the cinder cone. The coarser sands and ash, on the other hand, are the materials from which the cinder cone is largely constructed.

The manner of formation and the structure of cinder cones may be illustrated by use of a simple laboratory apparatus (Fig. 119). Through an opening in a board, first white and then colored sand is sent up in a light current of air or gas supplied from suitable apparatus. The alternating layers of the sand form in the attitudes shown; that is to say, dipping inward or toward the chimney of the volcano at all points within the crater rim, and outward or away from it at all points outside (Fig. 119). If the experiment is carried so far that at its termination sand slides down the crater walls into the chimney below, the inward dipping layers will be truncated, or even removed entirely, as shown in Fig. 119 *b*.

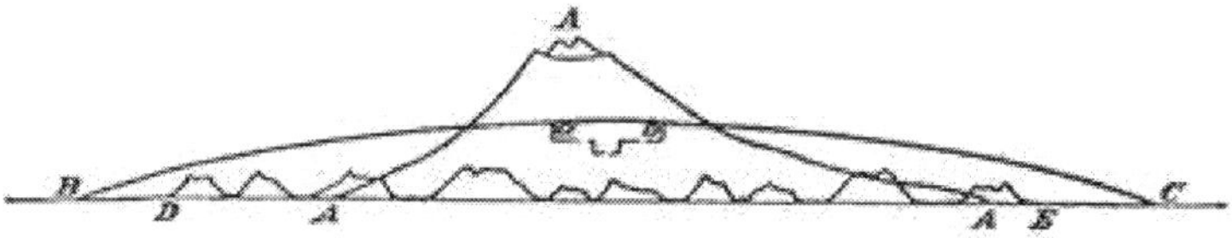

FIG. 120.—Diagram to show the contrast between a lava dome and a cinder cone. *AAA*, cinder cone; *BabC*, lava dome; *DE*, line of low cinder cones above a fissure (after Thoroddsen).

The profile lines of cinder cones.—The shapes of cinder cones are notably different from those of lava mountains. While the latter are domes, the mountains constructed of cinder are conical and have curves of profile that are concave upward instead of convex (Fig. 120). In the earlier stages of its growth the cinder cone has a crater which in proportion to the height of the mountain is relatively broad (Fig. 99, p. 104).

FIG. 121.—Mayon volcano on the island of Luzon, P.I. A remarkably perfect high cinder cone.

Speaking broadly, the diameter of the crater is a measure of the violence of the explosions within the chimney. A single series of short and violent explosive eruptions builds a low and broad cinder cone. A long-continued succession of moderately violent explosions, on the other hand, builds a high cone with crater diameter small if compared with the mountain's altitude, and the profile afforded is a remarkably beautiful sweeping curve (Fig. 121). Toward the summit of such a cone the loose materials of which it is composed are at as steep an angle as they can lie, the so-called angle of repose of the material; whereas lower down the flatter slopes have been determined by the distribution of the cinder during its fall from the air. When one makes the ascent of such a mountain, he encounters continually steeper grades, with the most difficult slope just below the crest.

FIG. 122.—A series of breached cinder cones where the place of eruption has migrated along the underlying fissure. The Puys Noir, Solas, and La Vache in the Mont Dore Province of central France (after Scrope).

The composite cone.—The life histories of volcanoes are generally so varied that lava domes and the pure types of cinder cones are less common than volcanoes in which paroxysmal eruptions have alternated with explosions, and where, therefore, the structure of the mountain represents a composite of lava and cinder. Such composite cones possess a skeleton of solid rock upon which have been built up alternate sloping layers of cinder and lava. In most respects such cones stand in an intermediate position between lava domes and cinder cones.

FIG. 123.—The *bocca* or mouth upon the inner cone of Mount Vesuvius from which flowed the lava stream of 1872. This lava stream appears in the foreground with its characteristic "ropy" surface.

Regarded as a retaining wall for the lava which mounts in the chimney, the cinder cone is obviously the weakest of all. Should lava rise in a cinder cone without an explosion occurring, the cone is at once broken through upon one side by the outwelling of the lava near the base. Thus arises the characteristic *breached* cone of horseshoe form (Fig. 122).

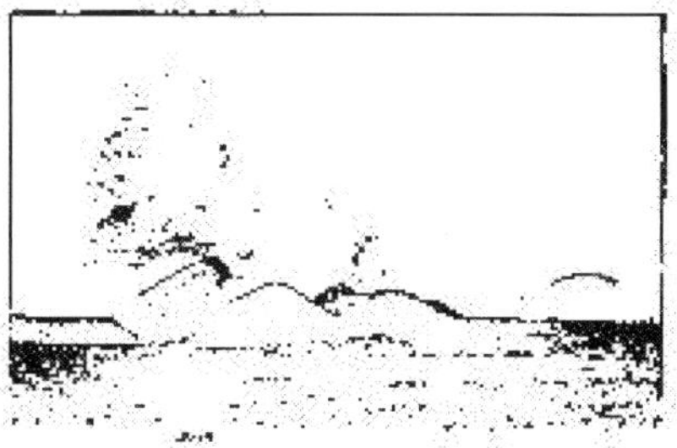

FIG. 124.—A row of parasitic cones raised above a fissure which was opened upon the flanks of Mount Etna during the eruption of 1892 (after De Lorenzo).

Quite in contrast with the weak cinder cone is the lava dome with its rock walls and relatively flat slopes. Considered as a retaining wall for lava it is much the strongest type of volcanic mountain, and it is likely that the hydrostatic pressure of the lava within the crater would seldom suffice to rupture the walls, were it not that the molten rock first fuses its way into old stream tunnels buried under the mountain slopes (see *ante*, p. 112). Composite cones have a strength as retaining walls for lava which is intermediate between that of the other types. Their Vulcanian eruptions of the convulsive type are initiated by the formation of a rent or fissure upon the mountain flanks at elevations well above the base, the opening of the fissure being generally accompanied by a local earthquake of greater or less violence.

From one or more such fissures the lava issues usually with sufficient violence at the place of outflow to build up over it either an enlarged type of driblet cone, referred to as a "mouth", or *bocca* (Fig. 123), or one or more cinder cones which from their position upon the flanks of the larger volcano are referred to as *parasitic cones* (Fig. 124). The lava of Vesuvius more frequently yields *bocchi* at the place of outflow, whereas the flanks of Etna are pimpled with great numbers of parasitic cinder cones, each the monument to some earlier eruption (Fig. 125).

FIG. 125.—View looking toward the summit of Etna from a position upon the southern flank near the village of Nicolosi. The two breached parasitic cones seen behind this village are the Monti Rossi which were thrown up in 1669 and from which flowed the lava which overran Catania (after a photograph by Sommer).

It is generally the case that a single eruption makes but a relatively small contribution to the bulk of the mountain. From each new cone or *bocca* there proceeds a stream of lava spread in a relatively narrow stream extending down the slopes (Fig. 126).

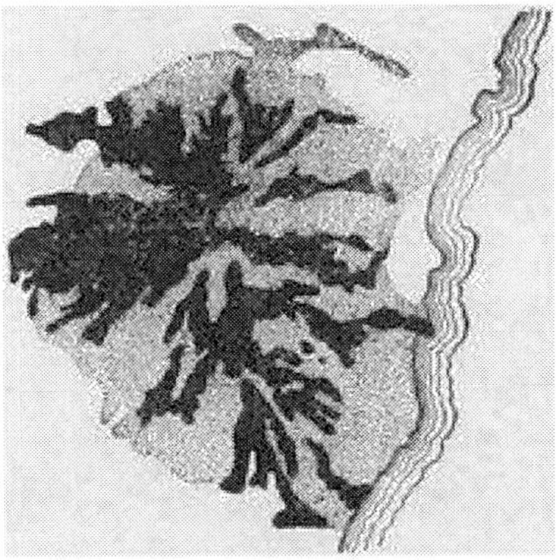

FIG. 126.—Sketch map of Etna, showing the individual surface lava streams (in black) and the tuff covered surface (stippled).

The caldera of composite cones.—Because of the varied episodes in the history of composite cones, they lack the regular lines characteristic of the two simpler types. The larger number of the more important composite cones have been built up within an outer crater of relatively large diameter, the Somma cone or *caldera*, which surrounds them like a gigantic ruff or collar. This caldera is clearly in most cases at least the relic of an earlier explosive crater, after which successive eruptions of lesser violence have built a more sharply conical structure. This can only be interpreted to mean that most larger and long-active volcanoes have been born in the grandest throes of their life history, and that a larger or smaller lateral migration of the vent has been responsible for the partial destruction of the explosion crater. Upon Vesuvius we find the crescent-like rim of Monte Somma; on Etna it is the Val del Bove, etc. It is this caldera of composite cones which gave rise to the theory of the "elevation crater" of von Buch (see *ante*, p. 95, and Fig. 127).

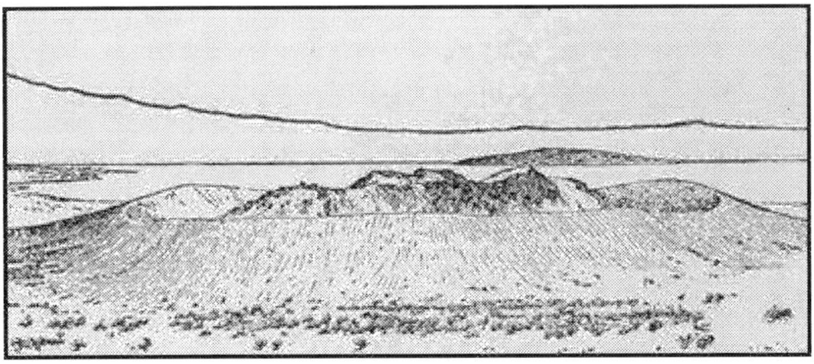

FIG. 127.—Panum crater, showing the caldera and the later interior cones (after Russell).

The eruption of Vesuvius in 1906.—The volcano Vesuvius rises on the shores of the beautiful bay of Naples only about ten miles distant from the city of Naples. The mountain consists of the remnant of an earlier broad-mouthed explosion crater, the Monte Somma, and an inner, more conical elevation, the Monte Vesuvio. Before the eruption of 1906 this central cone was sharply conical and rose to a height of about 4300 feet above the surface of the bay, or above the highest point of the ancient caldera. The base of this inner cone is at an elevation of something less than half that of the entire mass, and is separated from the encircling ring wall of the old crater by the *atrio*, to which corresponds in height a perceptible shelf or *piano* upon the slope toward the bay of Naples (Fig. 128).

FIG. 128.—View of Mount Vesuvius as it appeared from the Bay of Naples shortly before the eruption of 1906. The horn to the left is Monte Somma.

An active composite cone like that of Vesuvius is for the greater part of the time in the Strombolian condition; that is to say, light crater explosions continue with varying intensity and interval, except when the mountain has been excited to the periodic Vulcanian outbreaks with which its history has been punctuated. The Strombolian explosions have sufficient violence to eject small fragments of hot lava, which, falling about the crater, slowly built up a rather sharp cone. The period of Strombolian activity has, therefore, been called the *cone-producing period.* Just before each new outbreak of the Vulcanian type, the altitude of the mountain has, therefore, reached a maximum, and since the larger explosive eruptions remove portions of this cone at the same time that they increase the dimensions of the crater, the Vulcanian stage in contrast to the other has been called the *crater-producing period.* In this period, then, the material ejected during the explosions does not consist solely of fresh lava cakes, but in part of the older débris derived from the crater walls, whence it is avalanched upon the chimney after each larger explosion. The overhanging cloud, which during the Strombolian period has consisted largely of steam and is noticeably white, now assumes a darker tone, the "smoke" which characterizes the Vulcanian eruption.

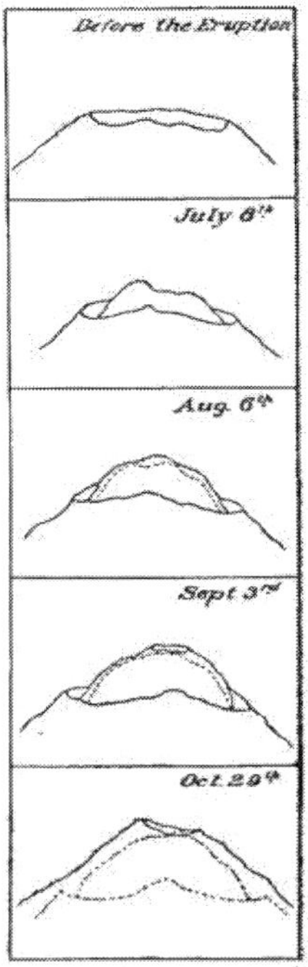

FIG. 129.—A series of consecutive sketches of the summit of the Vesuvian cone, showing the modifications in its outline (after Sir William Hamilton).

On several historical occasions the cone of Vesuvius has been lowered by several hundred feet, the greatest of relatively recent truncations having occurred in 1822 and in 1906. Between Vulcanian eruptions the Strombolian activity is by no means uniform, and so the upward growth of the cone is subject to lesser interruptions and truncations (Fig. 129).

The Vesuvian eruption of 1906 has been selected as a type of the larger Vulcanian eruption of composite cones, because it combined the explosive and paroxysmal elements, and because it has been observed and studied with greater thoroughness than any other. The latest previous eruption of the Vulcanian order had occurred in 1872. Some two years later the period of active cone building began and proceeded with such rapidity that by 1880 the new cone began to appear above the rim of the crater of 1872. From this time on occasional light eruptions interrupted the upbuilding process, and as the repairs were not in all cases completed before a new interruption, a nest of cones, each smaller than the last, arose in series like the outdrawn sections of an old-time spyglass. At one time no less than five concentric craters were to be seen.

For a brief period in the fall of 1904 Vesuvius had been in almost absolute repose, but soon thereafter the Strombolian crater explosions were resumed. On May 25, 1905, a small stream of lava began to issue from a fissure high up upon the central cone, and from this time on the lava continued to flow down to the valley or *atrio*, separating the inner cone from the caldera remnant of Monte Somma. Seen in the night, this stream of lava appeared from Naples like a red hot wire laid against the mountain's side (Fig. 130). With gradual augmentation of Strombolian explosions and increase in volume of the flowing lava stream, the same condition continued until the first days of April in 1906. The flowing lava had then overrun the tracks of the mountain railway and accumulated in considerable quantity within the *atrio* (Fig. 131).

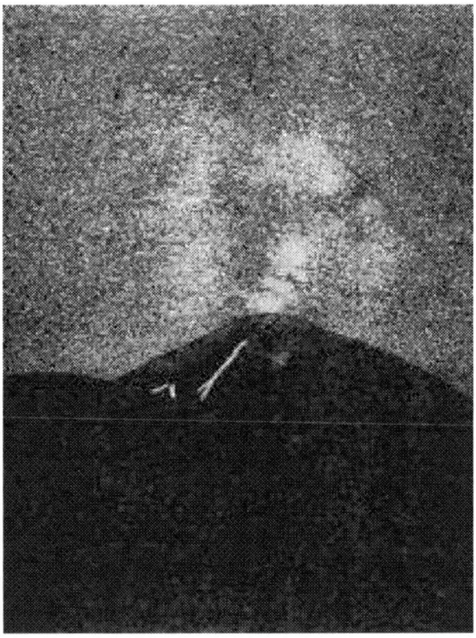

FIG. 130.—Night view of Vesuvius from Naples before the outbreak of 1906. A small lava stream is seen descending from a high point upon the central cone (after Mercalli).

On the morning of April 4, a preliminary stage of the eruption was inaugurated by the opening of a new radial fissure about 500 feet below the summit of the cone (Fig. 132 *a*), and by early afternoon the cone-destroying stage began with the rise of a dark "cauliflower cloud" or *pino* to replace the lighter colored steam cloud. The cone was beginning to fall into the crater, and old lava débris was mingled in the ejections with the lava clots blown from the still fluid material within the chimney. From now on short and snappy lightning flashes played about the black cloud, giving out a sharp staccato "tack-a-tack." The volume and density of the cloud and the intensity of the crater explosions continued to increase until the culmination on April 7. On April 5 at midnight a new lava mouth appeared upon the same fissure which had opened near the summit, but now some 300 feet lower (Fig. 132 *b*). The lava now welled out in larger volume corresponding to its greater head, and the stream which for ten months had been flowing from the highest outlet upon the cone now ceased to flow. The next morning, April 6, at about 8 o'clock, lava broke out at several points some distance east of the opening *b*, and evidently upon another fissure transverse to the first (Fig. 132 *c*). The lava surface within the chimney must still have remained near its old level,—effective draining had not yet begun,—since early upon the following morning a small outflow began nearly at the top of the cone upon the opposite side and at least a thousand feet higher.

FIG. 131.—Scoriaceous lava encroaching upon the tracks of the Vesuvian railway (after a photograph by Sommer).

FIG. 132.—Map of Vesuvius, showing the position and order of formation of the lava mouths upon its flanks during the eruption of 1906 (after Johnston-Lavis).

The culmination of the eruption came in the evening of April 7, when, to the accompaniment of light earthquakes felt as far as Naples, lava issued for the first time in great volume from a mouth more than halfway down the mountain side (Fig. 132 *f*), and thus began the drainage of the chimney. At about the same time with loud detonations a huge black cloud rose above the crater in connection with heavy explosions, and a rain of cinder was general in the region about the mountain but especially within the northeast quadrant. Those who were so fortunate as to be in Pompeii had a clear view of the mountain's summit where red hot masses of lava were thrown far into the air. The direction of these projections was reported to have been not directly upward, but inclined toward the northeast quadrant of the mountain; but since with a northeast surface wind the heaviest deposit of ash and dust should have been upon the southwestern quadrant of the mountain, it is evident that the material was carried upward until it reached the contrary upper currents of the atmosphere, to be by them distributed.

FIG. 133.—The ash curtain which had overhung Vesuvius lifting and disclosing the outlines of the mountain on April 10, 1911 (after De Lorenzo).

FIG. 134.—The central cone of Vesuvius as it appeared after the eruption of 1906, but with the earlier profile indicated. The truncation represents a lowering of the summit by some five hundred feet, with corresponding increase in the diameter of the crater (after Johnston-Lavis).

When the heavy curtain of ash, which now for a number of succeeding days overhung all the circum-Vesuvian country, began to lift (Fig. 133), it was seen that the summit of the cone had been truncated an average of some 500 feet (Fig. 134). All the slopes and much of the surrounding country had the aspect of being buried beneath a cocoa-colored snow of a depth to the northeastward of several feet, where it had drifted into all the hollow ways so as almost to efface them (Fig. 135). More than thrice as heavy as water, the weak roof timbers of the houses at the base of the mountain gave way beneath the added load upon them, thus making many victims. Inasmuch, however, as the ash-fall partakes of the same general characters as in eruptions from cinder cones, we may here give our attention especially to the streams of lava which issued upon the opposite flank of the mountain (Fig. 136).

FIG. 135.—A sunken road filled with indrifted cocoa-colored ash from the Vesuvian eruption of 1906.

FIG. 136.—View of Vesuvius taken from the southwest during the waning stages of the eruption of 1906. In the middle distance may be discerned the several lava mouths aligned upon a fissure, and the courses of the streams which descend from them. In the foreground is the main lava stream with scoriaceous surface (after W. Prinz).

The main lava stream descended the first steep slopes with the velocity of a mile in twenty-five minutes, about the strolling speed of a pedestrian, but this rate was gradually reduced as the stream advanced farther from the mouth. Taking advantage of each depression of the surface, the black stream advanced slowly but relentlessly toward the cities at the southwest base of the mountain. With a motion not unlike that of a heap of coal falling over itself down a slope, the block lava advances without burning the objects in its path (Fig. 137).

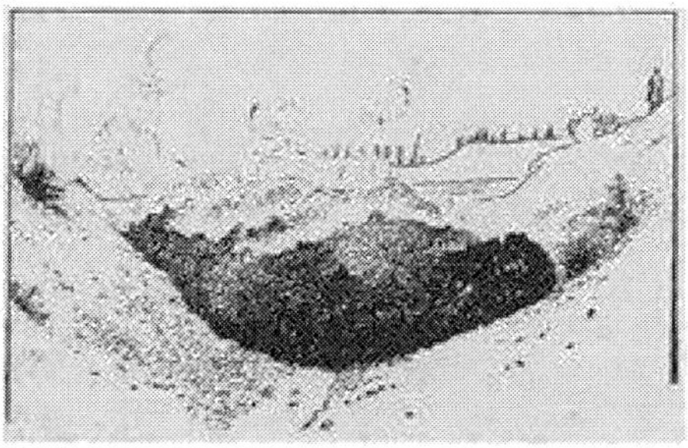

FIG. 137.—The main lava stream of 1906 advancing upon the village of Boscotrecase.

FIG. 138.—An Italian pine snapped off by the lava and carried forward upon its surface as a passenger (after Haug).

The beautiful pines are merely charred where snapped off and are carried forward upon the surface of the stream (Fig. 138). When a real obstruction, such as a bridge or a villa, is encountered, the stream is at first halted, but the rear crowding upon the van, unless a passage is found at the side, the lava front rises higher and higher until by its weight the obstruction is forced to give way (Figs. 139 and 140).

FIG. 139.—Lava front both pushing over and running around a wall which lies athwart its course (after Johnston-Lavis).

FIG. 140.—One of the villas in Boscotrecase which was ruined by the Vesuvian lava flow of 1906. The fragments of masonry from the ruined walls traveled upon the lava current, where they sometimes became incased in lava.

The sequence of events within the chimney.—The thorough study of this Vesuvian eruption has placed us in a position to infer with some confidence in our conclusions the sequence of events within the chimney and crater of the volcano, both before and during the eruption. Anticipating some conclusions derived from the observed dissection of volcanoes, which will be discussed below, it may be stated that what might be termed the core of the composite cone—the chimney—is a more or less cylindrical plug of cooled lava which during the active period of the vent has an interior bore of probably variable caliber. This plug in its lower section appears in solid black in all the diagrams of Fig. 141. During the cone-building period (Fig. 141 *a* and *b*) the plug is obviously built upward along with the cone, for lava often flows out at a level a few hundred feet only below the crater rim. By what process this chimney building goes on is not well understood, though some light is thrown upon it by the post-eruption stage of Mont Pelé in 1902-1903 (see below).

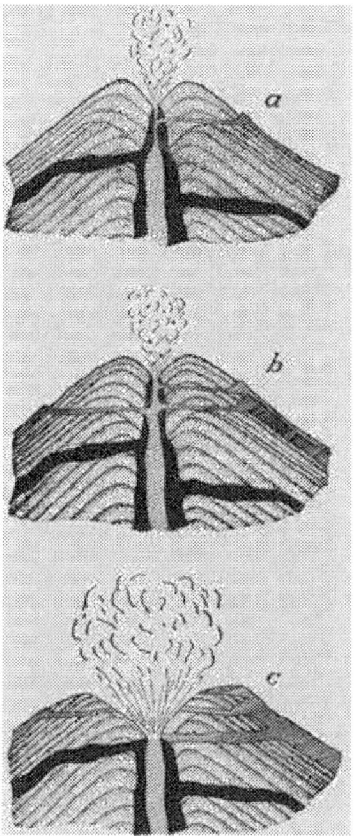

FIG. 141.—Three diagrams to illustrate the sequence of events within the crater of a composite cone during the cone-building and crater-producing periods. *a* and *b*, two successive stages of the cone building or Strombolian period; *c*, enlargement of the crater, truncation of the cone, and destruction of the upper chimney during the relatively brief crater-producing or Vulcanian period.

Both the older and newer sections of this plug or chimney are furnished some support against the outward pressure of the contained lava by the surrounding wall of tuff; and they are, therefore, in a condition not unlike that of the inner barrel of a great gun over which sleeves of metal have been shrunk so as to give support against bursting pressures. On the other hand, when not sustaining the hydrostatic pressure of the liquid lava within, the chimney would tend to be crushed in by the pressure of the surrounding tuff. Its strength to withstand bursting pressures is dependent not alone upon the thickness of its rock walls, but also upon its internal diameter or caliber. A steam cylinder of given thickness of wall, as is well known, can resist bursting pressures in proportion as its internal diameter is small. So in the volcanic chimney, any tendency to remelt from within the chimney walls must weaken them in a twofold ratio.

We are yet without accurate temperature observations upon the lava in volcanic chimneys, but it seems almost certain that these temperatures rise as the Vulcanian stage is approaching, and such elevation of temperature must be followed by a greater or less re-fusion of the chimney walls. The sequence of events during the late Vesuvian eruption is, then, naturally explained by progressive re-fusion and consequent weakening of the chimney walls, thus permitting a radial fissure to open near the top and gradually extend downwards. Thus at first small and high outlets were opened insufficient to drain the chimney, but later, on April 7, after this fissure had been much extended and a new and larger one had opened at a lower level, the draining began and the surface of lava commenced rapidly to sink.

FIG. 142.—The spine of Pelé rising above the chimney of the volcano after the eruption of 1902 (after Hovey).

When the rapid sinking of the lava surface occurred, the lower lava layers were almost immediately relieved of pressure, thus causing a sudden expansion of the contained steam and resulting in grand crater explosions. The partially refused and fissured upper chimney, now unable to withstand the inward pressure of the surrounding tuff walls, since outward pressures no longer existed, crushed in and contributed its materials and those of the surrounding tuff to the fragments of fresh lava rising in volume in the grand explosions (Fig. 141 *c*). In outline, then, these seem to be the conditions which are indicated by the sequence of observed events in connection with the late Vesuvian outbreak.

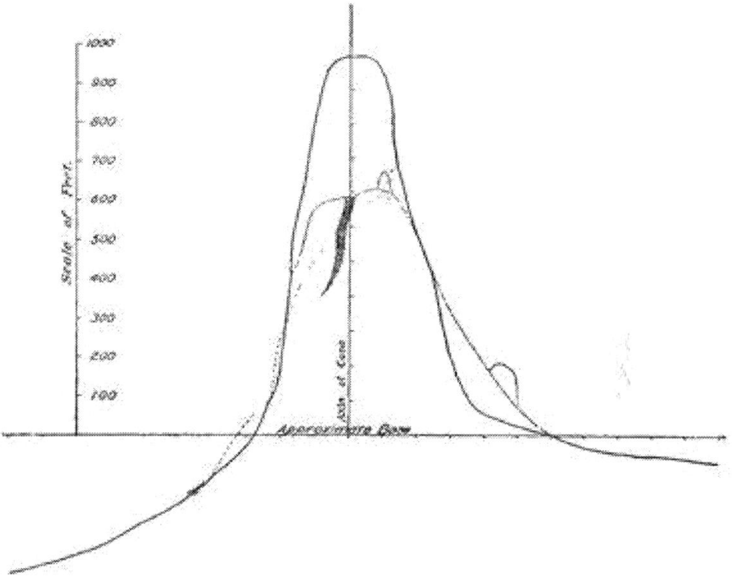

FIG. 143.—Outlines of the Pelé spine upon successive dates. The full line represents its outline on December 26, 1902; the dotted-dashed line is a profile of January 3, 1902; while the dotted line is that of January 9, 1903. The dark line is a fissure (after E. O. Hovey).

The spine of Pelé.—The disastrous eruption of Mont Pelé upon Martinique in the year 1902 is of importance in connection with the interesting problem of the upward growth of volcanic chimneys during the cone-building period of a volcano. After the conclusion of this great Vulcanian eruption, a spine of lava grew upward from the chimney of the main crater until it had reached an elevation of more then a thousand feet above its base, a figure of the same order of magnitude as the probable height of the upper section of the Vesuvian chimney previous to the eruption of 1906. The Pelé spine (Fig. 142) did not grow at a uniform rate, but was subject to smaller or larger truncations, but for a period of 18 days the upward growth was at the rate of about 41 feet per day. Later, the mass split upon a vertical plane revealing a concave inner surface, and was somewhat rapidly reduced in altitude to 600 feet (Fig. 143), only to rise again to its full height of about 1000 feet some three months later.

While apparently unique as an observed phenomenon, and not free from uncertainty as to its interpretation, the growth of this obelisk has at least shown us that a mass of rock can push its way up above the chimney of an active volcano even when there are no walls of tuff about it to sustain its outward pressures.

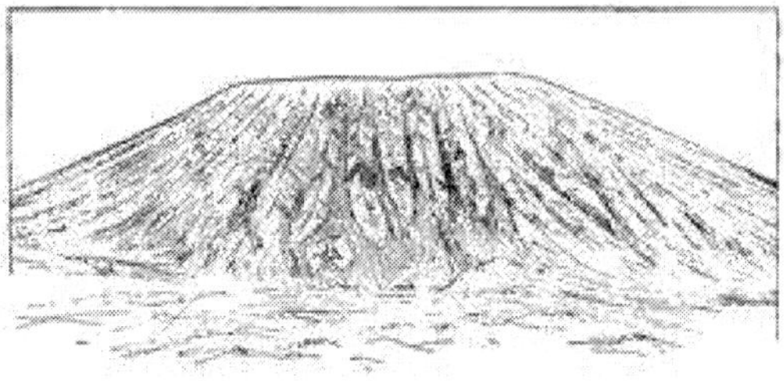

FIG. 144.—Corrugated surface of the Vesuvian cone after the mud flows which followed the eruption in 1906 (after Johnston-Lavis).

The aftermath of mud flows.—When the late Vulcanian explosions of Vesuvius had come to an end, all slopes of the mountain, but especially the higher ones, were buried in thick deposits of the cocoa-colored ash, included in which were larger and smaller projectiles. As this material is extremely porous, it greedily sucks up the water which falls during the first succeeding rains. When nearly saturated, it begins to descend the slopes of the mountain and soon develops a velocity quite in contrast with that of the slow-moving lava. The upper slopes are thus denuded, while the fields and even the houses about the base are invaded by these torrents of mud (*lava d'acqua*). Inasmuch as these mud flows are the inevitable aftermath of all grander explosive eruptions, the Italian government has of late spent large sums of money in the construction of dikes intended to arrest their progress in the future. It was streams of this sort that buried the city of Herculaneum after the explosive eruption of 79 A.D.

After the mud flows have occurred, the Vesuvian cone, like all similar volcanic cones under the same conditions, is found with deep radial corrugations (Fig. 144), such as were long ago described as "barrancoes" and supposed to support the "elevation crater" theory of volcano formation.

The dissection of volcanoes.—To the uninitiated it might appear a hopeless undertaking to attempt to learn by observation the internal structure of a volcano, and especially of a complex volcano of the composite type. The earliest successful attempt appears to have been made by Count Caspar von Sternberg in order to prove the correctness of the theory of his friend, the poet Goethe. Goethe had claimed that a little hill in the vicinity of Eger, on the borders of Bohemia, was an extinct volcano, though the foremost geologist of the time the famous Werner, had promulgated the doctrine that this hill, in common with others of similar aspect, originated in the combustion of a bed of coal. The elevation in question, which is known as the Kammerbühl, consists mainly of cinder, and Goethe had maintained that if a tunnel were to be driven horizontally into the mountain from one of its slopes, a core or plug of lava would be encountered beneath the summit. The excavations, which were completed in 1837, fully verified the poet's view, for a lava plug was found to occupy the center of the mass and to connect with a small lava stream upon the side of the hill (Fig. 145).

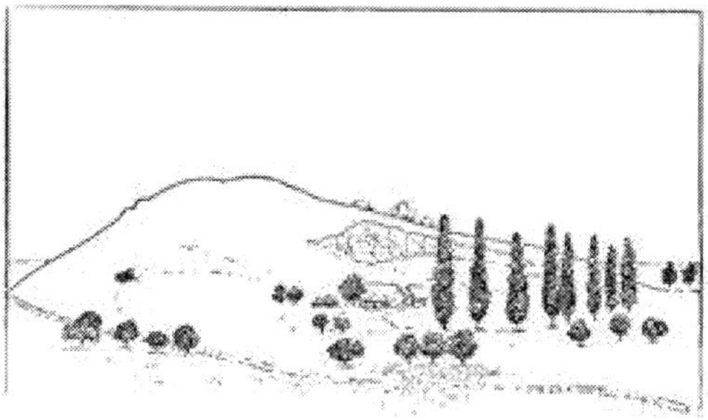

FIG. 145.—The Kammerbühl near Eger, showing the tunnel completed in 1837 which proved the volcanic nature of the mountain (after Judd).

It is not, however, to such expensive projects that reference is here made, but rather to processes which are continually going on in nature, and on a far grander scale. The most important dissecting agent for our purpose is running water, which is continually paring down the earth's surface and disclosing its buried structures. How much more convincing than any results of artificial excavation, as evidence of the internal structure of a volcano, is the monument represented in Fig. 146, since here the lava plug stands in relief like a gigantic thumb still surrounded by a remnant of cinder deposits. Such exposed chimneys of former volcanoes are found in many regions, and have become known as volcanic *necks*, *pipes*, or *plugs*.

FIG. 146.—Volcanic plug exposed by natural dissection of a volcanic cone in Colorado (U. S. G. S.).

FIG. 147.—A dike cutting beds of tuff in a partly dissected volcano of southwestern Colorado (after Howe, U. S. G. S.).

Not infrequently the beds of tuff composing the flanks of the volcano, upon dissection by the same process, bring to light walls of cooled lava standing in relief (Fig. 147)—the filling of the fissure which gave outlet to the flanks of the mountain at the time of the eruption. Study of exposed dikes formed in connection with recent eruptions of Vesuvius has shown that in many instances they are still hollow, the lava having drained from them before complete consolidation.

Another agent which is effective in uncovering the buried structures of volcanoes is the action of waves on shores. Always a relatively vigorous erosive agency, the softer structures of volcanic cones are removed with especial facility by this agent. On the shores of the island of Volcano, the little cone of Vulcanello has been nearly half carried away by the waves, so as to reveal with especial perfection the structure of the cinder beds as well as the internal rock skeleton of the mass. Here the characteristic dips of lava streams, intercalated as they now are between tuff deposits and the lava which consolidated in fissures, are both revealed.

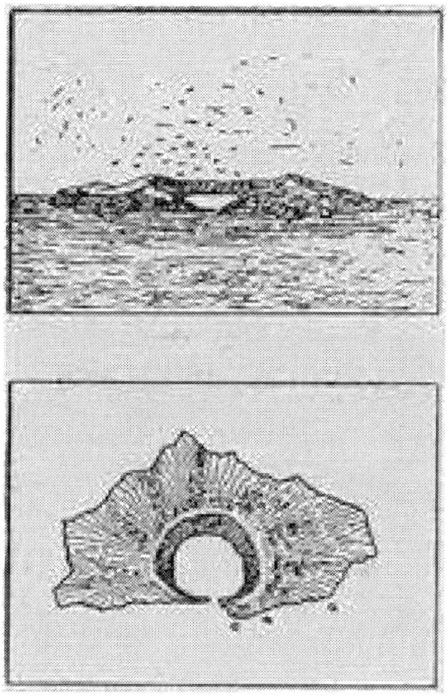

FIG. 148.—Map and general view of St. Paul's Rocks, a volcanic cone dissected by waves.

In mid-Atlantic a quite perfect crater, the St. Paul's Rocks, has been cut nearly in half so as to produce a natural harbor (Fig. 148).

In still other instances we may thank the volcano itself for opening up the interior of the mountain for our inspection. The eruption in 1888 of the Japanese volcano of Bandai-san, by removing a considerable part of the ancient cone, has afforded us a section completely through the mountain. The summit and one side of the small Bandai was carried completely away, and there was substituted a yawning crater eccentric to the former mountain and having its highest wall no less than 1500 feet in height (Fig. 149). In two hours from the first warning of the explosion the catastrophe was complete and the eruption over.

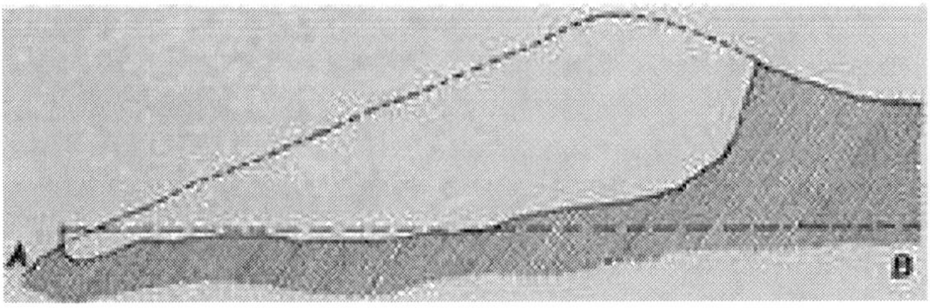

FIG. 149.—Dissection by explosion of Little Bandai-san in 1888 (after Sekiya).

The eruption of Krakatoa in 1883, probably the grandest observed volcanic explosion in historic times, left a volcanic cone divided almost in half and open to inspection (Fig. 150). Rakata, Danan, and Perbuatan had before constituted a line of cones built up round individual craters subsequent to the partial destruction of an earlier caldera, portions of which were still existent in the islands Verlaten and Lang. By the eruption of 1883 all the exposed parts and considerable submerged portions of the two smaller cones were entirely destroyed, and the larger one, known as Rakata, was divided just outside the plug so as to leave a precipitous wall rising directly from the sea and showing lava streams in alternation with somewhat thicker tuff layers, the whole knit together by numerous lava dikes.

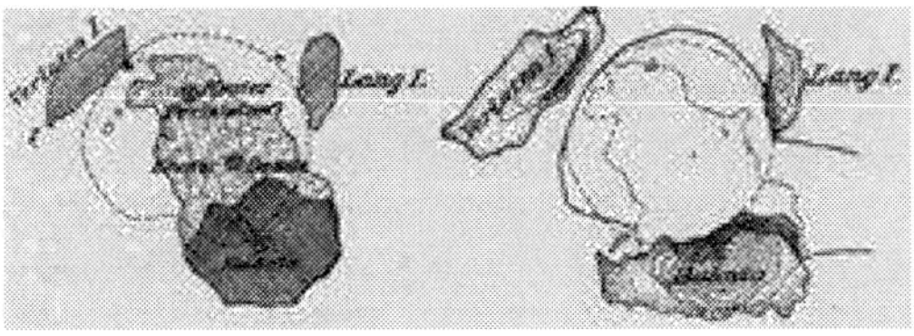

FIG. 150.—The half-submerged volcano of Krakatoa in the Sunda Straits before and after the eruption of 1883 (after Verbeek).

In order to carry our dissecting process down to levels below the base of the volcanic mountain, it is usually necessary to inspect the results of erosion by running water. Here the plug or chimney, instead of being surrounded by tuff, is inclosed by the country rock of the region, which is commonly a sedimentary formation. Such exposed lower sections of volcanic chimneys are numerous along the northwestern shores of the British Isles. Where aligned upon a dislocation or noteworthy fissure in the rocks, the group of plugs has been referred to as a scar or *cicatrice* (Fig. 151). Associated with the plugs of the cicatrice are not infrequently dikes, or, it may be, sheets of lava extended between layers of sediment and known as *sills*.

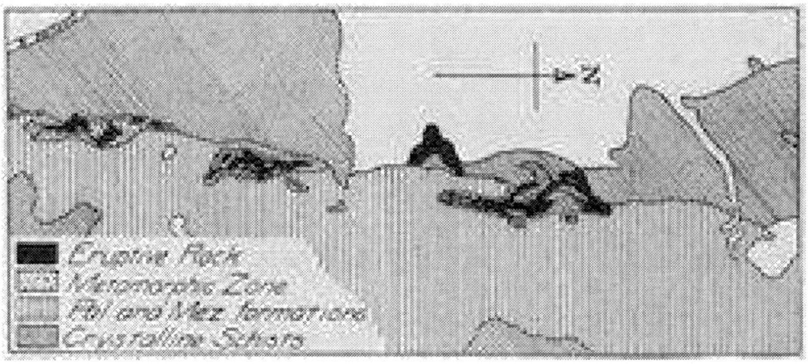

FIG. 151.—The cicatrice of the Banat (after Suess).

If we are able to continue the dissection process to still greater depths, we encounter at last igneous rock having a texture known as granitic and indicating that the process of consolidation was not only exceedingly slow but also uninterrupted. This rock is found in masses of larger dimensions, and though generally of more or less irregular form, no one dimension is of a different order of magnitude from the others. Such masses are commonly described as *bosses*, or, if especially large, as *batholites* (Fig. 152). Wherever the rock beds appear as though they had been forced up by the upward pressure of the igneous mass, the latter takes the form of a mushroom and has been described as a *laccolite* (Figs. 479-481, pp. 441-442). Evidence seems, however, to accumulate that in the greater number of cases the molten rock has fused its way upward, in part assimilating and in part inclosing the rock which it encountered. This process of upward fusion has been likened to the progress of a red hot iron burning its way through a board.

The formation of lava reservoirs.—The discarding of the earlier notion that the earth has a liquid interior makes it proper in discussing the subject of volcanoes to at least touch upon the origin of the molten rock material. As already pointed out, such reservoirs as exist must be local and temporary, or it would be difficult to see how the existing condition of earth rigidity could be maintained. From the rate at which rock temperatures rise, at increasing depths below the surface, it is clear that all rocks would be melted at very moderate depths only, if they were not kept in a solid state by the prodigious loads which they sustain. Any relief from this load should at once result in fusion of the rock.

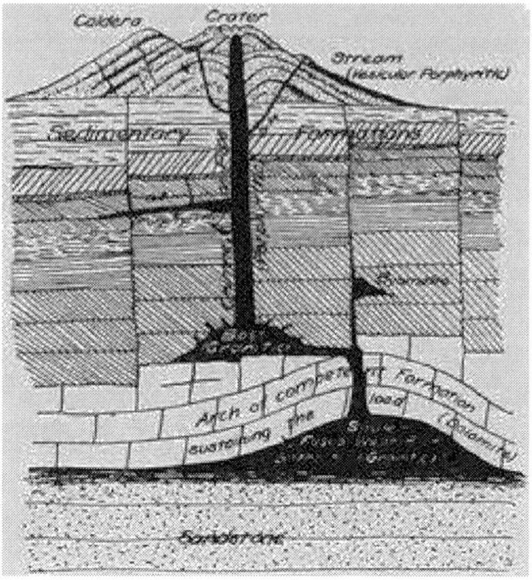

FIG. 152.—Diagram to illustrate a probable cause of formation of lava reservoirs, and to show the connection between such reservoirs and the volcanoes at the surface.

Now the restriction of active volcanoes to those zones of the earth's surface within which mountains are rising, and where in consequence earthquakes are felt, has furnished us at least a clew to the origin of the lava. Regarded as a structure capable of sustaining a load, the competency of an arch is something quite remarkable, so that the arching up of strong rock formations into anticlines within the upper layers of the zone of flow, or of combined fracture and flow, would be sufficient to remove the load from relatively weak underlying beds, which in consequence would be fused and form local reservoirs of lava (Figs. 152 and 153).

It has been further quite generally observed that lines of volcanoes, in so far as they betray any relation in position to neighboring mountain ranges, tend to appear upon the rear or flatter limb of unsymmetrical arches, or where local tension would favor the opening of channels toward the surface. Moreover, wherever recent block movements of surface portions of the earth's shell have been disclosed in the neighborhood of volcanoes, the latter appear to be connected with downthrown blocks, as though the lava had, so to speak, been squeezed out from beneath the depressed block or blocks.

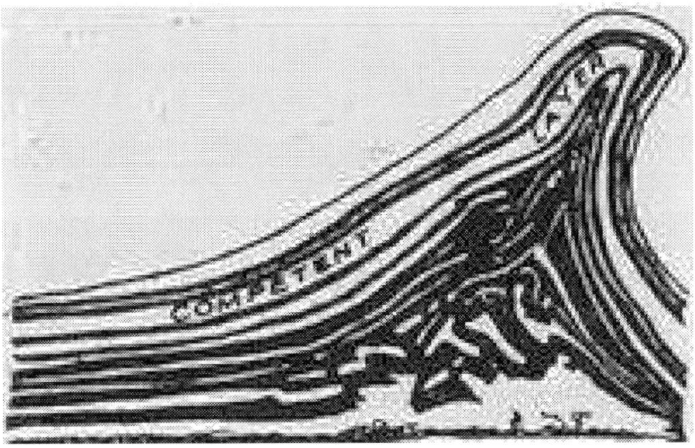

FIG. 153.—Result of experiment with layers of composition to illustrate the effect of relief of load upon rocks by arching of competent formation (after Willis).

We must not, however, forget that the igneous rocks are greatly restricted in the range of their chemical composition. No igneous rock type is known which could be formed by the fusion of any of the carbonate rocks such as limestone or dolomite, or of the more siliceous rocks, such as sandstone or quartzite. There remains only the argillaceous class of sediments, the shales and slates, and so soon as we examine the composition of these rocks we are struck by the remarkable resemblance to that of the class of igneous rocks. For purposes of comparison there is given below the composite or average constitution of igneous rocks in parallel column, with the average attained by combining the analyses of 56 slates and shales, the latter recalculated with water excluded:

	AVERAGE IGNEOUS ROCK				AVERAGE SHALE	
	(Clark)		(Washington)			
SiO_2	61.25		61.69		63.34	
Al_2O_3	15.81		15.94		15.56	
Fe_2O_3	2.70	} 6.31	1.88	} 4.53	4.41	} 7.89
FeO	3.61		2.65		3.48	
MgO	4.47		4.90		3.54	
CaO	5.03		5.02		3.33	
Na_2O	3.64		4.09		1.29	

K_2O	2.87	3.35	3.52
TiO_2	.62	.48	.53
	100.00	100.00	100.00

This close resemblance is probably of deep significance, for the reason that shales and slates are structurally the weakest of all rocks and for the further reason that they rather generally directly underlie the carbonate rocks, which are by contrast the strongest (see *ante*, p. 37). For these reasons shales and slates are the only rocks which are likely to be fused by relief from load through the formation of anticlinal arches within the earth's zone of flow. If this view is well founded, lavas and other igneous rocks are in large part fused argillaceous sediments formed in connection with the process of folding, or are refused rocks of igneous origin and similar composition.

Character profiles.—The character profiles of features connected in their origin with volcanoes are particularly easy to recognize, and in a few cases in which they might be confused with others of a different origin, an examination of the materials of the features should lead to a definitive judgment.

The lava plains which result from massive outflows of basalt might perhaps strictly be regarded as lack of feature, so great may be their continuous extent. Wherever definite vents exist, a broad flat dome is the usual result of the extravasation of a basaltic lava. The puys of France and many of the Kuppen of Germany, being formed from less fluid lava, have afforded profiles with relatively small radius of curvature.

In its youthful stage, the cinder cone usually presents a broad summit sag and relatively short side slopes, whereas the cone of later stages is apt to present long sweeping and upwardly concave curves with both the gradient and the radius of curvature increasing rapidly toward the summit. In contrast, too, with the earlier stage, the crest is relatively small. A marked reduction in the high symmetry of such profiles is noted wherever a breaching by lava outflow has occurred (Fig. 154).

With the composite cone, complexity and corresponding lack of symmetry is introduced, especially in the partially ruined caldera, and by the more or less accidental distribution of parasitic cones, as well as by migrations of the central cone. Peculiarly similar acuminated profiles result from spatter-cone formation, from the formation of a superchimney spine, and by the uncovering of the chimney through denudational processes—the volcanic neck.

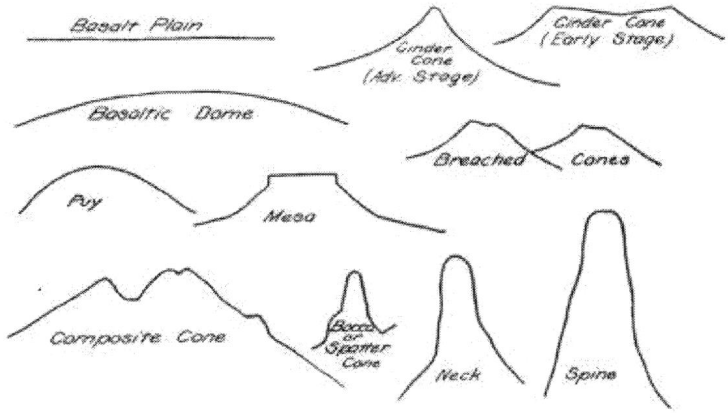

FIG. 154.—Character profiles connected with volcanoes.

Another important feature resulting from denudation is the Mesa or table mountain with its protecting basalt cap above softer rocks. Its profile most resembles that of table mountains due to differential erosion of alternately strong and weak horizontally bedded rocks, such as compose the upper portion of the section in the Grand Cañon of the Colorado. Here, however, in place of a single unusually strong top layer there are found several strong layers in alternation with weaker ones so as to produce additional steps in the profile.

READING REFERENCES TO CHAPTERS IX AND X

General works:—

PAULETT SCROPE. The Geology of the Extinct Volcanoes of Central France. John Murray, London, 1858, pp. 258. (An epoch-making work of early date which, like the following reference, may be studied to advantage to-day.)

SIR CHARLES LYELL. Principles of Geology, vol. 1, Chapters xxiii-xxv.

MELCHIOR NEUMAYR. Erdgeschichte, vol. 1, Allgemeine Geologie, revised edition by v. Uhlig, 1897, pp. 133-277 (a storehouse of valuable information clearly presented).

J. D. DANA. Characteristics of Volcanoes, with Contributions of Facts and Principles from the Hawaiian Islands. Dodd, Mead, and Company, New York, 1890, pp. 397.

TEMPEST ANDERSON. Volcanic Studies in Many Lands, being reproductions of photographs by the author with explanatory notes. John Murray, London, 1903, pp. 200, pls. 105.

T. G. BONNEY. Volcanoes, their Structure and Significance. John Murray, London, 1899, pp. 331.

I. C. RUSSELL. Volcanoes of North America. Macmillan, New York, 1897, pp. 346.

ELISÉE RÉCLUS. Les volcans de la terre, Belgian Society of Astronomy, Meteorology, and Physics of the Globe, 1906-1910 (a valuable descriptive geographical and bibliographical work of reference).

G. MERCALLI. I vulcani attivi della terre. Hoepli, Milan, 1907, pp. 421. (A most valuable work, beautifully illustrated, but in the Italian language.)

Arrangement of volcanic vents:—

TH. THORODDSEN. Die Bruchlinien und ihre Beziehungen zu den Vulkanen, Pet. Mitt., vol. 51, 1905, pp. 1-5, pl. 5.

R. D. M. VERBEEK. Various volumes and atlases of maps covering the Dutch East Indies and fully cited in the following reference (p. 21).

WILLIAM H. HOBBS. The Evolution and the Outlook of Seismic Geology, Proc. Am. Phil. Soc., vol. 48, 1909, pp. 17-27.

Birth of volcanoes:—

F. OMORI. The Usu-san Eruption and Earthquake and Elevation Phenomena, Bull. Earthq. Inv. Com., Japan, vol. 5, No. 1, 1911, pp. 1-37, pls. 1-13.

Fissure eruptions:—

TH. THORODDSEN. Island, IV, Vulkane, Pet. Mitt., Ergänzungsh. 153, 1906, pp. 108-111.

A. GEIKIE. Text-book of Geology, 4th ed., pp. 342-346.

Lava domes of Hawaii:—

J. D. DANA. Characteristics of Volcanoes (as above).

C. H. HITCHCOCK. Hawaii and Its Volcanoes. Honolulu, 1909, pp. 314.

Eruption of Matavanu volcano in 1906:—

KARL SAPPER. Der Matavanu-Ausbruch auf Savaii, 1905-1906, Zeit. d. Gesell. f. Erdk. z. Berlin, vol. 19, 1906, pp. 686-709, 4 pls.

H. J. JENSEN. The Geology of Samoa, and the Eruptions in Savaii, Proc. Linn. Soc., New South Wales, vol. 31, 1906, pp. 641-672, pls. 54-64.

TEMPEST ANDERSON. The Volcano of Matavanu in Savaii, Quart. Jour. Geol. Soc., London, vol. 66, 1910, pp. 621-639, pls. 45-52.

Eruption of Volcano in 1888:—

H. J. JOHNSTON-LAVIS. The South Italian Volcanoes. Naples, 1891, pp. 342, pls. 16.

Eruption of Taal volcano in 1911:—

W. E. PRATT. The Eruption of Taal Volcano, January 30, 1911, Phil. Jour. Sci., vol. 6, No. 2, Sec. A, 1911, pp. 63-86, pls. 1-14.

F. H. NOBLE. Taal Volcano, album of views of 1911 eruption, Manila, 1911, pp. 1-48.

The volcano of Etna:—

G. VOM RATH. Der Aetna. Bonn, 1872, pp. 1-33. (A beautiful piece of descriptive writing from both the geological and scenic standpoints.)

SARTORIUS VON WALTERSHAUSEN. Der Aetna. Leipzig, 1880, 2 quarto vols., pp. 371 and 548.

The eruption of Vesuvius in 1906:—

H. J. JOHNSTON-LAVIS. Geological Map of Monte Somma and Vesuvius, with a short and concise account, etc. Geo. Philip & Son, London, 1891.

H. J. JOHNSTON-LAVIS. The Eruption of Vesuvius in April, 1906, Trans. Roy. Dublin Soc., vol. 9, 1909, Pt. VIII, pp. 139-200, pls. 3-23 (the most authoritative work upon the subject).

T. A. JAGGAR, JR. The Volcano Vesuvius in 1906, Tech. Quart., vol. 19, 1906, pp. 105-115.

W. PRINZ. L'éruption du Vesuv d'avril, 1906, Ciel et Terre, 27e Année, 1906, pp. 1-49.

FRANK A. PERRET. Notes on the Electrical Phenomena of the Vesuvian Eruption, April, 1906, Sci. Bull., Brooklyn Inst. Arts and Sci., vol. 1, No. 11, pp. 307-312; Vesuvius, Characteristics and Phenomena of the Present Repose Period, Am. Jour. Sci., vol. 28, 1909, pp. 413-430.

WILLIAM H. HOBBS. The Grand Eruption of Vesuvius in 1906, Jour. Geol., vol. 14, 1906, pp. 636-655.

The spine of Pelée:—

E. O. HOVEY. The New Cone of Mont Pelée and the Gorge of the Rivière Blanche, Martinique, Am. Jour. Sci., vol. 16, 1903, pp. 269-281, pls. 11-14.

A. HEILPRIN. The Tower of Pelée. Philadelphia, 1904, pp. 62, pls. 22.

A. LACROIX. La montagne Pelée et ses éruptions, Acad. des Sciences, Paris, 1904, Chapter iii.

KARL SAPPER. In den Vulkangebieten Mittelamerikas und Westindiens, Stuttgart, 1905, pp. 172-178.

A. C. LANE. Absorbed Gases of Vulcanism, Science, N.S., vol. 18, 1903, p. 760.

G. K. GILBERT. The Mechanism of the Mont Pelée Spine, *ibid.*, vol. 19, 1904, pp. 927-928.

I. C. RUSSELL. Pelée Obelisk once More, *ibid.*, vol. 21, 1905, pp. 924-931.

The dissection of volcanoes:—

J. W. JUDD. Volcanoes, Chapter v.

S. SEKYA and Y. KIKUCHI. The Eruption of Bandai-San, Trans. Seis. Soc., Japan, vol. 13, Pt. 2, 1890, pp. 140-222, pls. 1-9.

R. D. M. VERBEEK. Krakatau. Batavia, 1885, pp. 557, pls. 25.

ROYAL SOCIETY. The Eruption of Krakatoa and Subsequent Phenomena. London, 1888, pp. 494.

G. K. GILBERT. Report on the Geology of the Henry Mountains, U.S. Geogr. and Geol. Surv., Rocky Mt. Region, Washington, 1877, pp. 22-60.

SIR A. GEIKIE. Ancient Volcanoes of Great Britain, vol. 2 especially.

D. W. JOHNSON. Volcanic Necks of the Mount Taylor Region, New Mexico, Bull. Geol. Soc. Am., vol. 18, 1907, pp. 303-324, pls. 25-30.

CHAPTER XI

THE ATTACK OF THE WEATHER

The two contrasted processes of weathering.—It has already been pointed out that change and not stability is the order of nature. Within the earth's outer shell and upon it rock alteration goes on continually, and from some portions of its surface the changed material is as constantly migrating to neighboring or even far distant regions. Before such transportation can begin the hard rock must first be broken down and reduced to fragments which the transporting agencies are competent to move.

To accomplish this breaking down, or *degeneration*, of the rock masses, either a wide range in temperature or chemical reaction is essential. In the atmosphere are found such active chemical agents as oxygen and carbon dioxide, the so-called carbonic acid gas; and these agents in the presence of water react chemically with the minerals of the rocks and form other minerals such as the hydrates and carbonates, which are lighter in weight and more soluble. This *chemical* attack upon the outer shell of the lithosphere is described as *decomposition.*

On the other hand the rock may succumb to changes which are purely mechanical and are due either to the stresses set up by differences between surface and interior temperatures, or to the prying action of the frost in the crevices. Such purely mechanical degeneration of the rocks is in contrast with decomposition and is described as *disintegration.* The two processes of decomposition and disintegration may, however, go on together; and the changes of volume that are caused by decomposition may result directly in considerable disintegration, as we are to see.

The rôle of the percolating water.—In order to effect chemical change or reaction, it is essential that the substances which are to react must be brought into such intimate contact with each other as it is seldom possible to attain except by solution. The chemical reactions which go on between the gaseous atmosphere and the solid lithosphere are accomplished through solution of the gases in water. This water, derived from rain or snow, percolates into the ground or descends along the crevices in the rocks, carrying with it a certain measure of dissolved air. This air differs from that of the surrounding atmospheric envelope by containing relatively large amounts of oxygen and of the other active element carbon dioxide. It follows from the important rôle thus performed by the percolating water that the process of decomposition will be relatively important in humid regions where the atmospheric precipitation is sufficient for the purpose.

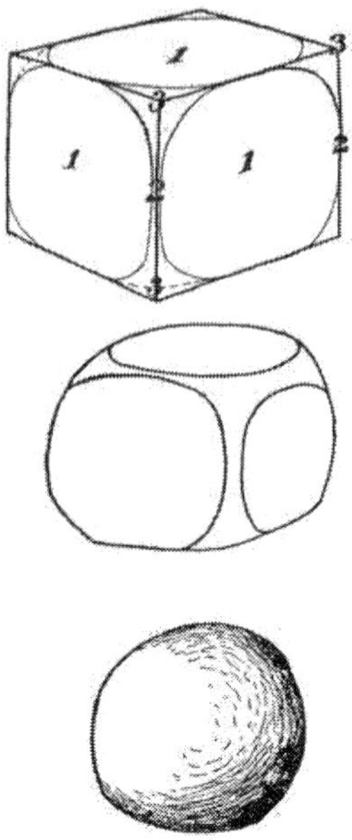

FIG. 155.—Successive diagrams to show the effect of decomposition and resulting disintegration upon joint blocks so as to produce spheroidal bowlders by weathering.

Within hot and dry regions there is a larger measure of rock disintegration, and distinct chemical changes unlike those of humid regions take place in the higher temperatures and with the more concentrated saline solutions. The discussion of such changes will be deferred until desert conditions are treated in another chapter.

Mechanical results of decomposition—spheroidal weathering.—From an earlier chapter it has been learned that the rocks of the earth's outermost shell are generally intersected by a system of vertical fissures which at each locality tend to divide the rock into parallel and upright rectangular prisms. It is these joints which offer relatively easy paths for the descent of the water into the rocks. In rocks of sedimentary origin there are found, in addition to the vertical joints, planes of bedding originally horizontal, and in the intrusive and volcanic rocks a somewhat similar parting, likewise parallel to the surface of the ground. The combined effect of the joints and the additional parting planes is thus to separate the rock mass into more or less perfect squared blocks (Fig. 155, upper figure) which stand in vertical columns.

The water which percolates downward upon the joints, finds its way laterally along the parting planes, and so subjects the entire surface of each block to simultaneous attack by its reagents. Though all parts of the surface of each block are alike subject to attack, it is the angles and the edges which are most vigorously acted upon. In the narrow crevices the solutions move but sluggishly, and as they are soon impoverished of their reagents in the attack upon the rock, fresh solution can reach the middle of the faces from relatively few directions. The edges are at the same time being reached from many more directions, and the corners from a still larger number.

The minerals newly formed by these chemical processes of hydration and carbonization are notably lighter, and hence more bulky than the minerals from whose constituents they have been largely formed. Strains are thus set up which tend to separate the bulkier new material from the core of unaltered rock below. As the process continues, distinct channels for the moving waters are developed favorable to action at the edges and corners of the blocks. Eventually, the squared block is by this process transformed into a spheroidal core of still unaltered rock wrapped in layers of decomposed material, like the outer wrappings of an onion. These in turn are usually imbedded in more thoroughly disintegrated material from which the shell structure has disappeared (Fig. 156).

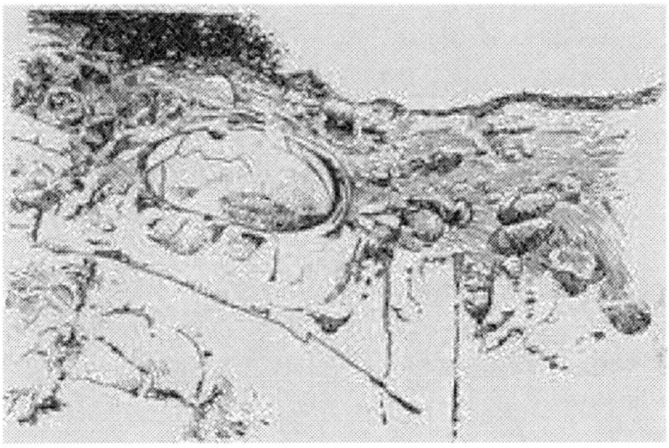

FIG. 156.—Spheroidal weathering of an igneous rock.

Exfoliation or scaling.—A fact of much importance to geologists, but one far too often overlooked, is that rocks are but poor conductors for heat. It results from this that in the bright sun of a summer's day a thin skin, as it were, upon the rock surface may be heated to a relatively high temperature, although the layer immediately below it is practically unaffected. The consequent expansion of the surface layer causes stresses that tend to scale it off from the layer below, which, uncovered in its turn, develops new strains of the same sort. This process of exfoliation acquires exceptional importance in desert regions where the rock surfaces are daily elevated to excessively high temperatures (see Chapter XV).

FIG. 157.—Dome structure in granite mass, Yosemite valley, California (after a photograph by Sinclair).

Dome structure in granite masses.—In large granite masses, such as are to be found in the ranges of the Sierra Nevada of California, a peculiar dome structure is sometimes found developed upon a large scale, and has had an important influence upon the breaking down of the rock and upon the shaping of the mountain (Fig. 157). Such a structure, made up as it is of prodigious layers, can have little in common with the veneers of weathered minerals which are the result of exfoliation, and it is quite likely that the dome structure is in some way connected with the relief of these massive rocks from their load—the rock which once rested upon them, but has been carried away by erosion since the uplift of the range.

The prying work of frost.—In all countries where winter temperatures range below the freezing point of water, a most potent agent of rock disintegration is the frost which pries at every crevice and cranny of the surface rock. Important in the temperate zones, in the polar regions it becomes almost the sole effective agent of rock weathering. There, as elsewhere, its efficiency as a disintegrating agent is directly dependent upon the nature of the crevices within the rock, so that the omnipresent joints are able to exercise a degree of control over the sculpturing of the surface features which is hardly to be looked for elsewhere (see plate 10 A).

FIG. 158.—Talus slope beneath a cliff.

Talus.—Wherever the earth's surface rises in steep cliffs, the rock fragments derived from frost action, or by other processes of disintegration, as they become detached either fall or slide rapidly downward until arrested upon a flatter slope. Upon the earlier accumulations of this kind, the later ones are deposited, until their surface slopes up to the cliff face as steeply as the material will lie—the angle of repose. Such débris accumulations at the base of a cliff (Fig. 158) are known as *talus*, and the slope is described as a talus slope, or in Scotland as a "scree."

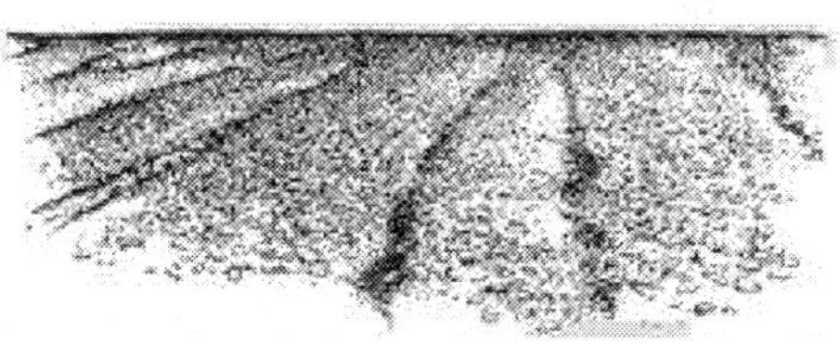

FIG. 159.—Striped ground from soil flow of chipped rock fragments upon a slope, Snow Hill Island, West Antarctica (after Otto Nordenskiöld).

Soil flow in the continued presence of thaw water.—So soon as the rocks are broken down by the weathering processes, they are easily moved, usually to lower levels. In part this transportation may be accomplished by gravity slowly acting upon the disintegrated rock and causing it to creep down the slope. Yet even in such cases water is usually present in quantity sufficient to fill the spaces between the grains, and so act as a lubricant to facilitate the migration.

Upon a large scale rocks which were either originally incoherent or have been made so by weathering, after they have become saturated with water, may start into sudden motion as great landslides or avalanches, which in the space of a few moments materially change the face of the country, and by burying the bottom lands leave disaster and misery in their wake.

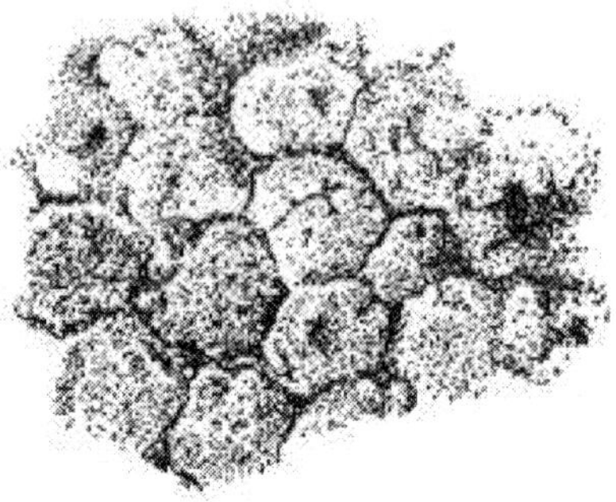

FIG. 160.—Pavement of horizontal surface due to soil flow, Spitzbergen (after Otto Nordenskiöld).

Within the subpolar regions, where a large part of the surface is for much of the year covered with snow, the underlying rocks are for long periods saturated with thaw water, and in alternation are repeatedly frozen and thawed. Essentially similar conditions are met with in the high, snow-capped mountains of temperate or torrid regions. For the subpolar regions particularly it is now generally recognized that somewhat special processes of soil flow, described under the name *solifluction*, are characteristic. The exact nature of these processes is as yet imperfectly understood, but there can be little doubt concerning the large rôle which they have played in the transportation of surface materials. Such soil flow is clearly manifested under different aspects, and it is likely that by this comprehensive term distinct processes have been brought together.

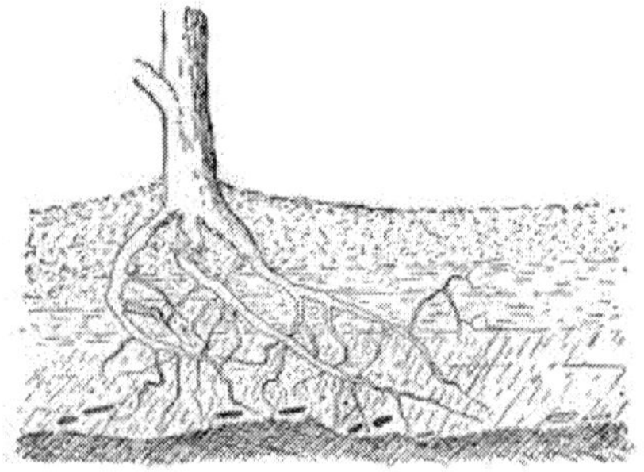

FIG. 161.—Tree roots entering fissured rock and prying its sections apart.

Possibly the most striking aspect of the soil flow in subpolar regions is furnished by the remarkable "stone rivers" and "rock glaciers"; though the more generally characteristic are peculiar stripings or other markings which appear upon the surface of the ground and thus betray the movements of the underlying materials. Upon slopes it is not uncommon for the surface to be composed of angular rock fragments riven by the frost and crossed by broad parallel furrows as though a gigantic plow had gone over it (Fig. 159). The direction of the furrows is always up and down the slope, and the striping is marked in proportion as the slope is steep. Where the bottom is reached, the furrows are replaced by a sort of mosaic pavement of hexagonal repeating figures, each of which may be an area of the surface six feet or more across (Fig. 160, and Fig. 390, p. 368). The depressions which separate the "blocks" of the pavement are often filled with clay, while the inclosed surfaces are made up of coarsely chipped stone.

The splitting wedges of roots and trees.—In the mechanical breakdown of the rocks within humid regions a not unimportant part is sometimes taken by the trees, which insinuate the tenuous extremities of their rootlets into the smallest cracks and by continued growth slowly wedge even the firmer rocks apart (Fig. 161). In a similar manner the small tree trunk growing within a crevice of the rock may in time split its parts asunder (Fig. 162).

FIG. 162.—A large glacial bowlder split by a growing tree near East Lansing, Michigan (after a photograph by Bertha Thompson).

The rock mantle and its shield in the mat of vegetation.—Through the action of weathering, the rocks, as we have seen, lose their integrity within a surface layer, which, though it may be as much as a hundred feet or more in thickness, must still be accounted a mere film above the underlying bed rock. The mechanical agents of the breakdown operate only within a few feet of the surface, and the agents of rock decomposition, derived as they are from the atmosphere, become inert before they have descended to any considerable depth. The surface layer of incoherent rock is usually referred to as the *rock mantle* (Fig. 163). Where the rock mantle is relatively deep, as it is in the states south of the Ohio in the eastern United States, there is found, deep below the outer layer of soil, a partially decomposed and disintegrated rock, of which the unaltered minerals lie unchanged in position but separated by the new minerals which have resulted from the breakdown of their more susceptible associates. While thus in a certain sense possessing the original structure, this altered material is essentially incoherent and easily succumbs to attack by the pick and spade, so that it is only at considerably greater depths that the unaltered rock is encountered.

FIG. 163.—Rock mantle consisting of broken rock, above which is soil and a vegetable mat. Coast of California (after a photograph by Fairbanks).

Because of the tendency of mantle rock to creep down upon slopes it is generally found thicker upon the crests and at the bases of hills and thinnest upon their slopes (Fig. 164).

In the transformation of the upper portion of the mantle rock into soil, additional chemical processes to those of weathering are carried through by the agency of earthworms, bacteria, and other organisms, and by the action of humus and other acids derived from the decomposition of vegetation. The bacteria particularly play a part in the formation of carbonates, as they do also in changing the nitrogen of the air into nitrates which become available as plant food. Within the humid tropical regions ants and other insects enter as a large factor in rock decomposition, as they do also in producing not unimportant surface irregularities.

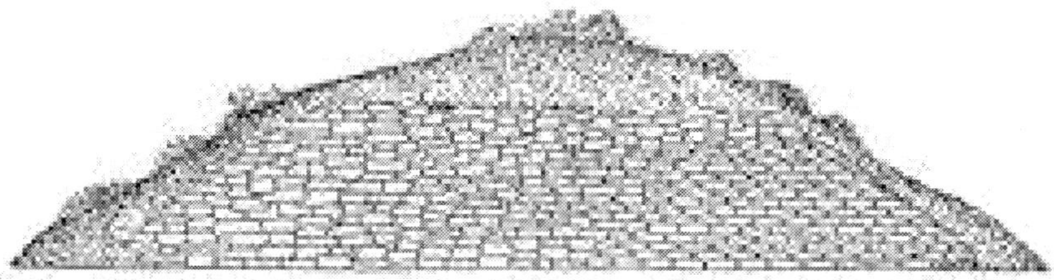

FIG. 164.—Diagram to show the varying thickness of mantle rock upon the different portions of a hill surface (after Chamberlin and Salisbury).

How important is the cover of vegetation in retaining the rock mantle and the upper soil layer in their respective positions, as required for agricultural purposes, may be best illustrated by the disastrous consequences of allowing it to be destroyed. Wherever, by the destruction of forests, by the excessive grazing of animals, or by other causes, the mat of turf has been destroyed, the surface is opened in gullies by the first hard rain, and the fertile layer of soil is carried from the slopes and distributed with the coarser mantle upon the bottom lands. Thus the face of the country is completely transformed from fertile hills into the most desolate of deserts where no spear of grass is to be seen and no animal food to be obtained (plate 5 A). The soil once washed away is not again renewed, for the continuation of the gullying process now effectively prevents its accumulation.

PLATE 5.

A. **Once wooded region in China now reduced to desert through deforestation (after Willis).**

B. **"Bad Lands" in the Colorado Desert (after Mendenhall).**

Decomposition and disintegration:—

GEORGE P. MERRILL. The Principles of Rock Weathering, Jour. Geol., vol. 4, 1896, pp. 704-724, 850-871. Rocks, Rock Weathering, and Soils. Macmillan, New York, 1897, Pt. iii, pp. 172-411.

ALEXIS A. JULIEN. On the Geological Action of the Humus Acids, Proc. Am. Assoc. Adv. Sci., vol. 28, 1879, pp. 311-410.

Corrosion of rocks:—

C. W. HAYES. Solution of Silica under Atmospheric Conditions, Bull. Geol. Soc. Am., vol. 8, 1897, pp. 213-220, pls. 17-19.

M. L. FULLER. Etching of Quartz in the Interior of Conglomerates, Jour. Geol., vol. 10, 1902, pp. 815-821.

C. H. SMYTH, JR. Replacement of Quartz by Pyrites and Corrosion of Quartz Pebbles, Am. Jour. Sci. (4), vol. 19, 1905, pp. 282-285.

Dome structure of granite masses:—

G. K. GILBERT. Domes and Dome Structure of the High Sierra, Bull. Geol. Soc. Am., vol. 15, 1904, pp. 29-36, pls. 1-4.

RALPH ARNOLD. Dome Structure in Conglomerate, *ibid.*, vol. 18, 1907, pp. 615-616.

Soil flow:—

J. GUNNAR ANDERSSON. Solifluction, a Component of Subaërial Denudation, Jour. Geol., vol. 14, 1906, pp. 91-112.

OTTO NORDENSKIÖLD. Die Polarwelt und ihre Nachbarländer, Leipzig, 1909, pp. 60-65.

ERNEST HOWE. Landslides in the San Juan Mountains, Colorado, etc., Prof. Pap., 67 U. S. Geol. Surv., 1909, pp. 1-58, pls. 1-20.

G. E. MITCHELL. Landslides and Rock Avalanches, Nat. Geogr. Mag., vol. 21, 1910, pp. 277-287.

WILLIAM H. HOBBS. Soil Stripes in Cold Humid Regions and a Kindred Phenomenon, 12th Rept. Mich. Acad. Sci., 1910, pp. 51-53, pls. 1-2.

Relation of deforestation to erosion:—

N. S. SHALER. Origin and Nature of Soils, 12th Ann. Rept. U. S. Geol. Surv., 1891, Pt. 1, pp. 268-287.

W. J. MCGEE. The Lafayette Formation, *ibid.*, pp. 430-448.

F. H. KING. Soils. Macmillan, New York, 1908, pp. 50-54.

BAILEY WILLIS. Water Circulation and Its Control, Rept. Nat. Conserv. Com., 1909, vol. 2, pp. 687-710.

W. J. MCGEE. Soil erosion, Bull. 71, U. S. Bureau of Soils, 1911, pp. 60, pls. 33.

CHAPTER XII

THE LIFE HISTORIES OF RIVERS

The intricate pattern of river etchings.—The attack of the weather upon the solid lithosphere destroys the integrity of its surface layer, and through reducing it to rock débris makes it the natural prey of any agent competent to carry it along the surface. We have seen how, for short distances, gravity unaided may pile up the débris in accumulations of talus, and how, when assisted by thaw water which has soaked into the material, it may accomplish a slow migration by a peculiar type of soil flow. Yet far more potent transporting agencies are at work, and of these the one of first importance is running water. Only in the hearts of great deserts or in the equally remote white deserts of the polar regions is the sound of its murmurings never heard. Every other part of the earth's surface has at some time its running water coursing in valleys which it has itself etched into the surface. It is this etching out of the continents in an intricate pattern of anastomosing valleys which constitutes the chief difference between the land surface and the relatively even floor of the oceans.

The motive power of rivers.—Every river is born in throes of Mother Earth by which the land is uplifted and left at a higher level than it was before. It is the difference of elevation thus brought about between separated portions of the land areas that makes it possible for the water which falls upon the higher portions to descend by gravity to the lower. This natural "head" due to differences of elevation is the motive power of the local streams, and for each increase in elevation there is an immediate response in renewed vigor of the streams. The elevated area off which the rivers flow is here termed an upland.

The velocity of a stream will be dependent not only upon the difference in altitude between its source and its mouth, but upon the distance which separates them, since this will determine the grade. The level of the mouth being the lowest which the stream can reach is termed the *base level*, and the current is fixed by the slope or declivity. The capacity to lift and transport rock débris is augmented at a quite surprising rate with every increase in current velocity, the law being that the weight of the heaviest transportable fragment varies with the sixth power of the velocity of the current. Thus if one stream flows twice as rapidly as another, it can transport fragments which are sixty-four times as heavy.

Old land and new land.—The uplifts of the continents may proceed without changes in the position of the shore lines, in which case areas, already carved by streams but no longer actively modified by them, are worked upon by tools freshly sharpened and driven by greater power. The land thus subjected to active stream cutting is described as old land, and has already had engraved upon it the characteristic pattern of river etchings, albeit the design has been in part effaced.

If, upon the other hand, the shore line migrates seaward with the uplift, a portion of the relatively even sea floor, or new land, is elevated and laid under the action of the running water. As we are to see, stream cutting is to some extent modified when a river pattern is inherited from the uplift. The uplift, whether of old land only or of both old land and new land, marks the starting point of a new river history, usually described as an *erosion cycle.*

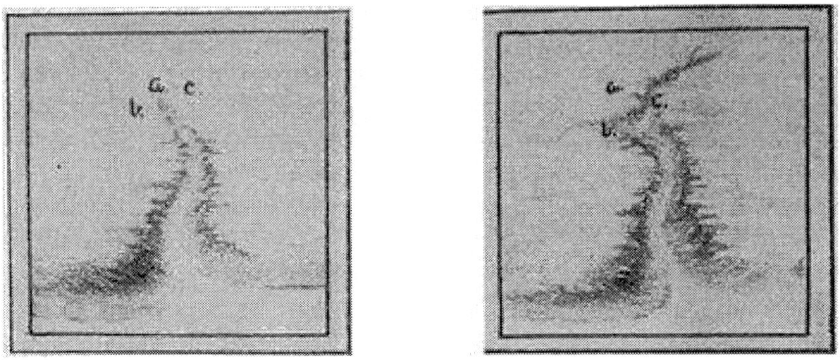

FIG. 165.—Two successive forms of gullies from the earliest stage of a river's life (after Salisbury and Atwood).

The earlier aspects of rivers.—Though geologists have sometimes regarded the uplift of the continents as a sort of upwarping in a continuous curved surface, the discussions of river histories and the pictorial illustrations of them have alike clearly assumed that the uplift has been essentially in blocks and that the elevated area meets the lower lying country or the sea in a more or less definite escarpment. The first rivers to develop after the uplift may be described as gullies shaped by the sudden downrush of storm waters and spaced more or less regularly along the margin of the escarpment (Fig. 165). These gullies are relatively short, straight, and steep; they have precipitous walls and few, if any, tributaries.

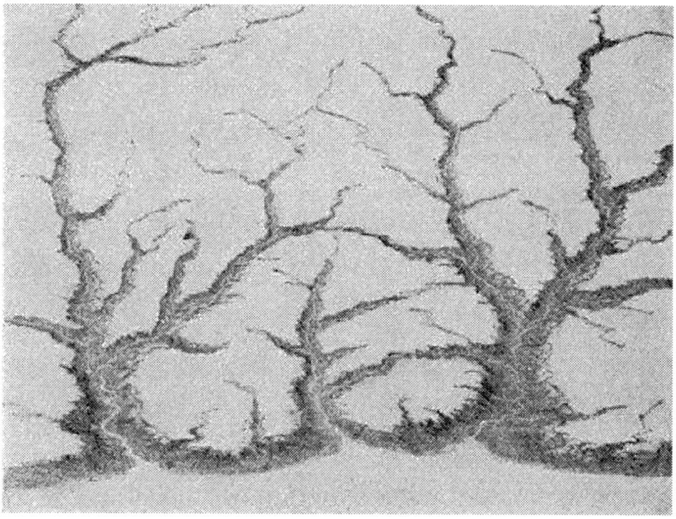

FIG. 166.—Partially dissected upland (after Salisbury and Atwood).

With time the gully heads advance into the upland as they take on tributaries; and so at length they in part invest it and dissect it into numerous irregularly bounded and flat-topped tables which are separated by cañons (Fig. 166). At the same time the grade of the channel is becoming flatter, and its precipitous walls are being replaced by curving slopes, as will be more fully described in the sequel. It is because of this progressive reduction of grades with increasing age that the early stages of a river's life are much the most turbulent of its history. The water then rushes down the steep grades in rapids, and is often at times opened out in some basin to form a lake where differences of uplift have been characteristic of neighboring sections. For several reasons such basins in the course of a stream are relatively short lived (Chapter XXX), and they disappear with the earlier stages of the river history.

The meshes of the river network.—From the continued throwing out of new tributaries by the streams, the meshes in the river network draw more closely together as the stages of its history advance. The closeness of texture which is at last developed upon the upland is in part determined by the quantity of rainfall, so that in New Jersey with heavy annual precipitation the meshes in the network are much smaller than they are, for example, upon the semiarid or arid plains of the western United States. Its design will, however, in either case more or less clearly express the plan of rock architecture which is hidden beneath the surface (Chapter XVII).

The upper and lower reaches of a river contrasted.—From the fact that the river progressively invades new portions of the upland and lays the acquired sections under more and more thorough investment, it has near its headwaters for a long time a frontier district which may be regarded as youthful even though the sections near its mouth have reached a somewhat advanced stage. The newly acquired sections of river valley may thus possess the steep grade and precipitous walls which are characteristic of early gullies and cañons and are in contrast with the more rounded and flat-bottomed sections below. Lateral streams, from the fact that they are newer than the main or trunk stream to which they are tributary, likewise descend upon somewhat steeper grades (Fig. 167).

FIG. 167.—Characteristic longitudinal sections of the upper portion of a river valley and its tributaries (after scaled sections by Nussbaum).

The balance between degradation and aggradation.—We have seen that the power to transport rock fragments is augmented at a most surprising rate with every increase in the current velocity. While the lighter particles of rock may be carried as high up as the surface of the water, the heavier ones are moved forward upon the bottom with a combined rolling and hopping motion aided by local eddies. Those particles which come in contact with the bottom or sides of the channel abrade its surface so as ever to deepen and widen the valley. This cutting accomplished by partially suspended débris in rapidly moving currents of water is known as *corrasion* and the stream is said to be *incising* its valley.

As the current is checked upon the lower and flatter grades, some of its load of sediment, and especially the coarser portion, will be deposited and so partially fill in the channel. A nice balance is thus established between *degradation* and the contrasted process known as *aggradation.* The older the river valley the flatter become the grades at any section of its course, and thus the point which separates the lower zone of aggradation from the upper one of degradation moves steadily upstream with the lapse of time.

The accordance of tributary valleys.—It is a consequence of the great sensitiveness of stream corrasion to current velocity that no side stream may enter the trunk valley at a level above that of the main stream—the tributary streams enter the trunk stream *accordantly.* Each has carved its own valley, and any abrupt increase in gradient of the side streams near where they enter the main stream would have increased the local corrasion at an accelerated rate and so have cut down the channel to the level of the trunk stream.

The grading of the flood plain.—All rivers are subject to seasonal variations in the volume of their waters. Where there are wet and dry seasons these differences are greatest, and for a large part of the year the valleys in such regions may be empty of water, and are in fact often utilized for thoroughfares. In the temperate climates of middle latitudes rivers are generally flooded in the spring when the winter snows are melted, though they may dwindle to comparatively small streams during the late summer. In the upper reaches of the river the current velocities are such that the usual river channel may carry all the water of flood time; but lower down and in the zone of aggradation, where the current has been checked, the level of the water rises in flood above the banks of its usual channel and spreads over the surrounding lowlands. As a deposit of sediment is spread upon the surface, the succession of the annual deposits from this source raises the general level as a broad floor described as the *flood plain* of the river.

The cycles of stream meanders.—The annual flooding with water and simultaneous deposition of silt is not, however, the only grading process which is in operation upon the flood plain. It is characteristic of swift currents that their course is maintained in relatively straight lines because of the inertia of the rapidly moving water. In proportion as their currents become sluggish, rivers are turned aside by the smallest of obstructions; and once diverted from their straight course, a law of nature becomes operative which increases the curvature of the stream at an accelerated rate up to a critical point, when by a change, sudden and catastrophic, a new and direct course is taken, to be in its turn carried through a similar cycle of changes. This so-called *meandering* of a stream is accompanied by a transfer of sediment from one bend or meander of the river to those below and from one bank to the other. Inasmuch as the later meanders cross the earlier ones and in time occupy all portions of the plain to the same average extent, a process of rough grading is accomplished to which the annual overflow deposit is supplementary.

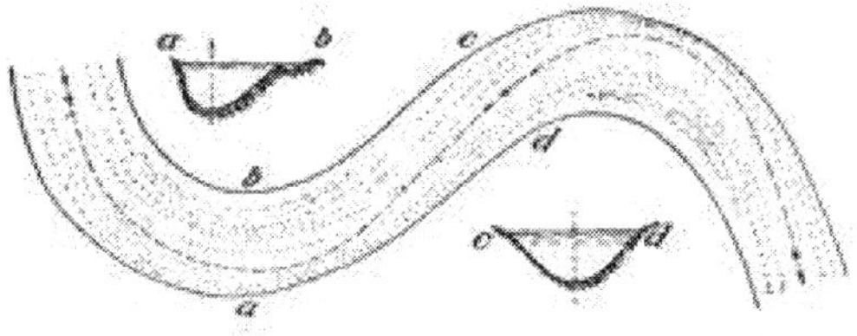

FIG. 168.—Map and sections of a stream meander. The course of the main current is indicated by the dashed line.

The course of the current in consecutive meanders and the cross sections of the channel which result directly from the meandering process will be made clear from examination of Fig. 168. So soon as diverted from its direct course, the current, by its inertia of motion, is thrown against the outer or convex side so as to scour or corrade that bank. Upon the concave or inner side of the curve there is in consequence an area of slack water, and here the silt scoured from higher meanders is deposited. The scouring of the current upon the outer bank and the filling upon the inner thus gives to the cross section of the stream a generally unsymmetrical character (Fig. 168 *ab*). Between meanders near the point of inflection of the curve, and there only, the current is centered in the middle of the channel and the cross section is symmetrical (Fig. 168 *cd*).

FIG. 169.—Tree in part undermined upon the outer bank of a meander.

The scour upon the convex side of a meander causes the river to swing ever farther in that direction, and through invasion of the silted flood plain to migrate across it. Trees which lie in its path are undermined and fall outward in the stream with tops directed with the current (Fig. 169). Whenever the flood plain is forested, the fallen trees may be so numerous as to lie in ranks along the shore, and at the time of the next flood they are carried downstream to jam in narrow places along the channel and give the erroneous impression that the flood has itself uprooted a section of forest (see p. 418).

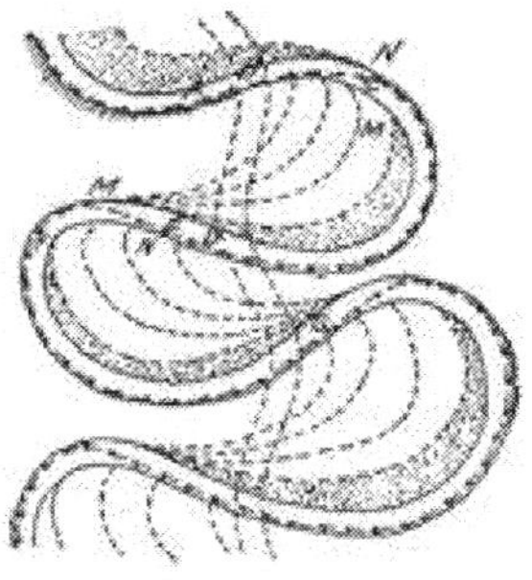

FIG. 170.—Diagrams to show the successive positions of stream meanders and the relatively stationary point near the sharpest curvature.

The cut-off of the meander.—As the meander swings toward its extreme position it becomes more and more closely looped. Adjacent loops thus approach nearer and nearer to each other, but in the successive positions a nearly stationary point is established near where the river makes its sharpest turn (Fig. 170, *G*, and Fig. 454, p. 417). At length the neck of land which separates meanders is so narrow that in the next freshet a temporary jamming of logs within the channel may direct the waters across the neck, and once started in the new direction a channel is scoured out in the soft silt. Thus by a breaking through of the bank of the stream, a so-called "crevasse", the river suddenly straightens its course, though up to this time it has steadily become more and more sharply serpentine. After the cut-off has occurred, the old channel may for a time continue to be used by the stream in common with the new one, but the advantage in velocity of current being with the cut-off, the old channel contains slacker water and so begins to fill with silt both at the beginning and the end of the loop. Eventually closed up at both ends, this loop or "ox-bow" is entirely separated from the new channel, and once abandoned of the stream is transformed into an ox-bow lake (Fig. 171 and p. 415).

FIG. 171.—An ox-bow lake in the flood plain of a river.

Meander scars.—Swinging as it occasionally does in its meanderings quite across the flood plain and against the bank of the earlier degrading river in this section, the meander at times scours the high bank which bounds the flood plain, and undermining it in the same manner, it excavates a recess of amphitheatral form which is known as a meander *scar* (Fig. 172). At length the entire bank is scarred in this manner so as to present to the stream a series of concave scallops separated by sharp intermediate salients of cuspate form.

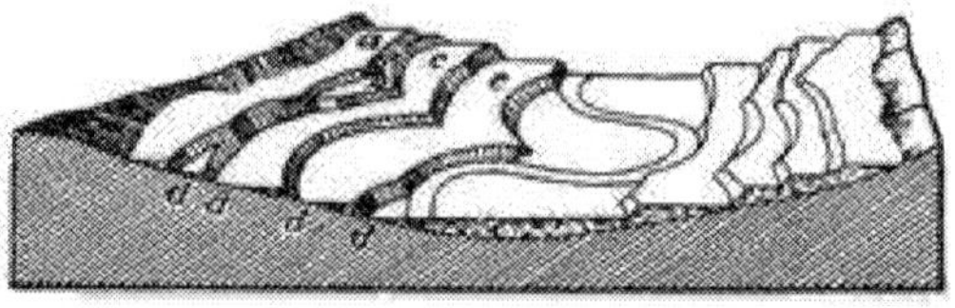

FIG. 172.—Schematic representation of a series of river terraces. *a*, *b*, *c*, *e*, successive terraces in order of age. *d*, *d*, *d*, *d*, terrace slopes formed of meander scars.

River terraces.—Whenever the river's history is interrupted by a small uplift, or the base level is for any reason lowered, the stream at once begins to sink its channel into the flood plain. Once more flowing upon a low grade, it again meanders, and so produces new walls at a lower level, but formed, like the first, of intersecting meander scars. Thus there is produced a new flood plain with cliff and terrace above, which is known as a *river terrace*. A succession of uplifts or of depressions of the base level yields terraces in series, as they appear schematically represented in Fig. 172. Such terraces are to be found well developed upon most of our larger rivers to the northward of the Ohio and Missouri. The highest terrace is obviously the remnant of the earliest flood plain, as the lowest represents the latest.

The delta of the river.—As it approaches its mouth the river moves more and more sluggishly over the flat grades, and swings in broader meanders as it flows. Yet it still carries a quantity of silt which is only laid down after its current has been stopped on meeting the body of standing water into which it discharges. If this be the ocean, the salinity of the sea water greatly aids in a quick precipitation of the finest material. This clarifying effect upon the water of the dissolved salt may be strikingly illustrated by taking two similar jars, the one filled with fresh and the other with salt water, and stirring the same quantity of fine clay into each. The clay in the salt water is deposited and the water cleared long before the murkiness of the other has disappeared.

By the laying down of the residue of its burden of sediment where it meets the sea, the river builds up vast plains of silt and clay which are known as deltas and which often form large local extensions of the continents into the sea. Whereas in its upper reaches the river with its tributary streams appears in the plan like a tree and its branches, in the delta region the stream, by dividing into diverging channels called distributaries (Fig. 458, p. 420), completes the resemblance to the tree by adding the roots. From the divergence of the distributaries upon the delta plain the Greek capital letter Δ is suggested and has supplied the name for these deposits. Of great fertility, the delta plains of rivers have become the densely populated regions of the globe, among which it is necessary to mention only the delta of the Nile in Egypt, those of the Ganges and Brahmaputra in India, and those of the Hoang and Yangtse rivers in China.

The levee.—When the snows thaw upon the mountains at the headwaters of large rivers, freshets result and the delta regions are flooded. At such times heavily charged with sediment, a thin deposit of fertile soil is left upon the surface of the delta plain, and in Egypt particularly this is depended upon for the annual enrichment of the cultivated fields. Though at this time the waters spread broadly over the plain, the current still continues to flow largely within the normal channel, so that the slack water upon either side becomes the locus for the main deposit of the sediment. There is thus built up on either side of the channel a ridge of silt which is known as a *levee*, and this bank is steadily increased in height from year to year (Fig. 452).

To prevent the danger of floods upon the inhabited plains, artificial levees are usually raised upon the natural ones, and in a country like Holland, such levees (dikes) involve a large expenditure of money and no small degree of engineering skill and experience to construct. So important to the life of the nation is the proper management of its dikes, that in the past history of China each weak administration has been marked by the development of graft in this important department and by floods which have destroyed the lives of hundreds of thousands of people.

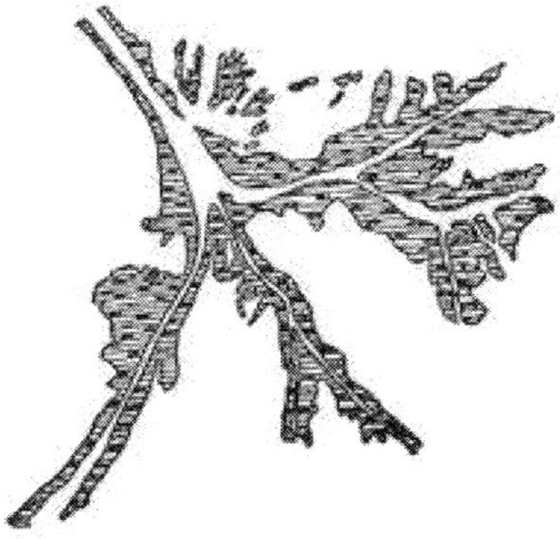

FIG. 173.—"Bird-foot" delta of the Mississippi River.

Wherever there has been a markedly rapid sinking upon a delta region, and depressions are common in delta territory, no doubt as a result of the loading down of the crust, the river may present the paradoxical condition of flowing at a higher level than the surrounding country. Between the levees of neighboring distributaries there are peculiar saucer-shaped depressions of the country which easily become filled with water. At the extremity of the delta the levee may be the only land which shows above the ocean surface, and so present the peculiar "bird-foot" outline which is characteristic of the extremity of the Mississippi delta, though other processes than the mere sinking of the deposits may contribute to this result (Fig. 173).

The sections of delta deposits.—If now we leave the plan of the delta to consider the section of its deposits, we find them so characteristic as to be easily recognized. Considered broadly, the delta advances seaward after the manner of a railroad embankment which is being carried across a lake. Though the greater portion of the deposit is unloaded upon a steep slope at the front, a smaller amount of material is dropped along the way, and a layer of extremely fine material settles in advance as the water clears of its finely suspended particles (Fig. 174). Simultaneous deposits within a delta thus comprise a nearly horizontal layer of coarser materials, the so-called top-set bed; the bulk of the deposit in a forward sloping layer, the so-called fore-set bed; and a thin film of clay which is extended far in advance, the bottom-set bed (Fig. 174, 2). If at any point a vertical section is made through the deposits, beds deposited in different periods are encountered; the oldest at the bottom in a horizontal position, the next younger above them and with forward dip, and the youngest and coarsest upon the top in nearly horizontal position (Fig. 174, 3).

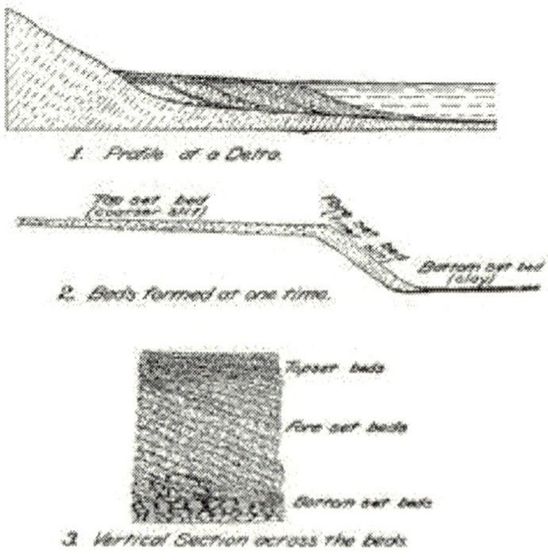

FIG. 174.—Diagrams to show the nature of delta deposits as exhibited in section.

It has been estimated that the surface of the United States is now being pared down by erosion at the average rate of an inch in 760 years. The derived material is being deposited in the flood plain and delta regions of its principal rivers. Some 513 million tons of suspended matter is in the United States carried to tidewater each year, and about half as much more goes out to sea as dissolved matter. If this material were removed from the Panama Canal cutting, an 85-foot sea-level canal would be excavated in about 73 days. The Mississippi River alone carries annually to the sea 340 million tons of suspended matter, or two thirds of the entire amount removed from the area of the United States as a whole. It is thus little wonder that great deltas have extended their boundaries so rapidly and that the crust is so generally sinking beneath the load.

CHAPTER XIII

EARTH FEATURES SHAPED BY RUNNING WATER

The newly incised upland and its sharp salients.—The successive stages of incising, sculpturing, and finally of reducing an uplifted land area, are each of them possessed of distinctive characters which are all to be read either from the map or in the lines of the landscape. Upon the newly uplifted plain the incising by the young rivers is to be found chiefly in the neighborhood of the margins. In this stage the valleys are described as V-shaped cañons, for the valley wall meets the upland surface in sharp salients (plate 12 A), and the lines of the landscape are throughout made up from straight elements. Though the landscapes of this stage present the grandest scenery that is known and may be cut out in massive proportions, often with rushing river or placid lake to enhance the effect of crag and gorge, they lack the softness and grace of outline which belong only to the maturer erosion stages. The grand cañon of the Colorado presents the features characteristic of this stage in the grandest and most sublime of all examples, and the castled Rhine is a gorge of rugged beauty, carved out from the newly elevated plateau of western Prussia, through which the water swirls in eddying rapids (Fig. 175).

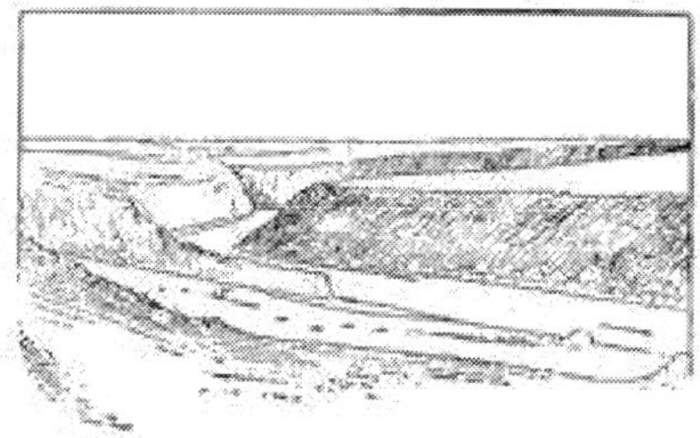

FIG. 175.—Gorge of the River Rhine near St. Goars, incised within an uplifted plain which forms the hill tops.

The stage of adolescence.—As the upland becomes more largely invaded as a consequence of the headward advance of the cañons and their sending out of tributary side cañons, the sharp angles in which the cañon walls intersect the plain become gradually replaced by well-rounded shoulders. Thus the lines in the landscape of this stage are a combination of the straight line with a simple curve convex toward the sky (Fig. 176). In this stage large sections of the original plateau remain, though cut into small areas by the extensions of the tributary valleys.

FIG. 176.—V-shaped valley with well-rounded shoulders characteristic of the stage of adolescence. Allegheny plateau in West Virginia.

The maturely dissected upland.—Continued ramifications by the rivers eventually divide the entire upland area into separated parts, and the rounding of the shoulders of valleys proceeds simultaneously until of the original upland no easily recognizable compartments are to be found. Where before were flat hilltops are now ridges or watersheds, the well-known *divides*. The upland is now said to be completely dissected or to have arrived at *maturity*. The streams are still vigorous, for they make the full descent from the upland level to base level, and yet a critical turning point of their history has been reached, and from now on they are to show a steady falling off in efficiency as sculpturing agents.

FIG. 177.—View of a maturely dissected upland from one of its hilltops, Klamath Mountains, California (after a photograph by Fairbanks).

Viewed from one of the hilltops, the landscape of this stage bears a marked resemblance to a sea in which the numberless divides are the crests of billows, and these, as distance reduces their importance in the landscape, fade away into the even line of the horizon (Fig. 177).

The Hogarthian line of beauty.—Since the youthful stage of the upland, when the lines of its landscape were straight, its character rugged, and its rivers wild and turbulent, there has been effected a complete transformation. The only straight line to be seen is the distant horizon, for the landscape is now molded in softened outlines, among which there is a repeated recurrence of the line of beauty made famous by Hogarth in his "Analysis of Beauty." As well known to all art students, this is a sinuous line of reversed or double curvature—a curve which passes insensibly at a point of inflection from convex to concave (Fig. 178). The curve of beauty is now found in every section of the hills, and it imparts to the landscape a gracefulness and a measure of restfulness as well, which are not to be found in the landscapes of earlier stages in the erosion cycle. In the bottoms of the valleys also the initial windings of the rivers within their narrow flood plains add silver beauty lines which stand out prominently from the more somber background of the hills.

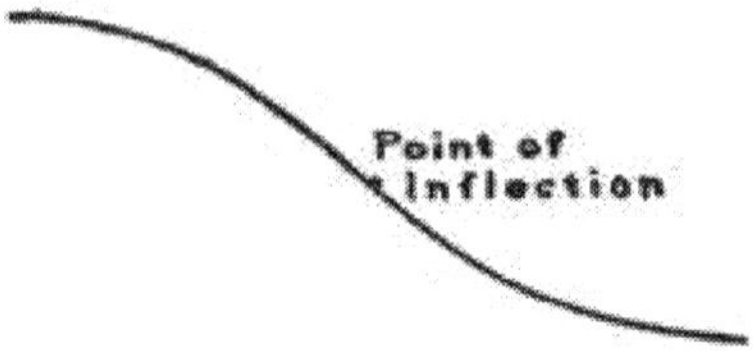

FIG. 178.—Hogarth's line of beauty.

Considered from the commercial viewpoint, the mature upland is one of the least adaptable as a habitation for highly civilized man. Direct lines of communication run up hill and down dale in monotonous alternation, and almost the only way of carrying a railroad through the region, without an expenditure for trestles which would be prohibitive, is to follow the tortuous crest of a main divide or the equally winding bed of one of the larger valleys.

FIG. 179.—View of the old land of New England, with Mount Monadnock rising in the distance.

The final product of river sculpture—the peneplain.—When maturity has been reached in the history of a river, its energies are devoted to a paring down of the valley slopes and crests so as to reduce the general level. From this time on hill summits no longer fall into a common level—that of the original upland—for some mount notably higher than others, and with increasing age such differences become accentuated. There is now also a larger aggradation of the valleys to form the level floors of flood plains, out of which at length the now slight elevations rise upon such gentle slopes that the process of land sculpture approaches its end. Gradually the vigor of the stream has faded away, and can now only be renewed through a fresh uplift of the land, or, what would amount to the same thing, a depression of the base level. Upland and river have reached old age together, and the approximation to a new plain but little elevated above base level is so marked that the name *peneplain* is applied to it. Scattered elevations, which because of some favoring circumstance rise to greater heights above the general level of the peneplain, are known as *monadnocks* after the type example of Mount Monadnock in New Hampshire (Fig. 179).

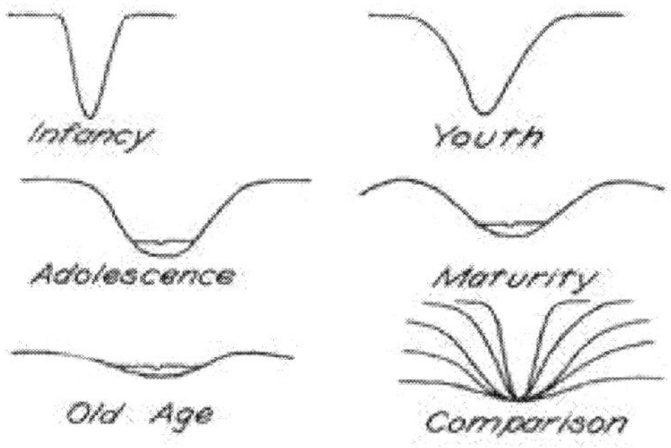

FIG. 180.—Comparison of the cross sections of river valleys for the different stages of the erosion cycle.

The river cross sections of successive stages.—To the successive stages of a river's life it has been common to carry over the names from the well-marked periods of a human life. If neglecting for the moment the general aspect of the upland, we fix our attention upon the characteristic cross sections of the river valley, we find that here also there are clearly marked characters to distinguish each stage of the river's life (Fig. 180). In infancy the steep, narrow, and sharp-angled cañon is a characteristic; with youth the wider V-form has already developed; in adolescence the angles of the cañon are transformed into well-rounded shoulders, and the valley broadens so as in the lower reaches to lay down a flood plain; in maturity the divides and the double curves of the line of beauty appear; while in the decline of old age the valleys are extremely broad and flat and are floored by an extended flood plain.

The entrenchment of meanders with renewed uplift.—Upon the reduced grades which are characteristic of the declining stage of a river's life, the current has little power to modify the surface configuration. On the old land of this stage a renewed uplift starts the streams again into action. This infusion of driving power into moving water, regarded as a machine capable of accomplishing certain work, is like winding up a clock that has run down. Once more the streams acquire a velocity sufficient to enable them to cut their valleys into the land surface, and so a new erosional cycle may be inaugurated upon the old land surface—the peneplain. After such an uplift has been accomplished and the rivers have sunk their early valleys within the new upland, we may look out from this now elevated surface and the eye take in but a single horizontal line, since we view the plain along its edge.

FIG. 181.—The Beavertail Bend of the Yakima Cañon in central Washington (after George Otis Smith).

By the uplift the meanders of the earlier rivers may become entrenched in the new upland, the wide lobes of the individual meanders being now separated by mountains where before had been plains of silt only. The New River of the Cumberland plateau and the Yakima River of central Washington (Fig. 181) furnish excellent American examples of intrenched meanders, as the Moselle River does in Europe. Upon the course of the latter river near the town of Zell a tunnel of the railroad a quarter of a mile in length pierces a mountain in the neck of a meander lobe in which the river itself travels a distance of more than six miles in order to make the same advance. The Kaiser Wilhelm tunnel in the same district penetrates a larger mountain included in a double meander of the river. Although intrenched, river meanders are still competent to scour and so undermine the outer bank, and with favoring conditions they may by this process erode extended "bottoms" out of the plateau. (See Lockport quadrangle, U. S. G. S.)

The valley of the rejuvenated river.—Whenever a new uplift occurs before an erosional cycle has been completed, the rivers become intrenched, not in a peneplain, but in the bottoms of broad valleys. The sweeping curves which characterize mature landscapes may thus be brought into striking contrast with the straight lines of youthful cañons which with V-sections descend from their lowest levels (Fig. 182). The full cross section of such a valley shows a central V whose sharp shoulders are extended outward and upward in the softened curves of later erosion stages.

FIG. 182.—A rejuvenated river valley (after a photograph by Fairbanks).

The arrest of stream erosion by the more resistant rocks.—The capacity of a river to erode and carry away the rock material that lies along its course is dependent not only upon the velocity of the current, but also upon the hardness, the firmness of texture, and the solubility of the material. Particularly in arid and semiarid regions, where no mantle of vegetation is at hand to mask the surfaces of the firmer rock masses, differences of this kind are stamped deeply upon the landscape. The rock terraces in the Grand Cañon of the Colorado together represent the stronger rock formations of the region, while sloping talus accumulations bury the weaker beds from sight.

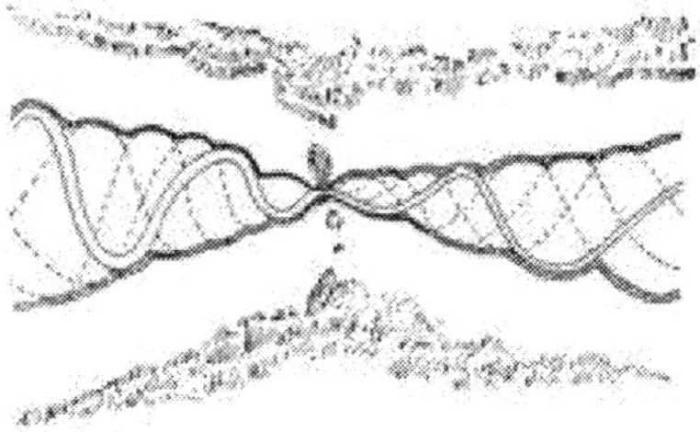

FIG. 183.—Plan of a river narrows.

Each area of harder rock which rises athwart the course of a stream causes a temporary arrest in the process of valley erosion and is responsible for a noteworthy local contraction of the river valley. The valley is carved less widely as well as less deeply, and since a river can never corrade below its base, a "temporary base level" is for a time established above the area of harder rock. Owing to the contraction of the valley under these conditions, the locality is described as a river *narrows* (Fig. 183). The narrows upon the Hudson River occur in the Highlands where the river leaves a broad expanse occupied by softer sediments to traverse an island-like area of hard crystalline rocks. Within the narrows of a river the steep walls, characteristic of youth and the turbulent current as well, are often retained long after other portions of the river have acquired the more restful lines of river maturity. The picturesque crag and the generally rugged character of river narrows render them points of special interest upon every navigable river.

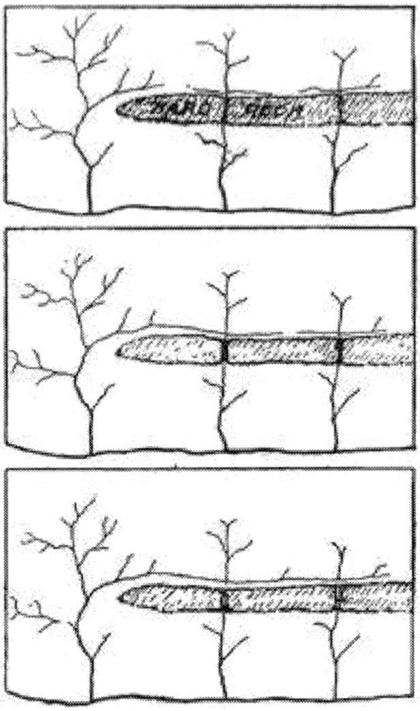

FIG. 184.—Successive diagrams to illustrate repeated river piracy and the development of "trellis drainage", (after Russell).

The capture of one river's territory by another.—The effect of a hard layer of rock interposed in the course of a stream is thus always to delay the advance of the erosional process at all levels above the obstruction. When a stream in incising its valley degrades its channel through a veneer of softer rocks into harder materials below, it is technically described as having *discovered* the harder layer. Where several neighboring streams flow by similar routes to their common base level, those which discover a harder rock will advance their headwaters less rapidly into the upland and so will be at a disadvantage in extending their drainage territory. A stream which is not thus hindered will in the course of time rob the others of a portion of their territory, for it is able to erode its lower reaches nearer to base level and thus acquire for its upper reaches, where erosion is chiefly accomplished, an advantage in declivity. The divide which separates its headwaters from those of its less favored neighbor will in consequence migrate steadily into the neighbor's territory. The divide is thus a sort of boundary wall separating the drainage basins of neighboring streams, and any migration must extend the territory of the one at the expense of the other. As more and more territory is brought under the dominion of the more favored stream, there will come a time when the divide in its migration will arrive at the channel of the stream that is being robbed, and so by a sudden act of annexation draw off all the upper waters into its own basin. By this *capture* the stream whose territory has been invaded is said to have been *beheaded*. By this act of *piracy* the stronger stream now develops exceptional activity because of the local steep grades near the point of capture, and with this newly acquired cutting power the invader is competent to advance still further and enter the territory of the stream that lies next beyond. The type of drainage network which results from repeated captures of this kind is known as "trellis drainage" (Fig. 184), a type well illustrated by the rivers of the southern Appalachians.

In general it may be said that, other conditions being the same, of two neighboring streams which have a common base level, that one which takes the longest route will lose territory to the other, since it must have the flatter average slope. Stream capture may thus come about without the discovery of hard rock layers which are more unfavorable to one stream than another.

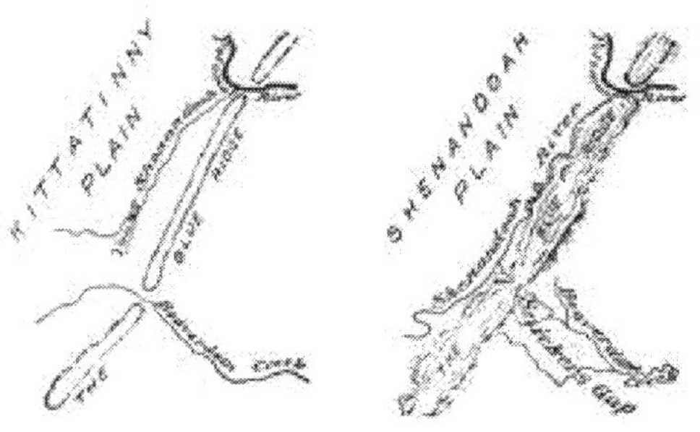

FIG. 185.—Sketch maps to show the earlier and the present drainage condition about the Blue Ridge near Harper's Ferry.

Water and wind gaps.—In the Allegheny plateau rivers cross, the range of harder rocks in deep mountain narrows which upon the horizon appear as gateways through the barrier of the mountain wall. Such gateways are sometimes referred to as "water gaps", of which the Delaware Water Gap is perhaps the best known example, though the Potomac crosses the Blue Ridge at the historic Harper's Ferry through a similar portal. The valley of the tributary Shenandoah has been the scene of an interesting episode in the struggle of rival streams which is typical of others in the same upland region. The records which may be made out from the landscapes show clearly that in an earlier but recent period, when the general surface stood at a higher level which has been called the Kittatinny Plain, the younger Potomac of that time and a younger but larger ancestor of Beaverdam Creek each crossed the

Blue Ridge of the time through similar water gaps (Fig. 185, map, and Fig. 186). The Potomac of that time was, however, the more deeply intrenched, and possessing an advantage in slope it was able to advance the divide at the head of its tributary, the Shenandoah, into the territory of Beaverdam Creek. Thus the beheading of the Beaverdam by the Shenandoah was accomplished (Fig. 185, second map) and its upper waters annexed to the Potomac system. With the subsequent lowering of the general level of the country which yielded the present Shenandoah Plain, the former water gap of Beaverdam Creek was abandoned of its stream at a high level in the range. Known as Snickers Gap, it may serve as a type of the "wind gaps" of similar origin which are not altogether uncommon in the Appalachian Mountain system (Fig. 186).

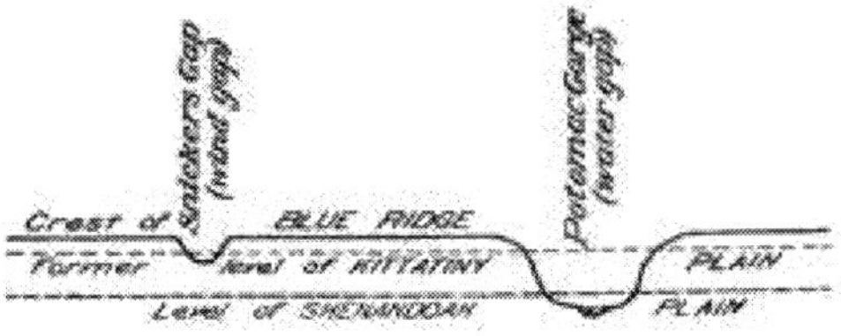

FIG. 186.—Section to illustrate the history of Snickers Gap.

Character profiles.—For humid regions the landscapes possess characters which, speaking broadly, depend upon the stage of the erosion cycle. For the earliest stages the straight line enters as almost the only element in the design; as the cycle advances to adolescence the rounded forms begin to replace the angles of the immature stages, and with full maturity the lines of beauty alone are characteristic. As this critical stage is passed irregularity of feature and ever more flattened curves are found to correspond to the decline of the river's vital energies. There are thus marks of senility in the work of rivers (Fig. 187).

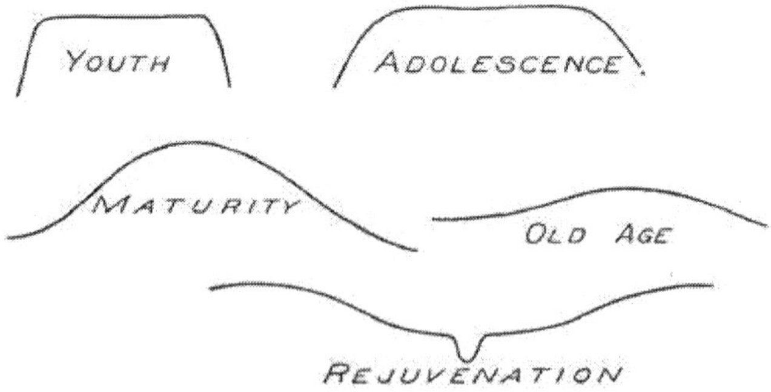

FIG. 187.—Character profiles of landscapes shaped by stream erosion in humid climates.

READING REFERENCES FOR CHAPTERS XII AND XIII

General:—

SIR JOHN PLAYFAIR. Illustrations of the Huttonian Theory of the Earth. Edinburgh, 1802, pp. 350-371.

J. W. POWELL. Exploration of the Colorado River of the West and its Tributaries. Washington, 1875, pp. 149-214.

G. K. GILBERT. Report on the Geology of the Henry Mountains. Washington, 1877, pp. 99-150. (A classic upon the work of rivers.)

C. E. DUTTON. Tertiary History of the Grand Cañon District (with atlas), Mon. 2, U. S. Geol. Surv., 1882, pp. 264.

W. M. DAVIS. The Rivers and Valleys of Pennsylvania, Nat. Geogr. Mag. vol. 1, 1889, pp. 203-219; The Triassic Formation of Connecticut, 18th Ann. Rept. U. S. Geol. Surv., Pt. ii, 1898, pp. 144-153.

SIR A. GEIKIE. The Scenery of Scotland. London, 1901, pp. 1-12.

I. C. RUSSELL. Rivers of North America. Putnam. New York, 1898, pp. 327.

M. R. CAMPBELL. Drainage Modifications and their Interpretation, Jour. Geol., vol. 4, 1896, pp. 567-581, 657-678.

HENRY GANNETT. Physiographic Types, U. S. Geol. Surv., Topographic Atlas, Folios 1-2, 1896, 1900.

W. M. DAVIS. The Geographical Cycle, Geogr. Jour., vol. 14, 1899, pp. 481-504.

The flood plain:—

HENRY GANNETT. The Flood of April, 1897, in the Lower Mississippi, Scot. Geogr. Mag., vol. 13, 1897, pp. 419-421.

W. M. DAVIS. The Development of River Meanders, Geol. Mag., Decade iv, vol. 10, 1903, pp. 145-148.

W. S. TOWER. The Development of Cut-off Meanders, Bull. Am. Geogr. Soc., vol. 36, 1904, pp. 589-599.

River terraces:—

W. M. DAVIS. The Terraces of the Westfield River, Massachusetts, Am. Jour. Sci., vol. 14, 1902, pp. 77-94, pl. 4; River Terraces in New England, Bull. Mus. Comp. Zoöl., vol. 38, 1902, pp. 281-346.

River deltas:—

G. K. GILBERT. The Topographic Features of Lake Shores, 5th Ann. Rept. U. S. Geol. Surv., 1885, pp. 104-108; Lake Bonneville, Mon. I, U. S. Geol. Surv., 1890, pp. 153-167.

Charts of Mississippi River Commission.

G. R. CREDNER. Die Deltas, ihre Morphologie, geographische Verbreitung und Entstehungsbedingungen, Pet. Mitt. Ergh. 56, 1878, pp. 1-74, pls. 1-3.

The peneplain:—

W. M. DAVIS. Plains of Marine and Subaërial Denudation, Bull. Geol. Soc. Am., vol. 7, 1896, pp. 377-398; The Peneplain, Am. Geol., vol. 23, 1899, pp. 207-239.

Intrenchment of meanders:—

W. M. DAVIS. The Seine, the Meuse, and the Moselle, Nat. Geogr. Mag., vol. 7, 1896, pp. 189-202.

Stream capture:—

N. H. DARTON. Examples of Stream Robbing in the Catskill Mountains, Bull. Geol. Soc. Am., vol. 7, 1896, pp. 505-507, pl. 23.

COLLIER COBB. A Recapture from a River Pirate, Science, vol. 22, 1893, p. 195.

WILLIAM H. HOBBS. The Still Rivers of Western Connecticut, Bull. Geol. Soc. Am., vol. 13, 1902, pp. 17-22, pl. 1.

ISAIAH BOWMAN. A Typical Case of Stream Capture in Michigan, Jour. Geol., vol. 12, 1904, pp. 326-334.

CHAPTER XIV

THE TRAVELS OF THE UNDERGROUND WATER

The descent within the unsaturated zone.—Of the moisture precipitated from the atmosphere, that portion which neither evaporates into the air nor runs off upon the surface, sinks into the ground and is described as the *ground water*. Here it descends by gravity through the pores and open spaces, and at a quite moderate depth arrives at a zone which is completely saturated with water. The depth of the upper surface of this saturated zone varies with the humidity of the climate, with the altitude of the earth's surface, and with many other similarly varying factors. Within humid regions its depth may vary from a few feet to a few hundred feet, while in desert areas the surface may lie as low as a thousand feet or more.

The surface of the zone of the lithosphere that is saturated with water is called the *water table*, and though less accentuated it conforms in general to the relief of the country (Fig. 188). Its depth at any point is found from the levels of all perennial streams and from the levels at which water stands in wells.

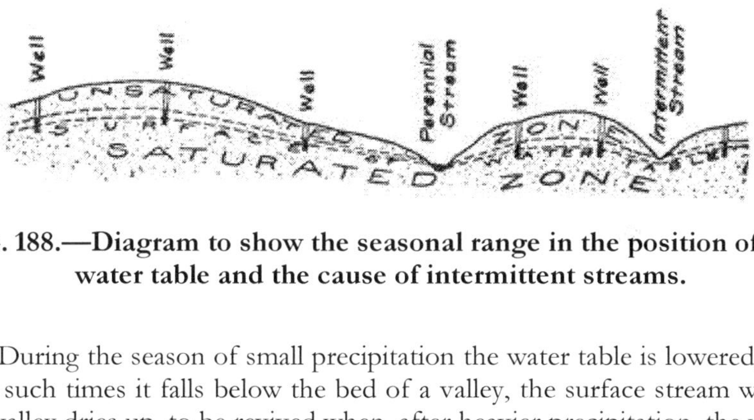

FIG. 188.—Diagram to show the seasonal range in the position of the water table and the cause of intermittent streams.

During the season of small precipitation the water table is lowered, and if at such times it falls below the bed of a valley, the surface stream within the valley dries up, to be revived when, after heavier precipitation, the water table has in turn been raised. Such streams are said to be *intermittent*, and are especially characteristic of semiarid regions (Fig. 188).

Wherever in descending from the surface an impervious layer, such as clay, is encountered, the further downward progress of the water is arrested. Now conducted in a lateral direction it issues at the surface as a spring at the line of emergence of the upper surface of the impervious layer (Fig. 189).

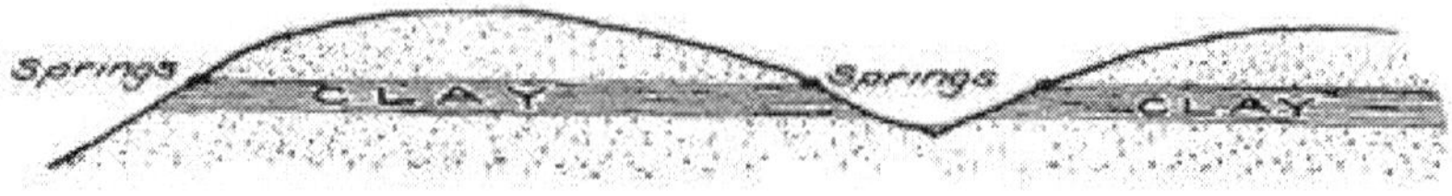

FIG. 189.—Diagram to show how an impervious layer conducts the descending water in a lateral direction to issue in surface springs.

The trunk channels of descending water.—While within the unconsolidated rock materials near the surface of the earth, it is clear that water can circulate in proportion as the materials are porous and so relatively pervious. As the pore spaces become minute and capillary, the difficulty of permeation through the materials becomes very great. Thus in the noncoherent rocks it is the coarse gravel and the layers of sand which serve as the underground channels, while the fine clays have the effect of an impervious wall upon the circulating waters. In coarse sand as much as a third of the volume of the material is pore space for the absorption and transmission of water. Even under these favorable conditions the movement of the water is exceedingly slow and usually less than a fifth of a mile a year.

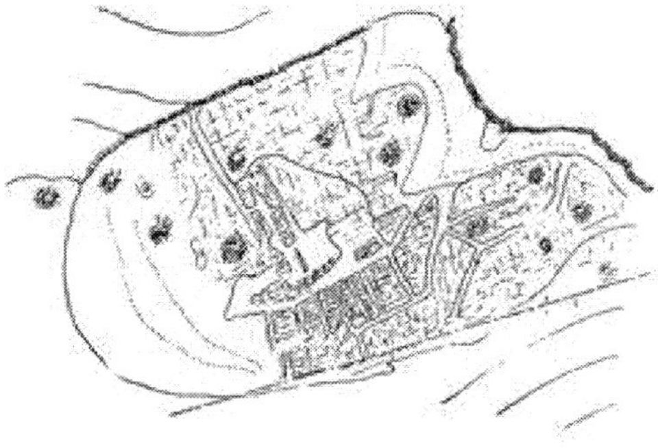

FIG. 190.—Sketch map of the Oucane de Chabrières near Chorges in the High Alps, to illustrate the corrosion of limestone along two series of vertical joints (after Martel).

Within the hard rocks it is the sandstones which have the largest pore spaces, but in nearly all consolidated rocks there are additional spaces along certain of the bedding planes, the joint openings (Fig. 190), and the crushed zones of displacement, so that these parting planes become the trunk channels, so to speak, of the circulating water. It is along such crevices that in the course of time the mineral matter carried in solution by the water is deposited to produce the ore veins and the associated crystallized minerals.

The caverns of limestones.—Where limestone formations have a nearly flat upper surface, a large part of the surface water enters the rock by way of the joint spaces, which it soon widens by solution into broad crevices with well-rounded shoulders. At joint intersections solution of the limestone is so favored that the water may here descend in a sort of vertical shaft until it meets a bedding plane extending laterally and offering more favorable conditions for corrosion. Its journey now begins in a lateral direction, and solution of the rock continuing, a tunnel may be etched out and extended until another joint is encountered which is favorable to its further descent into the formation. By this process on alternating shafts and galleries the water descends to near the surface of the water table by a series of steps, and is eventually discharged into the river system of the district (Fig. 191). Within the larger caverns the water at the lowest level usually flows as a subterranean river to emerge later into the light from beneath a rock arch.

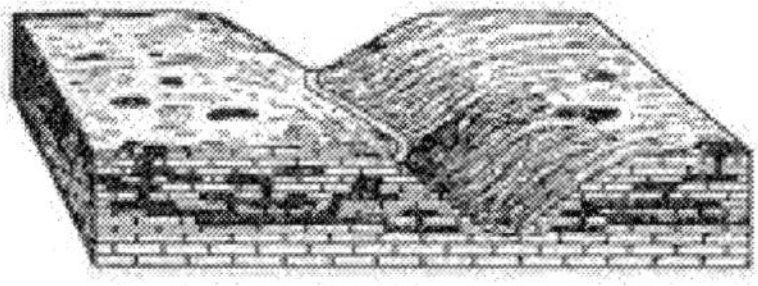

FIG. 191.—Diagram to show the relation of caverns in limestone to the river system of the district and to the "swallow holes" upon the surface.

From the plan of a system of connecting caverns it may often be observed that the galleries of the several levels are alike directed along two rectangular directions which indicate the master joint directions within the limestone formation. This is especially clear from the map of the galleries in the explored portions of the Mammoth Cave (Fig. 192).

Swallow holes and limestone sinks.—Above the caverns of limestone formations there are selected points where the water has descended in the largest volume, and here funnel-shaped depressions have been dissolved out from the surface of the rock. In different districts such depressions have become known as "sinks", "swallow holes", *entonnoirs*, and *Orgeln*. Wherever the depressions have a characteristic circular outline, there can be little doubt that they are the product of solution by the descending water, and have relatively small connections only with the subterranean caverns. They have thus naturally collected upon their bottoms the insoluble clay which was contained in the impure limestone as well as a certain amount of slope wash from the surface. Inasmuch as the clays are impervious to water, the bottoms of these swallow holes are better supplied with moisture than the surrounding rock surfaces, and by nourishing a more vigorous plant growth are strongly impressed upon the landscape (Fig. 193).

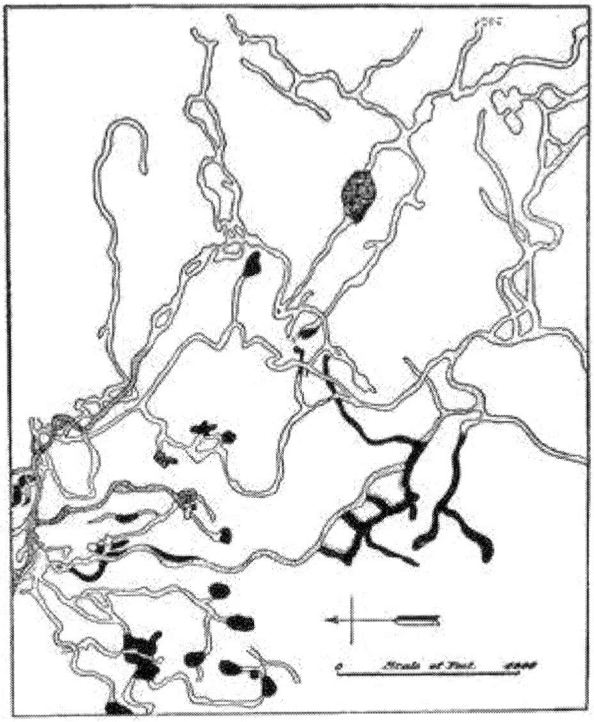

FIG. 192.—Plan of a portion of Mammoth Cave, Kentucky (after H. C. Hovey).

FIG. 193.—Trees and shrubs growing luxuriantly upon the bottoms of sinks within a limestone country (after a photograph by H. T. A. de L. Hus).

Certain of the depressions above caverns are, however, less regular in outline, and their bottoms are occupied by a mass of limestone rubble. In some instances, at least, these depressions appear to be the result of local incaving of the cavern roofs. An incaving of this nature may close up an earlier gallery in the cavern and divert the cave waters to a new course. The destruction of the roofs of caverns through this process of incaving may continue until only relatively small remnants are left. From long subterranean tunnels the caves are thus transformed into subaërial rock bridges that have become known as "natural bridges." The best-known American example is the Natural Bridge near Lexington, Virginia. Much grander natural bridges have been formed in sandstone by a totally different process, and must not be confused with these limestone remnants of caverns.

The sinter deposits.—Just as water can dissolve the calcareous rocks with the formation of caverns, it can under other conditions deposit the material which has thus been taken into solution. Its power to hold carbonate of lime in solution is dependent upon the presence of carbonic acid gas within the water. Water charged with gas and dissolved lime carbonate is said to be "hard", and if the gas be driven off by boiling or otherwise, the dissolved lime is thrown out of solution and deposited in a form well known to all housekeepers.

Hard water flowing in a surface stream, if dashed into spray at a cascade, may deposit its lime carbonate in an ever thickening veneer wherever the spray is dashed about the falls. This material, when cut in section, has waving parallel layers and is known as *travertine* or *calcareous sinter*. Some of the most remarkable deposits of this nature may be seen at the cascade of Tivoli near Rome, and most of the Roman buildings have been constructed from travertine that has been quarried in the vicinity.

The growth of stalactites.—Water, after percolating slowly through the crevices of limestone, where it becomes charged with the carbonic acid gas and with dissolved carbonate of lime, may trickle from the roof of a cavern. Emerging from the narrow crevice, it may give off some of its contained gas and is usually subject to evaporation, with the result that the lime carbonate is left adhering to the rock surface from which evaporation took place. If the water collects upon the cavern roof so slowly that it can entirely evaporate before a drop can form, the entire content of carbonate will be left adhering to the roof. Evaporation is most rapid near the margins and over the center of each drop as it develops, and the deposit which is left thus takes the form of tiny white rings at those points upon the crevice where there is the easiest passage for the trickling water. To the outer surface of these rings water will first adhere and then evaporate, as it will also slowly ooze through the passage in the ring, but here without evaporation until it reaches the lower surface. A pendant structure will, therefore, develop, growing outward in all directions by the deposition of concentric layers which are thickest near the roof, and downward into the form of a rock "icicle" through evaporation of the water which collects near the tip. These pendant sinter formations are known as stalactites and are thus formed of concentric layers arranged like a series of nested cornucopias with a perforation of nearly uniform caliber along the axis of the structure (Fig. 194).

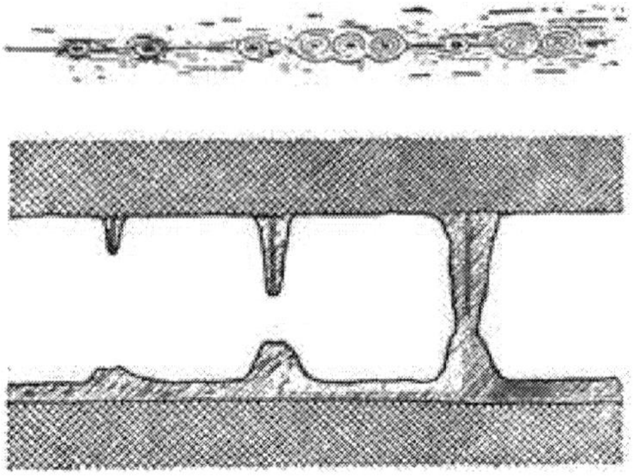

FIG. 194.—Diagrams to show the manner of formation of stalactites, stalagmites, and sinter columns beneath parallel crevices upon the roofs of caverns (in part after von Knebel).

Formation of stalagmites.—Wherever the water percolates through the roof of the cavern so rapidly that it cannot entirely evaporate upon the roof, a portion falls to the floor, and, spattering as it strikes, builds up a relatively thick cone of sinter known as a stalagmite, and this is accurately centered beneath a stalactite upon the roof. In proportion as the cavern is high, the dropping water is widely dispersed as it strikes the floor, with the formation of a correspondingly thick and blunt stalagmite. As this rises by growth toward the roof, it often develops upon its summit a distinct crater-like depression (Fig. 194, lower figure). When the process is long continued, stalactites and stalagmites may grow together to form columns which may be ranged with their neighbors like the pipes of an organ, and like them they give out clear tones when struck lightly with a mallet. At other times the columns are joined to their neighbors to form hangings and draperies of the most fantastic and beautiful design (Fig. 195).

FIG. 195.—Sinter formations in the Luray caverns, Virginia.

In remote antiquity limestone caverns afforded a refuge to many species of predatory birds and animals as well as to our earliest ancestors. The bones of all these denizens of the caves lie entombed within the clays and the sinter formations upon the cavern floors, and they tell the story of a fierce and long-continued warfare for the possession of these natural strongholds. The evidence is clear that these cave men with their primitive weapons were able at times to drive away the cave bears, lions, and hyenas, and to set up in the cavern their simple hearths, only in their turn to be conquered by the ferocity of their enemies. Some of the European caves have yielded many wagonloads of the skeletons of these fierce predatory animals, together with the simple weapons of the primitive man.

The Karst and its features.—Most so-called limestones have a large admixture of argillaceous materials (clays) and of siliceous or sandy particles. Such impurities make up the bulk of the clays and muds which are left behind when the soluble portions of the limestone have been dissolved.

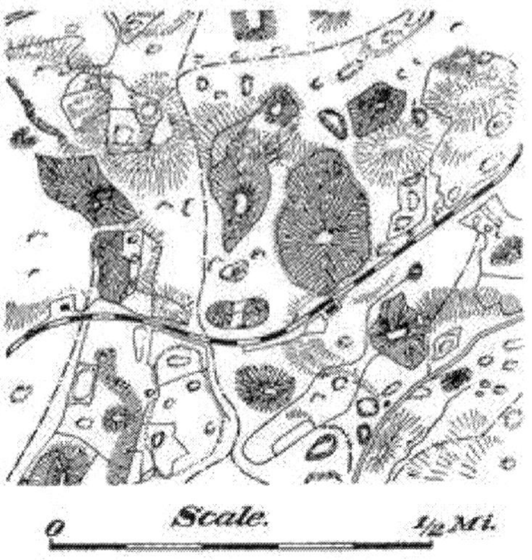

FIG. 196.—Map of the dolines of the Karst region near Divača.

Swallow holes we have found to be characteristic features within such districts. When limestones are more nearly pure, as in the Karst region east of the Adriatic Sea, similar features are developed, but upon a grander scale, and certain additional forms are encountered. In place of the sink or swallow hole, there appears the "karst funnel" or *doline*, a deep, bowl-shaped depression having a flat bottom. Such funnels may be 30 to 3000 feet across and from 6 to 300 feet in depth (Fig. 196). Though in one or two instances known to be the result of the break down of cavern roofs (Fig. 197), yet like the swallow holes of other regions these larger funnels appear generally to be the work of solution by the descending waters. Where they have been opened in artificial cuttings along railroads or in mines, the original rock is found intact at the bottom, with small crevices only going down to lower levels. Over the bottoms of the dolines there is spread a layer of fertile red clay, the *terra rossa*, like that which is obtained as a residue when a fragment of the limestone has been dissolved in laboratory experiments.

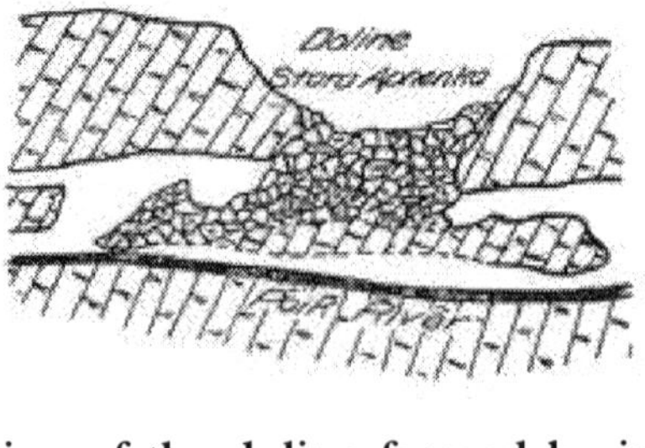

FIG. 197.—Cross section of the doline formed by inbreak of a cavern roof. The Stara Apnenka doline in Carinthia (after Martel).

A desert from the destruction of forests.—Between the dolines is found a veritable desert with jutting limestone angles and little if any vegetation. The water which falls upon the surface either runs off quickly or goes down to the subterranean caverns by which so much of the country is undermined. Hence it is that the gardens which furnish the sustenance for the scattered population are all included within the narrow limits of the doline bottoms. Although to-day so largely a barren waste, we know that the Karst upon the Adriatic was in remote antiquity a heavily forested region and that it supplied the myriads of wooden piles upon which the city of Venice is supported. The vessels which brought to this port upon the Adriatic its ancient prosperity were built from wood brought from this tract of modern desert. In the days of Venetian grandeur the fertile terra rossa formed a veneer upon the rock surface of the Karst and so retained the surface waters for the support of the luxuriant forest cover. After deforestation this veneer of rich soil was washed by the rains into the dolines or into the few stream courses of the region, thus leaving a barren tract which it will be all but impossible to reclaim (plate 6 A).

FIG. 198.—Sharp *Karren* of the Ifenplatte Allgäu (after Eckert).

Upon the steeper slopes over the purer limestones, the rain water runs away, guided by the joints within the rock. There is thus etched out a more or less complete network of narrow channels (Fig. 190, p. 181), between which the remnants rise in sharp blades to produce a structure often simulated upon the fissured surface of a glacier that has been melted in the sun's rays (Fig. 401). These almost impassable areas of karst country are described as *Schratten* or *Karrenfelder* (Fig. 198).

The ponore and the polje.—To-day large areas of the Karst are devoid of surface streams, nearly all the surface water finding its way down the crevices of the limestone into caverns, and there flowing in subterranean courses. The foot traveler in the Karst country is sometimes suddenly arrested to find a precipice yawning at his feet, and looking down a well-like opening to the depth of a hundred feet or more, he may see at the bottom a large river which emerges from beneath the one wall to disappear beneath the other. These well-like shafts are in the Austrian Karst known as *Ponores*, while to the southward in Greece they are called *Katavothren.*

PLATE 6.

A. **Barren Karst landscape near the famous Adelsberg grottoes. (*Photograph by I. D. Scott.*)**

B. **Surface of a limestone ledge where joints have been widened through solution. Syracuse, N.Y. (*Photograph by I. D. Scott.*)**

Elsewhere the karst river may emerge from its subterranean course in a broader depressed area bounded by vertical cliffs, from which it later disappears beneath the limestone wall. Such depressions of the karst are known as *poljen*, and appear in most cases to be above the downthrown blocks in the intricate fault mosaic of the region. Some of these steeply walled inclosures have an area of several hundred square miles, and especially at the time of the spring snow melting they are flooded with water and so transformed into seasonal lakes (Fig. 199 and p. 422). It appears that at such times the cave galleries of the region with their local narrows are not able to carry off all the water which is conducted to them; and in consequence there is a temporary impounding of the flood waters in those portions of the river's course which are open to the sky and more extended. The rush of water at such times may bring the red clay into the subterranean channels in sufficient quantity to clog the passages. The Zirknitz Lake usually has high water two or three times a year, and exceptionally the flooding has continued for a number of years. It has thus in some districts been necessary to afford relief to the population through the construction of expensive drainage tunnels.

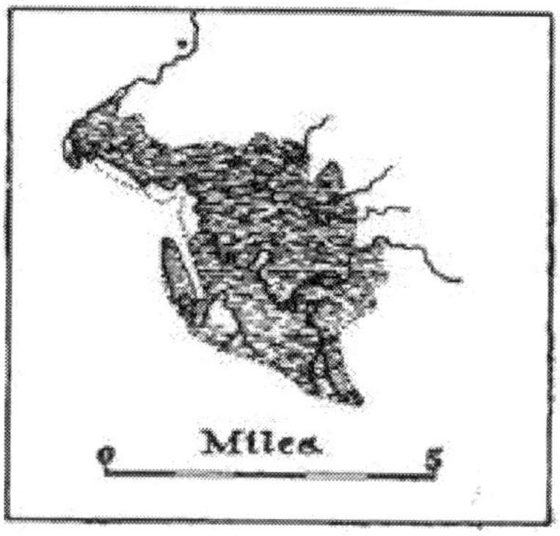

FIG. 199.—The Zirknitz seasonal lake within a polje of the Karst (after Berghaus).

The conditions which are typified in the Karst area to the east of the Adriatic Sea are encountered also in many other lands; as, for example, in the Vorarlberg and Swiss Alps, in Lebanon, and in Sicily.

The return of the water to the surface.—Water which has descended from the surface and been there held between impervious layers, may be under the pressure of its own weight or "head"; and will later find its way upward, it may be to the surface or higher, where a perforation is discovered in its otherwise impervious cover. Such local perforations are produced naturally by lines of fracture or faulting (widened at their intersections), and artificially through the sinking of deep wells. The water, which at ordinary times reaches the surface upon fissures, is usually concentrated locally at the intersections of the fracture network, where it issues in lines of fissure springs (Fig. 200); but at the time of earthquakes the water may rise above the surface in lines of fountains (p. 83), or occasionally as sheets of water which may mount some tens of feet into the air.

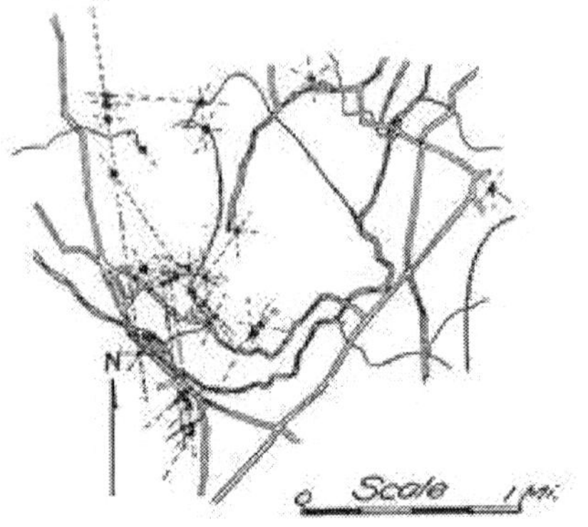

FIG. 200.—Fissure springs arranged upon lines of rock fracture at intersections, Pomperaug valley, Connecticut.

In contrast to the flow of surface springs, which varies with the season through wide ranges both in its volume and in temperature of the water, the volume of fissure springs is but slightly affected by the seasonal precipitation, and the water temperature is maintained relatively constant. Rock is but a poor heat conductor, and the seasonal temperature changes descend a few feet only into the ground. Thus water which rises from depths of a few hundred feet only is apt to be icy cold, while from greater depths the effect of the earth's internal heat is apparent in a uniform but relatively higher temperature of the water. Such "warm" or *thermal* springs are apt to contain considerable mineral matter in solution, both because the water is far traveled and because its higher temperature has considerably increased its solvent properties.

It has long been recognized that lines of junction of different rock formations at the base of mountain ranges are localities favorable for the occurrence of thermal springs. These junction lines are usually within zones where by movement upon fractures the widest openings in the rock have formed, and the catchment area of the neighboring mountain highland has supplied head for the ground water. A map of the hot springs within the Great Basin of the western United States would present in the main a map of its principal faults.

Artesian wells.—From the natural fissure spring an artesian well differs in the artificial character of the perforation of the impervious cover to the water layer. The water of artesian wells may flow out at the surface under pressure, or it may require pumping to raise it from some lower level. Ideal conditions are furnished where the geological structure of the district is that of a broad basin or syncline. The water which falls in a neighboring upland is here impounded between two parallel, saucer-like walls and will flow under its head if the upper wall be perforated at some low level (Fig. 201, 3).

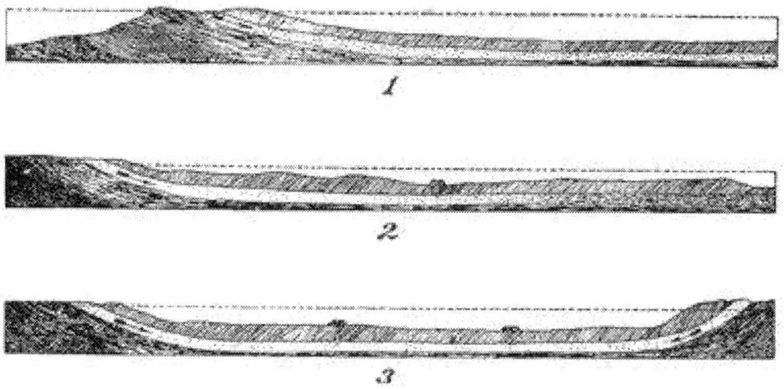

FIG. 201.—Schematic diagrams to illustrate the different types of artesian wells, (1) A non-flowing well; (2) flowing wells without basin structure caused by clogging of the pervious formation; (3) flowing wells in an artesian basin. The dotted lines are the water levels within the pervious layers (after Chamberlin).

A monoclinal structure may furnish artesian conditions when the generally pervious layer has become clogged at a low level so as to hold back the water (Fig. 248, 2). Pumping wells may be used successfully even when such clogging does not exist, for the slow-moving underground water flows readily in the direction of all free outlets (Fig. 201, 1).

Hot springs and geysers.—Thermal springs whose temperature approaches the boiling point of water are known as *hot springs.* A *geyser* is a hot spring which intermittently ejects a column of water and steam. Both hot springs and geysers are to be found only in volcanic regions, and appear to be connected with uncooled masses of siliceous lava. In two of the three known geyser regions, Iceland and New Zealand, the volcanoes of the neighborhood are still active, and the lavas of the Yellowstone National Park date from the quite recent geological period which immediately preceded the so-called "Ice Age."

Wherever found, geysers are in the low levels along lines of drainage where the underground water would most naturally reappear at the surface. Their water has penetrated to considerable depths below the surface, but has been chiefly heated by ascending steam or other vapors. The water journey has been chiefly made along fissures, as is shown by the cool springs which often issue near them. Though some hot springs and geysers may disappear from a district, others are found to be forming, and there is no good reason to think that geysers are rapidly dying out, as was at one time supposed.

The action of a geyser was first satisfactorily explained by the great German chemist Bunsen after he had made studies of the Icelandic geysers, and the mechanics of the eruption was later strikingly illustrated in the laboratory by an artificial geyser constructed by the Irish physicist Tyndall. In many respects this action is like that of the Strombolian eruption within a cinder cone, since it is connected with the viscosity of the fluid and the resistance which this opposes to the liberation of the developing vapor. In the case of the geyser, a column of heated water stands within a vertical tube and is heated near the bottom of the column.

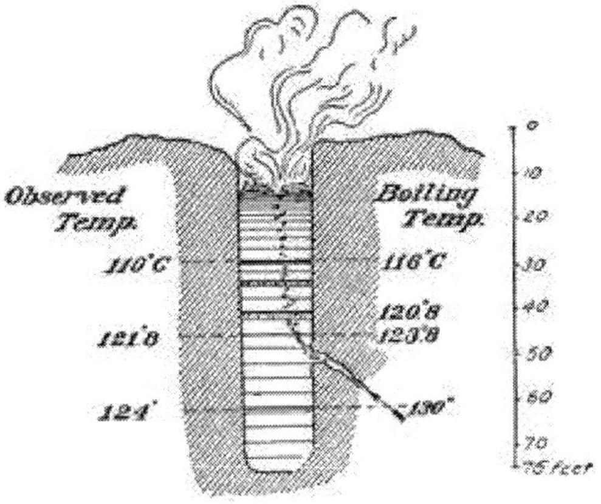

FIG. 202.—Cross section of Geysir, Iceland, with simultaneously observed temperatures recorded at the left, and the boiling temperatures for the same levels at the right (after Campbell).

Though the water may at its surface have the normal boiling temperature and be there in quiet ebullition, the boiling point for all lower levels is raised by the weight of the column of superincumbent liquid, and so for a time the formation of steam within the mass is prevented. In Fig. 202 is shown a cross section of the Icelandic *Geysir* from which our name for such phenomena has been derived, and to this section have been added the actual observed temperatures of the water at the different levels as well as the temperatures at which boiling can take place at these levels. From this it will be seen that at a depth of 45 feet the water is but 2° Centigrade below its boiling point. A slight increase of temperature at this level, due to the constantly ascending steam, will not only carry this layer above the boiling point, but the expansion of the steam within the mass will elevate the upper layers of the water into zones where the boiling points are lower, and thus bring about a sudden and violent ebullition of all these upper portions. Thus is explained the almost universal observation that just before geysers erupt the hot water rises in the bowls and generally overflows them.

FIG. 203.—Apparatus for simulating geyser action in the lecture room (by courtesy of Professor B. W. Snow).

The water ejected from the geyser is considerably cooled in the air; and after its return to the tube must be again heated by the ascending vapors before another eruption can occur. The measure of the cooling, the time necessary to fill the tube, and the supply of rising steam, all play a part in fixing the period which separates consecutive eruptions. If the top of the tube be narrowed from its average caliber, as is commonly observed to be true of the geysers within the Yellowstone National Park, the escape of the steam is further hindered, and frequent geyser eruption promoted.

An artificial geyser for demonstration of the phenomenon in the lecture room is represented in Fig. 203. The cut has been prepared from a photograph of an apparatus designed by Professor B. W. Snow of the University of Wisconsin. In this design the tube is contracted so as to have a top diameter one fourth only of what it is at the bottom, where heat is directly applied by multiple Bunsen lamps. The water once sufficiently heated, this artificial geyser erupts at regular intervals of time which are dependent upon the dimensions of the apparatus and the quantity of heat applied.

In case of natural geysers a considerable quantity of heat escapes between eruptions in steam which issues quietly from the bowl of the geyser. If this heat be retained by plugging the mouth of the tube with a barrowful of turf, as is sometimes done with the geyser *Strokr* in Iceland, eruption is promoted and so takes place earlier. Another method of securing the same result is to increase the viscosity of the water through the addition of soap, as was accidentally discovered by a Chinaman who was utilizing the geyser water in the Yellowstone Park for laundry operations. After this discovery it became a common custom to "soap" the Yellowstone geysers in order to make them play; but this method was prohibited under heavy penalty after the disastrous eruption of the Excelsior Geyser.

The deposition of siliceous sinter by plant growth.—Geysers are known only from areas of siliceous volcanic lava, and this may perhaps have its cause in the easier solution of the geyser tube from such materials. The silica dissolved in the heated waters is *again* deposited at the surface to form *siliceous sinter* or *geyserite*. This material forms terraces surrounding the geysers or is built up into mounds which are often quite symmetrical, such as those of the Bee Hive and Lone Star geysers of the Yellowstone Park (Fig. 204).

FIG. 204.—Cone of siliceous sinter built up about the mouth of the Lone Star Geyser in the Yellowstone National Park.

The greater part of this separation of silica from the heated geyser waters is due to the action of plants or algæ that are able to grow in the boiling waters and which produce the beautiful colors in the linings to the hot springs. The wonderful variety of the tints displayed is accounted for by the fact that the algæ take on different colors at different temperatures. The silica is deposited from the water in the gelatinous hydrated form, which, however, dries in the sun to a white sand. The growth within the pools goes on in a manner similar to that of a coral reef, the algæ dying below and there becoming encased in the rock lining while still continuing to grow upon the surface. Whereas sinter of this nature, when deposited by evaporation alone, can produce a maximum thickness of layer of a twentieth of an inch each year, the growth from alga deposition within limited areas may be as much as eight inches during the same period.

READING REFERENCES FOR CHAPTER XIV

General:—

F. H. KING. Principles and Conditions of the Movements of Ground Water, 19th Ann. Rept. U. S. Geol. Surv., 1899, Pt. ii, pp. 59-294, pls. 6-16.

C. S. SLICHTER. The Motions of the Underground Waters, Water Supply Paper No. 67, U. S. Geol. Surv., 1902, pp. 1-106, pls. 1-8; Field Measurements of the Rate of Movement of Underground Waters, *ibid.*, No. 140, 1905, pp. 1-122, pls. 1-15.

M. L. FULLER. Occurrence of Underground Water, *ibid.*. No. 114, 1905, pp. 18-40, pls. 4; Bibliographic review and index of papers relating to underground waters published by the United States Geological Survey, 1879-1904, *ibid.*, No. 120, 1905, pp. 1-128.

Caverns:—

E. A. MARTEL. Les abimes, les eaux souterraines, les cavernes, les sources, la spélæologie. Delagrave, Paris, pp. 578. (Lavishly illustrated.)

H. C. HOVEY. Celebrated American Caverns. Cincinnati, 1896, pp. 228; The Mammoth Cave of Kentucky. Louisville, 1897, pp. 111.

J. W. BEEDE. Cycle of Subterranean Drainage in the Bloomington Quadrangle, Proc. Ind. Acad. Sci., 1910, pp. 1-31.

Karst conditions:—

J. CVIJIC. Das Karstphänomen, Geogr. Abh., vol. 5, 1893.

ÉMILE CHAIX. La topographie du desert de platé (Hautes Savoie), Le Globe, vol. 34, 1895, pp. 1-44, pls. 1-16, pp. 217-330.

W. v. KNEBEL. Höhlenkunde mit Berücksichtigung der Karstphänomene. Vieweg, Braunschweig, 1906, pp. 222.

A. GRUND. Die Karsthydrographie, Studien aus Westbosnien, Geogr. Abh., vol. 7, No. 3, 1903, pp. 200.

ÉMILE CHAIX-DU BOIS et ANDRÉ CHAIX. Contribution a l'étude des lapies en Carniole et au Steinernes Meer, Le Globe, vol. 46, 1907, pp. 17-56, pls. 26.

P. ARBENZ. Die Karrenbildungen geschildert am Beispiele der Karrenfelder bei der Frutt in Kanton Obwalden (Schweiz). Deutsch. Alpenzeitung, Munich, 1909, pp. 1-9.

F. KATZER. Karst und Karsthydrographie. Sarejevo, 1909, pp. 95.

M. NEUMAYR. Erdgeschichte, vol. 1, pp. 500-510.

E. DE MARTONNE. Traité de Géographie Physique, pp. 462-472 (excellent summaries in this and the last reference).

E. A. MARTEL. The Land of the Causses, Appalachia, vol. 7, 1893, pp. 18-149, pls. 4-13.

Fissure springs:—

A. C. PEALE. Natural Mineral Waters of the United States, 14th Ann. Rept. U. S. Geol. Surv., Pt. ii, 1894, pp. 49-88.

WILLIAM H. HOBBS. The Newark System of the Pomperaug Valley. Connecticut, 21st Ann. Rept. U. S. Geol. Surv., Pt. iii, 1901, pp. 91-93.

Artesian wells:—

T. C. CHAMBERLIN. Requisite and Qualifying Conditions of Artesian Wells, 5th Ann. Rept. U. S. Geol. Surv., 1885, pp. 131-173.

Hot springs and geysers:—

A. C. PEALE. Yellowstone Park, Thermal Springs, 12th Ann. Rept. Geol. and Geogr. Surv. Ter. (Hayden), Pt. ii, Sec. ii, pp. 63-454 (many plates and maps).

W. H. WEED. Geysers, Rept. Smithson. Inst., 1891, pp. 163-178.

ARNOLD HAGUE and W. H. WEED (on hot springs and geysers of Yellowstone National Park), C. R. Cong. Géol. Intern., Washington, 1891, pp. 346-363.

W. H. WEED. Formation of Travertine and Siliceous Sinter by the Vegetation of Hot Springs, 9th Ann. Rept. U. S. Geol. Surv., 1889, pp. 613-676, pls. 78-87.

M. NEUMAYR. Erdgeschichte, vol. 1, pp. 500-510.

ARNOLD HAGUE. Soaping Geysers, Trans. Am. Inst. Min. Eng., vol. 17, 1889, pp. 546-553.

JOHN TYNDALL. Heat as a Mode of Motion, New York, 1873, pp. 115-121 (artificial geyser).

CHAPTER XV

SUN AND WIND IN THE LANDS OF INFREQUENT RAINS

The law of the desert.—It is well to keep ever in mind that there is no universal law which dominates Nature's processes in all the sections of her realm. Those changes which, because often observed, are most familiar, may not be of general application, for the reason that the areas habitually occupied by highly civilized races together comprise but a small portion of the earth's surface. In the dank tropical jungle, upon the vast arid sand plains, and in the cold white spaces near the poles, Nature has instituted peculiar and widely different processes.

The fundamental condition of the desert is aridity, and this necessitates an exclusion from it of all save the exceptional rain cloud. Thus deserts are walled in by mountain ranges which serve as barriers to intercept the moisture-bringing clouds. They are in consequence saucer-shaped depressions, often with short mountain ranges rising out of the bottoms, and such rain as falls within the inclosure is largely upon the borders. Of this rainfall none flows out from the desert, for the water is largely returned to the atmosphere through evaporation.

The desert history is thus begun in isolation from the sea from which the cloud moisture is derived, a balance being struck between inflow and evaporation. Yet if deserts have no outlets, it is not true that they have no rivers. These are occasionally permanent, often periodic, but generally ephemeral and violent. The characteristic drainage of deserts comes as the immediate result of sudden cloudburst. As a consequence, the desert stream flows from the mountain wall choked with sediment, and entering the depressed basin, is for the most part either sucked down into the floor or evaporated and returned to the atmosphere. The dissolved material which was carried in the water is eventually left in saline deposits, and the great burden of sediment accumulates in thick stratified masses which in magnitude outstrip the largest deltas in the ocean.

The self-registering gauge of past climates.—From the initiation of the desert in its isolation from the lands tributary to the sea, its history becomes an individual and independent one. An increasing quantity of rainfall will be marked by larger inflow to the basin, and the lakes which form in its lowest depression will, as a consequence, rise and expand over larger areas. A contrary climatic change will bring about a lowering of the lakes and leave behind the marks of former shorelines above the water level (Fig. 205). Deserts are thus in a sense self-registering climatic gauges whose records go back far beyond the historic past. From them it is learned that there have been alternating periods of larger and smaller precipitation, which are referred to as *pluvial* and *interpluvial* periods.

FIG. 205.—Former shore lines on the mountain wall surrounding the desert of the Great Basin. View from the temple in Salt Lake City (after Gilbert).

From such records it is learned that the Great Basin of the western United States was at one time occupied by two great desert lakes, the one in the eastern portion being known as Lake Bonneville (Fig. 206). With the desiccation which followed upon the series of pluvial periods, which in other latitudes resulted in great continental glaciers and has become known as the Glacial Period, this former desert lake dried up to the limits of Great Salt Lake and a few smaller isolated basins. Between 1850 and 1869 the waters of Great Salt Lake were rising, while from 1876 to 1890 their level was falling, though subject to periodic fluctuations, and in recent years the waters of the lake have risen so high as to pass all records since the occupation of the country. As a consequence the so-called Salt Lake "cut-off" of the Union Pacific Railway, constructed at great expense across a shallow portion of the lake, has been overflowed by its waters. The Sawa Lake in the Persian Desert, which disappeared some five hundred years ago, again came into existence in 1888 so as to cover the caravan route to Teheran.

FIG. 206.—Map of the former Lake Bonneville (dotted shores), and the boundaries of the Great Salt Lake of 1869 (smaller area) and that of the present (after Berghaus).

The record in the rocks of the distant past reveals the fact that in some former deserts barriers were, in the course of time, broken down, with the result that an invading sea entered through the breached wall. The result was the sudden destruction of land life, the remains of which are preserved in "bone beds", now covered by true marine deposits. A still later episode of the history was begun when the sea had disappeared and land animals again roamed above the earlier desert. Such an alternation of marine deposits with the remains of land plants and animals in the deposits of the Paris Basin, led the great Cuvier to his belief that geologic history was comprised of a succession of cataclysms in which life was alternately destroyed and re-created in new forms—a view which later, under the powerful influence of Lyell and Darwin, gave way to that of more gradual changes and the evolution of life forms.

Some characteristics of the desert wastes.—The great stretches of the arid lands have been often compared to the ocean, and the Bedouin's camel is known as "the ship of the desert." Though a deceptive resemblance for the most part, the comparison is not without its value. Both are closed basins, and it is in this respect that the desert and the ocean may be said to most resemble each other, for none of the water and none of the sediment is lost to either except as boundaries are, with the progress of time, transposed or destroyed. Flatness of surface and monotony of scenery both have in common, and the waters and the sand are in each case salt; yet the ocean, from the tropics to the poles, has the same salts in essentially the same proportions, while in the desert the widest variations are found both in the salts which are present and in their relative quantities.

Upon the borders of the ocean are found ridges of yellow sand heaped up by the wind, but these ramparts are small in comparison to those which in deserts are found upon the borders (plate 7 A).

The desert is a land of geographic paradoxes. As Walther has pointed out, we have rain in the desert which does not wet, springs which yield no brooks, rivers without mouths, forests preserved in stone, lakes without outlets, valleys without streams, lake basins without lakes, depressions below the level of the sea yet barren of water, intense weathering with no mantle of disintegrated rock, a decomposition of the rocks from within instead of from without, and valleys which branch sometimes upstream and sometimes down.

Within the deserts curious mushroom-like remnants of erosion afford a local relief from the searching rays of the desert sun. Pocket-like openings large enough for a hermit's habitation are hollowed out by the wind from the disintegrated rock masses. Amphitheaters open out from little erosion valleys or wadi, and isolated outliers of the mountains stand like sentinels before their massive fronts.

Because of the general absence of clouds above a desert, no shield such as is common in humid regions is provided against the blinding intensity of the sun's rays. Sun temperatures as high as 180° Fahrenheit have been registered over the deserts of western Africa. Every one is familiar with the fact that a blanket of thick clouds is a prevention of frosts at night, for, with the setting of the sun and the consequent radiation of heat from the earth, these rays are intercepted by the clouds, returned and re-returned in many successive exchanges. Over desert regions the absence of any such blanket of moisture is responsible for the remarkable falls of temperature at sunset. Though shortly before temperatures of 100° Fahrenheit or greater may have been measured, it is not uncommon for water to freeze during the following night. Much the same conditions of sudden temperature change with nightfall are experienced in high mountains when one has ascended above the blanketing clouds.

FIG. 207.—Borax deposits upon the floor of Death valley, California (after a photograph by Fairbanks).

Dry weathering—the red and brown desert varnish.—In desert lands the fierce rays of the sun suck up all the available moisture, and the water table may be hundreds of feet below the surface. Roots of trees a hundred feet or more in length have been found to testify to the fierce struggle of the desert plant with the arid conditions. In humid regions the meteoric water dissolves the more soluble sodium salts near the surface of the rock and carries them out to the ocean, where they add to the saltness of the sea. In the desert the rare precipitations prevent an outflow, but the sun's strong rays suck out with the moisture the salts from within the rock, and evaporating upon the surface, the salts are left as a coat of "alkali", which is in part carried away on the wind and in part washed off in one of the rare cloudbursts. In either case these constituents find their way to the lowest depressions of the basin, where they contribute to the saline deposits of the desert lakes (Fig. 207).

FIG. 208.—Hollowed forms of weathered granite in a desert of central Asia (after Walther).

Certain of the saline constituents of the rocks, as they are thus drawn out by the sun's rays, fuse with the rock at the surface to form a dense brown substance with smooth surface coat, known as *desert varnish.* Within the interior a portion of the salts crystallize within the capillary fissures, and like water freezing within a pipe, they rend the walls apart. As a direct consequence of this disintegrating process the interior of rock masses may crumble into sand; and if the hard shell of varnish be broken at any point, the wind makes its entrance and removes the interior portion so as to leave a hollow shell—the characteristic "pocket rock" (Fig. 208) of the desert. The nummulitic limestone of Mokkatan and many of the great hewn blocks of Egyptian limestone sound hollow under the tap of the hammer, and when broken, they reveal a shell a few inches only in thickness (Fig. 209).

FIG. 209.—Hollow hewn blocks in a wall in the Wadi Guerraui (after Walther).

The brown desert varnish is one of the most characteristic marks of an arid country. It is found in all deserts under much the same conditions, and is especially apt to be present in sandstone. When scratched, the surface of the rock becomes either cherry-red, indicating anhydrous ferric oxide, or it is yellowish, due to the hydrated iron oxide which we know as iron rust. Thus it is seen that the sands of deserts, in contrast to those yielded by other processes within humid regions, have a characteristic red color, and this may vary from brownish red upon the one hand to a rich carmine upon the other.

The mechanical breakdown of the desert rocks.—The chemical changes of decomposition within desert rocks are, as we have seen, largely due to the action of concentrated solutions of salts at high temperatures. That there is a certain mechanical rending of these rocks, due to the "freezing" of salts within the capillary fissures, has been already mentioned. A further strain effect arises in rocks like granite, which are a mixture of different minerals. Heated to a high temperature during the day and cooled through a considerable range at night, the different minerals alternately expand and contract at different rates and by different relative amounts, so that strains are set up, tending to tear them apart. The effect of these strains is thus a surface crumbling of rocks.

But rock is, as already pointed out, a relatively poor conductor of heat, and hence it is a relatively thin skin only which passes through the daily round of wide temperature range. This outer shell when heated is expanded, and so tends to peel off, or exfoliate, like the outer skin of an onion. The process is therefore described as *exfoliation.* In all rocks of homogeneous texture the continued action of this process results in convexly spherical surfaces, the material scaled off in the process remaining as a slope or talus which surrounds the projecting knob (Fig. 210). Naked, these projecting domes rise above the rim of débris at their bases. Not a particle of dust adheres to the fresh rock surface—no dirt interferes with its glaring whiteness. Yet close at hand lie masses of débris into which wells may be carried to depths of more than six hundred feet without encountering either solid rock or ground water. The bare walls of granite sometimes mount upwards for thousands of feet into the air, as steep and as inaccessible as the squared towers of the Tyrolean Dolomites.

FIG. 210.—Smooth granite domes shaped by exfoliation and surrounded by a rim of talus. Gebel Karsala, Nubian Desert (after Walther).

Rock is such a poor conductor of heat that special strains are set up at the margin of sunlight and shade. This localization of the disintegration on the margin of the shaded portions of rock masses is known as *shadow weathering* (see Fig. 215, p. 206).

There is, however, still another mechanical disintegrating process characteristic of the desert regions, which is likewise dependent upon the sudden changes of temperature. Rains, though they may not occur for a year or more, come as sudden downpours of great volume and violence. Rock masses, which are highly heated beneath the desert sun, if suddenly dashed with water, may be rent apart by the differential strains set up near the surface. That rocks may be easily rent as a result of sudden chilling is well known to our Northern farmers, who are accustomed to rid themselves of objectionable bowlders by first building a fire about them and then dashing water upon their surface. Thus split into fragments, even the larger bowlders may be handled and so removed from the farming land. The natural process of rock rending by the occasional cloudburst may be described as *diffission*. Blocks as much as twenty-five feet in diameter have been observed in the desert of western Texas, soon after being broken into several fragments at the time of a downpour of rain (Fig. 211).

FIG. 211.—Granite blocks in the Sierra de los Dolores of Texas, rent into several fragments by the dash of rain (after Walther).

The natural sand blast.—Because of the saucer-like shape, the vast expanse, and the absence of wind breaks, the potency of wind as a geological agent is in desert areas not easily overestimated. While most of its work is accomplished with the aid of tools, it has been proven that even without this help, considerable work is done through the friction of the wind alone, particularly when moving as powerful eddies in cracks and crannies. This wear of the wind, unaided by cutting tools, is known as *deflation*.

The greater work of the wind is, however, accomplished with the aid of larger or smaller rock particles, the sand and dust, with which it is so generally charged above the deserts. Unprotected by any mat of vegetation the materials of the desert surface are easily lifted and are constantly migrating with the wind. The finest dust is raised high into the air, and is carried beyond the marginal barriers, but none of the sand or coarser materials ever passes beyond the borders.

FIG. 212.—"Mushroom rock" from a desert in Wyoming (after Fairbanks).

The efficiency of this sand as a cutting tool when carried by the wind is directly proportioned to the size of the grain, since with larger fragments a heavier blow is struck when carried at any given velocity. These more effective grains are, however, not lifted far above the ground, but advance with a squirming or hopping motion, much as do the larger pebbles upon the bottom of a river at the time of a spring freshet. To quote Professor Walther: "Whoever has had the opportunity to travel over a surface of dune sand when a strong wind is blowing has found it easy to convince himself of the grinding action of the wind. At such times the ground becomes alive, everywhere the sand is creeping over the surface with snake-like squirmings, and the eye quickly tires of these writhing movements of the currents of sand and cannot long endure the scene."

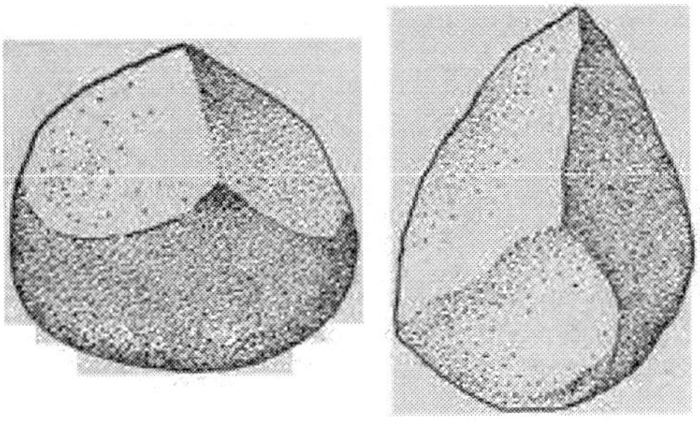

FIG. 213.—Windkanten shaped by the desert sand blast (after Chamberlin and Salisbury).

A direct consequence of this restriction of the more effective cutting tools to the layer of air just above the ground, is the strong tendency to cut away all projecting masses near their bases. The "mushroom rocks", which are so characteristic of desert landscapes, have been shaped in this manner (Fig. 212). Another product of the desert sand blast is the so-called *Windkante* (wind-edge) or *Dreikante* (three-edge), a pebble which is usually shaped in the form of a pyramid (Fig. 213).

Whenever a rock face, open to direct attack by the drifting sand, is constituted of parts which have different hardness, the blast of sand pecks away at the softer places and leaves the harder ones in relief. Thus is produced the well-known "stone lattice" of the desert (Fig. 214). Particularly upon the neck of the great Sphinx have the flying sand grains, by removing the softer layers, brought the sedimentary structures of the sandstone into strong relief.

FIG. 214.—The "stone lattice" of the desert, the work of the natural sand blast (after Walther).

When guided both by planes of sedimentation and planes of jointing, forms of a very high degree of ornamentation are developed. Some of the most remarkable forms are due to the protection afforded to the sun-exposed surfaces by the shell of desert varnish. In the shaded portions of projecting masses there is no such protection, and here the sand blast insinuates itself into every crack and cranny. In this it is aided by shadow weathering due to the differential strains set up at the border of the expanded sun-heated surface. As a result, projecting rock masses are sometimes etched away beneath and give the effect of a squatting animal. These forms, due to shadow erosion, have also been likened to projecting faucets. (Fig. 215).

FIG. 215.—Projecting rock carved by the drifting sand into the form of a couchant animal as a result of shadow weathering and erosion. Cut in granite on the north Indian Desert (after Walther).

Worn by its impact upon neighboring sand grains while in transport, but much more as it is thrown against the ground or hard rock surfaces, the wind-driven or *eolian* sand is at last worn into smoothly rounded granules which approach the form of a sphere. Compared to the surface which sea sand acquires by attrition, this shaping process is much the more efficient, since in the water the beach sand is buoyed up and is more effectively cushioned against its neighboring grains. The grains of beach sand when examined under a microscope are found to be much more irregular in form and usually display the original fracture surfaces only in part abraded.

FIG. 216.—Cliffs in loess 200 feet in height which exhibit the characteristic vertical jointing (after von Richtofen).

The dust carried out of the desert.—When, standing upon the mountain wall that surrounds a desert, the traveler gazes out to windward over the great depression, his field of view is generally obscured by the yellow haze of the dust clouds moving across the margins. Upon the mountain flanks and extending far outside the borders, this cloud of dust settles as a shrouding mantle of impalpable yellow powder, which is known as *loess*. These deposits are continually deepening, and have sometimes accumulated until they are hundreds or even thousands of feet in thickness. Before reaching its final resting place the dust of this deposit may have settled many times, and has certainly been in part redistributed by the streams near the desert margin. In it are the ingredients which are necessary for the nourishment of plants, and it constitutes the most important of natural soils. Continually fed by new deposits from the desert, and refertilized from below by a natural process so soon as the upper layers become impoverished, it requires no artificial fertilization. Without artificial aids the loess of northern China has been tilled for thousands of years without any signs of exhaustion.

FIG. 217.—A cañon in loess worn by traffic and wind. A highway in northern China (after von Richtofen).

Though easily pulverized between the fingers, loess is none the less characterized by a perfect vertical jointing and stands on vertical faces as does the solid rock (Fig. 216), but it is absolutely devoid of layers or bedding. Its capacity of standing in vertical cliffs the loess owes to a never failing content of lime carbonate which acts as a cement, and to a peculiar porous structure caused by capillary canals that run vertically through the mass, branching like rootlets and lined with carbonate of lime. This texture once destroyed, loess resolves itself into a common sticky clay.

By the feet of passing animals or by wheels of vehicles, the loess is crushed, and a portion is lifted and carried away by the wind. Thus in the course of time roadways sink deep into the mass as steep-walled cañons (Fig. 217). A portion of the now structureless clay remaining upon the roadway is at the time of the rains transformed into a thick mud which makes traveling all but impossible, though before its structure has been destroyed the loess is perfectly drained to the bottom of its deposits.

The particles which compose the loess are sharply angular quartz fragments, so fine that all but a few grains can be rubbed into the pores of the skin. Fine scales of mica, such as are easily lifted by the wind, are disseminated uniformly throughout the mass. The only inclosures which are arranged in layers consist of irregularly shaped concretions of clay. These show a striking resemblance to ginger roots and are called by the Chinese "stone ginger", though they are elsewhere more generally known by their German name of *Loessmännchen*, or loess dolls. These concretions are so disposed in the loess that their longer axes are vertical, and they were evidently separated from the mass and not deposited with it.

CHAPTER XVI

THE FEATURES IN DESERT LANDSCAPES

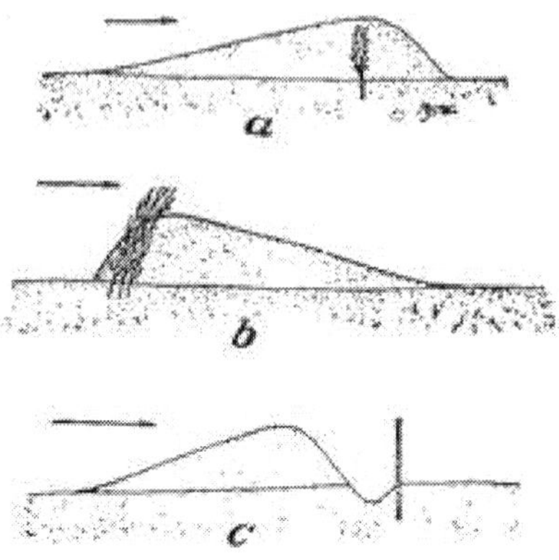

FIG. 218.—Diagrams to illustrate the effects of obstructions of different types in arresting wind-driven sand. *a*, An unyielding obstruction which permits the wind to pass through it; *b*, a flexible and perforated obstruction; *c*, an unyielding closed barrier (after Schulze).

The wandering dunes.—Over the broad expanse of the desert, sand and dust, and occasionally gypsum from the saline deposits, are ever migrating with the wind; on quiet days in the eddying "sand devils", but especially during the terrifying sand storms such as in the windy season darken the air of northern China and southern Manchuria. This drift of the sand is halted only when an obstruction is encountered—a projecting rock, a bush, or a bunch of grass, or again the buildings of a city or a town. The manner in which the sand is arrested by obstacles of different kinds is of great interest and importance, and is utilized in raising defenses against its encroachments. If the obstacle is unyielding but allows some of the wind to pass through it, no eddies are produced and the sand is deposited both to windward and to leeward of the obstruction to form a fairly symmetrical mound (Fig. 218 *a*). An obstruction which yields to the wind causes the sand to deposit in a mound which is largely to leeward of the obstruction (Fig. 218 *b*). A solid wall, on the other hand, by inducing eddies, is at first protected from the sand and mounds deposit both to windward and to leeward (Fig. 218 *c* and Fig. 219).

Except when held up by an obstruction, the drifting sand travels to leeward in slowly migrating mounds or ridges which are known as *dunes*. Their motion is due to the wind lifting the sand from the windward side and carrying it over the crest, from where it slides down the leeward slope and assumes a surface which is the angle of repose of the material. In contrast with this the windward slope is notably gradual, being shaped in conformity to the wind currents.

FIG. 219.—Sand accumulating both to windward and to leeward of a firm and impenetrable obstruction. The wind comes from the left (after a photograph by Bastin).

The dunes which are raised upon seashores, like those of the desert, are constantly migrating, those upon the shores of the North Sea at the average rate of about twenty feet per year. Relentlessly they advance, and despite all attempts to halt them, have many times overwhelmed the villages along the coast. Upon the great barrier beach known as the *Kurische Nehrung*, on the southeastern shore of the Baltic Sea, such a burial of villages has more than once occurred, but as in the course of time further migration of the dune has proceeded, the ruins of the buried villages have been exhumed by this natural excavating process (Fig. 220).

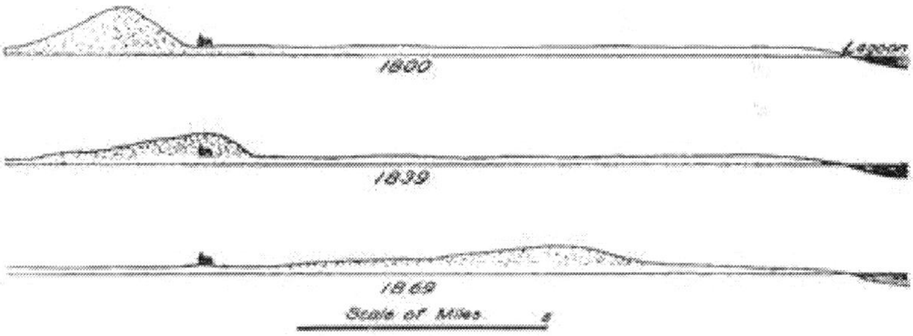

FIG. 220.—Successive diagrams to show how the town of Kunzen was buried, and subsequently exhumed in the continued migration of a great dune upon the Kurische Nehrung (after Behrendt).

PLATE 7.

A. **Ranges of dunes upon the margin of the Colorado Desert (after Mendenhall).**

B. **Sand dunes encroaching upon the oasis of Wed Souf. Algeria (after T. H. Kearney).**

The forms of dunes.—The forms assumed by dunes are dependent to a very large extent upon the strength of the wind and the available supply of sand. With small quantities of sand and with moderate winds, sickle-shaped dunes known as *barchans* (Fig. 221) are formed, whose convex and flatter slopes are toward the wind and whose steep concave leeward slopes are maintained at the angle of repose. The barchan is shaped by the wind going both over and around the dune, constantly removing sand from the windward side and depositing it to leeward. With larger supplies of sand and winds which are not too violent a series of barchans is built up, and these are arranged transversely to the wind direction (Fig. 222 *b*). If the winds are more violent, the minor depressions in the crests of the dunes become wind channels, and the sand is then trailed out along them until the arrangement of the ridges is parallel to the wind (Fig. 222 *c*). The surfaces of dunes are generally marked by beautiful ripples in the sand, which, seen from a little distance, may give the appearance of watered silk (plate 7 A).

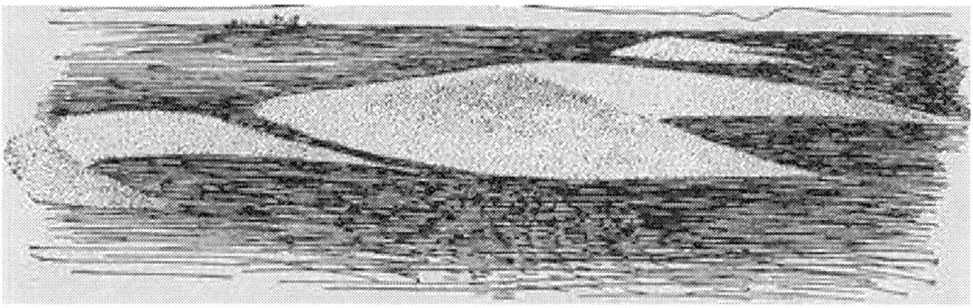

FIG. 221.—View of desert barchans (after Haug).

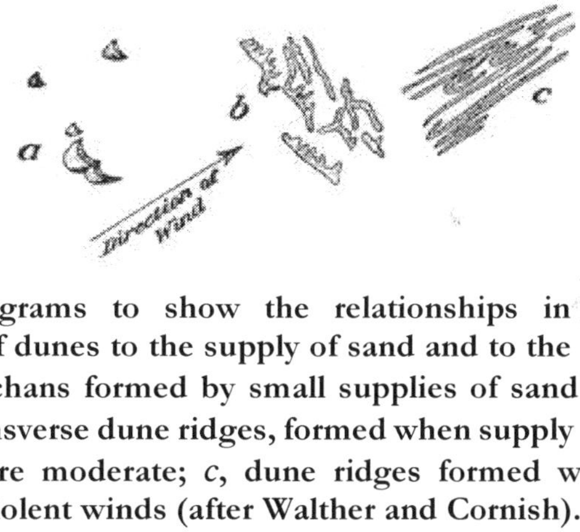

FIG. 222.—Diagrams to show the relationships in form and in orientation of dunes to the supply of sand and to the strength of the wind. *a*, barchans formed by small supplies of sand and moderate winds; *b*, transverse dune ridges, formed when supply of sand is large and winds are moderate; *c*, dune ridges formed with large sand supply and violent winds (after Walther and Cornish).

Under normal conditions dunes are not stationary but continue to wander with the prevailing winds until they have reached the outer edge of the zone of vegetation near the base of the foothills at the margin of the desert. Here the grasses and other desert plants arrest the first sand grains that reach them, and they continue to grow higher as the sands accumulate. Some of the desert plants, like the yuccas, have so adapted themselves to desert conditions that they may grow upward with the sand for many feet and so keep their crowns above its surface.

The cloudburst in the desert.—Such clouds as enter the desert through its mountain ramparts, and those derived from evaporation from the hot desert soil, usually precipitate their moisture before passing out of the basin. Above the highly heated floor the heavy rain clouds are unable to drop their burden. The rain can sometimes be seen descending, but long before it has reached the ground it has again passed into vapor, and through repetition of this process the clouds become so charged with moisture that when they encounter a mountain wall and are thus forced to rise, there is a sudden downpour not equaled in the humid regions. Desert rains are rare, but violent beyond comparison. Often for a year or more there is no rainfall upon the loose sand or porous clay, and the few plants which survive must push their roots deep down until they have reached the zone of ground water. When the clouds burst, each small cañon or *wed* (pl. *wadi*) within the mountain wall is quickly occupied by a swollen current which carries a thick paste of sediment and drowns everything before it. Ere it has flowed a mile, it may be that the water has disappeared entirely, leaving a layer of mud and sand which rapidly dries out with the reappearance of the sun.

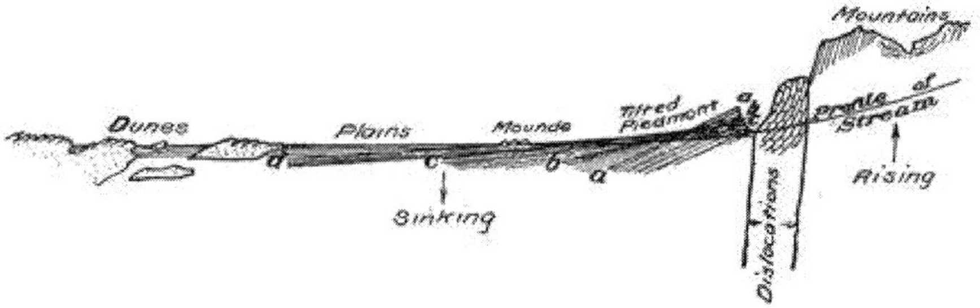

FIG. 223.—Ideal section across the rising mountain wall surrounding a desert and a part of the neighboring slope (after R. W. Pumpelly).

As the mountains upon seacoasts are generally rising with reference to the neighboring sea bottom, so the mountains which hem in the deserts are generally growing upward with reference to the inclosed desert floor. The marginal dislocations which separate the two are often in evidence at the foot of the steep slope (Fig. 223), and these may even appear as visible earthquake faults to indicate that the uplift is more accelerated than the deposition along the mountain front.

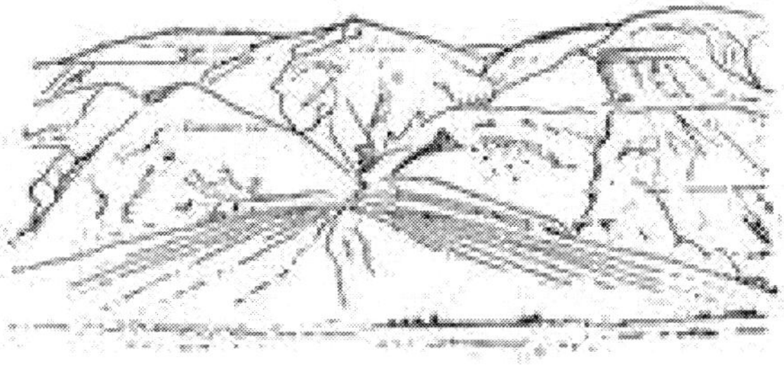

FIG. 224.—Dry delta or alluvial fan at the foot of a mountain range upon the borders of a desert.

The zone of the dwindling river.—The rapid uplift so generally characteristic of desert margins gives to the torrential streams which develop after each cloudburst such an unusual velocity that when they emerge from the mountain valleys on to the desert floor, the current is suddenly checked and the burden of sediment in large part deposited at the mouth of the valley so as to form a coarse delta deposit which is described as a *dry delta* (Fig. 224). Dependent upon its steepness of slope, this delta is variously referred to as an *alluvial fan* or *apron*, or as an *alluvial cone*. Over the conical slopes of the delta surface the stream is broken up into numerous distributaries which divide and subdivide as do the roots of a tree. In the Mohammedan countries described as *wadi*, these distributaries upon dry deltas are on the Pacific coast of the United States referred to as "washes" (Fig. 225).

FIG. 225.—Map of the distributaries of neighboring streams which emerge at the western base of the Sierra Nevadas in California (after W. D. Johnson).

Fast losing their velocity after emerging from the mountains, the various distributaries drop first of all the heavy bowlders, then the large pebbles and the sand, so that only the finer sand and the silt are carried to the margin of the delta. As they enlarge their boundaries, the neighboring deltas eventually coalesce and so form an *alluvial bench* or "gravel piedmont" at the foot of the range. Only the larger streams are able to entirely cross this bench of parched deposits with its coarsely porous structure, for the water is soon sucked up by the thirsty materials. Encountering in its descent more clayey layers, this water is conducted to the surface near the margin of the bench and may there appear as a line of springs. At this level there develops, therefore, a zone of vegetation, though there is no local rain.

The alluvial bench grows upward by accretion of layers which are thickest at the mountain end, so that the steepness of the bench increases with time.

Erosion in and about the desert.—The violent cloudburst that is characteristic of the arid lands is a most potent agent in modeling the surface of the ground wherever the rock materials are not too firmly coherent. Under the dash of the rain a peculiar type of "bad land" topography is developed (plate 5 B and Fig. 226). Such a rain-cut surface is a veritable maze of alternating gully and ridge, a country worthless for agricultural purposes and offering the greatest difficulty in the way of penetrating it. When composed of stiff clay with scattered pebbles and bowlders, the effect of the "rain erosion" is to fashion steep clay pillars each capped by a pebble and described as "demoiselles" (Fig. 226).

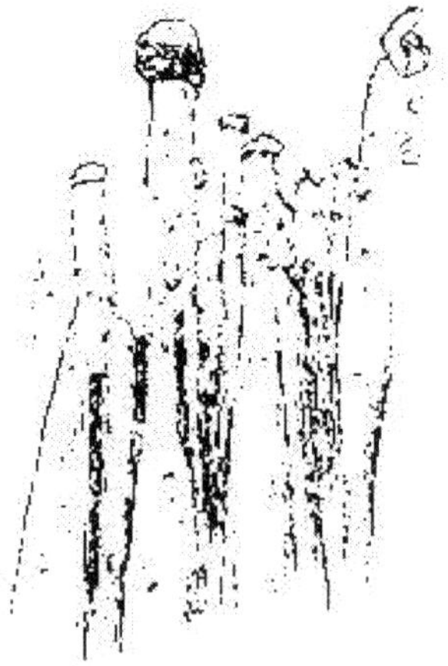

FIG. 226.—A group of "demoiselles" in the "bad lands" (after a photograph by Fairbanks).

Behind the mountain front the valleys out of which the torrents are discharged are usually short with steep side walls and a relatively flat bottom, ending headward in an amphitheater with precipitous walls (Fig. 227). In the western United States such valleys are referred to as "box cañons", but in Mohammedan countries the name "wed" applies to the river valley within the mountains and to the distributaries as well.

Characteristic features of the arid lands.—It is characteristic of erosion and deposition within humid regions that all outlines become softened into flowing curves, due to the protective mat of vegetation. In arid lands those massive rocks which are without structural planes of separation, partly as a consequence of exfoliation, develop broad domes which are projected upon the horizon as great semicircles, broken in half it may be by displacement. The same massive rocks where intersected by vertical joint planes yield, on the contrary, sharp granite needles like those of Harney Peak (plate 8 A). Similarly, schistose or bedded rocks, when tilted at a high angle, may yield forms which are almost identical. Examples of such needles are found in the Garden of the Gods in Colorado.

At lower levels, where the flying sand becomes effective as an eroding agent, flat bedded rocks become etched into shelves and cornices, and if intersected by joints, the shelves and cornices are transformed into groups of castellated towers and pinnacles of a high degree of ornamentation. These fantastic erosion remnants are usually referred to as "chimneys" and may be seen in numbers in the bad lands of Dakota, as they may in Colorado either in Monument Park or in the new Monolithic National Park (plate 8 B).

FIG. 227.—Amphitheater at the head of the Wed Beni Sur (after Walther).

Where wind erosion plays a smaller rôle in the sculpture, but where after an uplift a river has made its way, horizontally bedded rocks are apt to be carved into broad *rock terraces*, nowhere shown upon so grand a scale as about the Grand Cañon of the Colorado. Each harder layer has here produced a floor or terrace which ends in a vertical escarpment, and this is separated from the next lower layer of more resistant rock by a slope of talus which largely hides the softer intermediate beds. The great Desert of Sahara is shaped in a series of rock terraces or steppes which descend toward the interior of the basin.

A single harder layer of resistant rock comes often to form the flat capping of a plateau, and is then known as a *mesa*, or table mountain. Along its front, detached outliers usually stand like sentinels before the larger mass, and according to their relative proportions, these are referred to either as small mesas or as the smaller *buttes* (Fig. 228).

FIG. 228.—Mesa and outlying butte in the Leucite Hills of Wyoming (after Whitman Cross, U. S. G. S.).

The war of dune and oasis.—In every desert the deposits are arranged in consecutive belts or zones which are alternately the work of wind and water. Surrounding the desert and upon the flanks of the mountain wall there is found (1) the deposit of loess derived from the dust that is carried out of the desert by the wind. Immediately within the desert border at the base of the mountains is (2) the zone of the dwindling river with its sloping bench of coarse rubble and gravel.

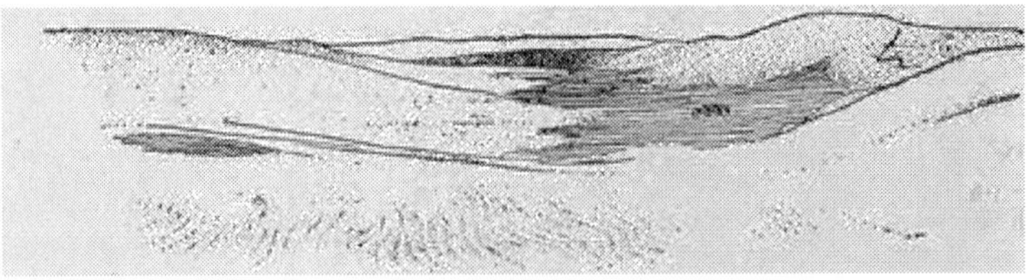

FIG. 229.—Flat-bottomed basin separating dunes—*bajir* or *takyr* (after Ellsworth Huntington).

Next in order is (3) the belt of the flying sand, a zone of dune ridges often separated by narrow, flat-bottomed basins (Fig. 229) into which the strongest streams bring the finer sands and silt from the mountains. Lastly, there is (4) the central sink or sinks, into which all water not at once absorbed within the zone of alluviation or in the zone of dunes is finally collected. Here are the true lacustrine deposits of clay and separated salts (Fig. 230 and Fig. 207, p. 201). The lake deposits fill in all the original irregularities of the desert floor, out of which the tops of isolated ranges of mountains now project like islands out of the surface of the sea. The several zones of deposits in their order from the margin to the center of the desert are given schematically in Fig. 231.

PLATE 8.

A. **The granite needles of Harney Peak in the Black Hills of South Dakota (after Darton).**

B. **Castellated erosion chimneys in El Cobra Cañon, New Mexico.** *(Photograph by E. C. Case.)*

FIG. 230.—Billowy surface of the salt crust on the central sink in the Lop Desert of central Asia (after Ellsworth Huntington).

The zone of vegetation, as already stated, lies near the foot of the alluvial bench, so that here are found the oases about which have clustered the cities of the desert from the earliest records of antiquity until now. Just without the line of oases is the wall of dunes held back from further advance only by the vegetation which in turn is dependent upon the rains in the neighboring mountains. With every diminution in the water supply, the dunes advance and encroach upon the oases (plate 7 B); while with every considerable increase in this supply of moisture the alluvial bench advances over the dunes and acquires a strip of their territory. Thus with varying fortunes a war is continually waged between the withering river and the flying sand, and the alternations of climate are later recorded in the dovetailing together of the eolian and alluvial deposits at their common junction (Fig. 231).

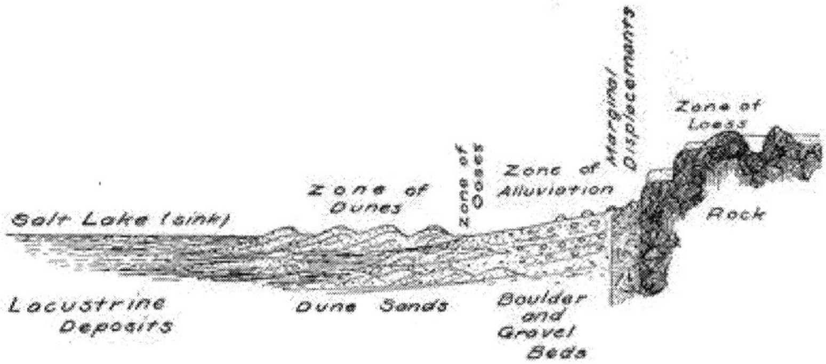

FIG. 231.—Schematic diagram to show the zones of deposition in their order from the margin to the center of a desert.

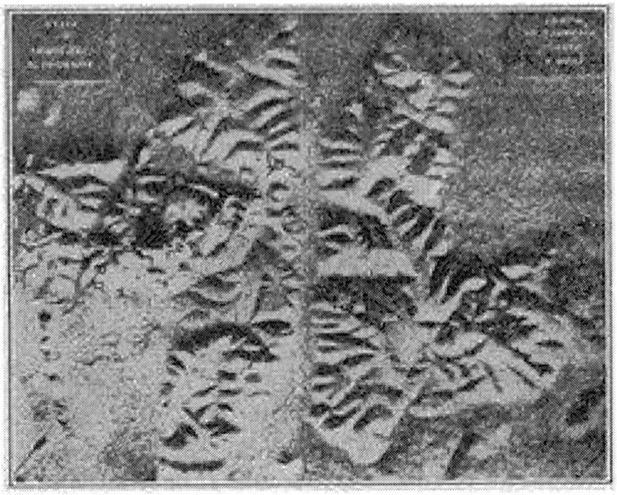

FIG. 232.—Mounds upon the site of the buried city of Nippur (after the cast by Muret).

In addition to the smaller periodic alternations of pluvial and interpluvial climate—the "pulse of Asia"—the record of the Asiatic deserts indicates a progressive desiccation of the entire region, which has now given the victory to the dune. The ancient history of the cities of the plains supplies the records of many that have been buried in the dunes. To-day, where once were prosperous cities, nothing is to be seen at the surface but a group of mounds (Fig. 232). Exhumed after much painstaking labor, the walls and palaces of these ancient cities have once more been brought to the light of day, and much has thus been learned of the civilization of these early times (Fig. 233). Quite recently the mounds which cover between one and two hundred buried villages have been found upon the borders of the Tarim basin of central Asia, where they were lost to history when they were overwhelmed in the early centuries of the Christian Era.

FIG. 233.—Exhumed structures in the buried city of Nippur (after Hilprecht).

The origin of the high plains which front the Rocky Mountains.—To the eastward of the great backbone of the North American continent stretches a vast plain gently inclined away from the range and generally known as the High Plains region (plate 9). The tourist who travels westward by train ascends this slope so gradually that when he has reached the mountain front it is difficult to realize that he has climbed to an altitude of five thousand feet above the level of the sea. That he has also passed through several climatic zones—a humid, a semiarid, and an arid—and has now entered a semiarid district, is more easily appreciated from study of the vegetation (Fig. 234). The surface of the High Plains, where not cut into by rivers, is remarkably even, so that it might be compared to the quiet surface of a great lake.

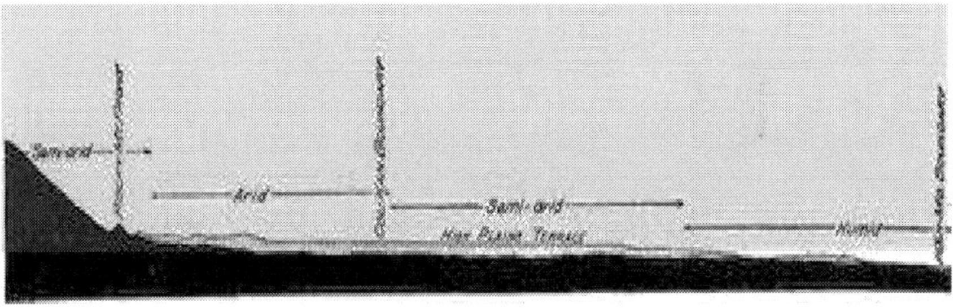

FIG. 234.—Section across the High Plains, showing the position of the terrace and the climatic zones above it (after W. D. Johnson).

The materials which compose the surface veneer of these plains are coarse conglomerates, gravels, and sands, and the so-called "mortar beds", which are nothing but sands cemented into sandstone by carbonate of lime. The pebbles in all these deposits are far-traveled and appear to have been derived from erosion of those crystalline rocks which compose the eastern front of the Rocky Mountains. These different materials are not arranged in strictly parallel beds, as are the deposits of a lake or sea; but the beds are made up of long threads of lenticular cross section which are interlaced in the most intricate fashion and which extend down the slope, or outward from the mountain front (Fig. 235). It is thus shown that the High Plains are a bench or plain of alluviation formed at the front of the Rocky Mountains during an earlier series of pluvial periods, and that subsequent uplift has produced the modern river valleys which are cut out of the plain. The plexus of long threads of the coarser materials are the courses of dwindling rivers which interlaced over the former plain, and which in time were buried under other channel deposits of the same nature but in different positions (Fig.

236). The pluvial periods in which this bench was formed produced in other latitudes the great continental glaciers which wrought such marvelous changes in northern North America and in northern Europe.

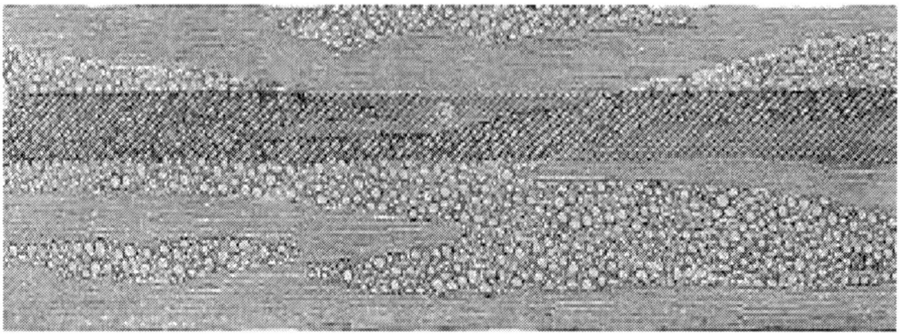

FIG. 235.—Section across the great lenticular threads of alluvial deposits which compose the veneer of the High Plains (after W. D. Johnson).

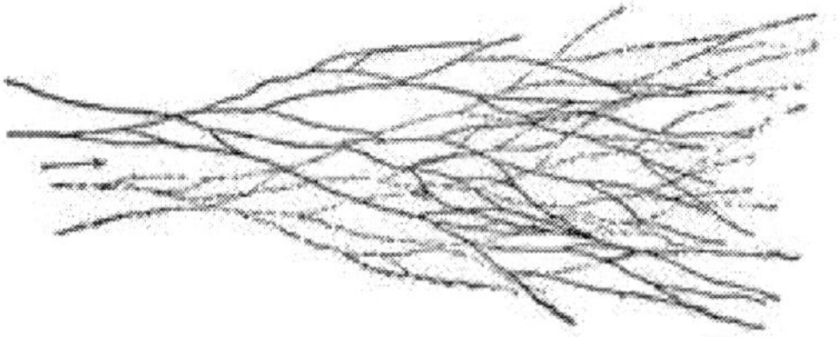

FIG. 236.—Distributaries of the foothills superimposed upon an earlier series (after W. D. Johnson).

Character profiles.—In contrast with the profiles in the landscapes of humid regions (see Fig. 187, p. 177), those of arid lands are marked by straighter elements (Fig. 237). Almost the only exception of importance is furnished by the domes of massive granite monoliths, which are sometimes broken in half by great displacements. Below the horizon the secondary lines in the landscape betray the same straightness of the component elements by the gabled slopes of talus which are many times repeated so as almost to reproduce the lines in a house of cards, since the sloping lines are maintained at the angle of repose of the materials (Fig. 482, p. 443). Wherever the waves of desert lakes have made an attack upon the rocks and have retired the projecting spurs, other gables characterized by slightly different slopes are introduced into the landscape.

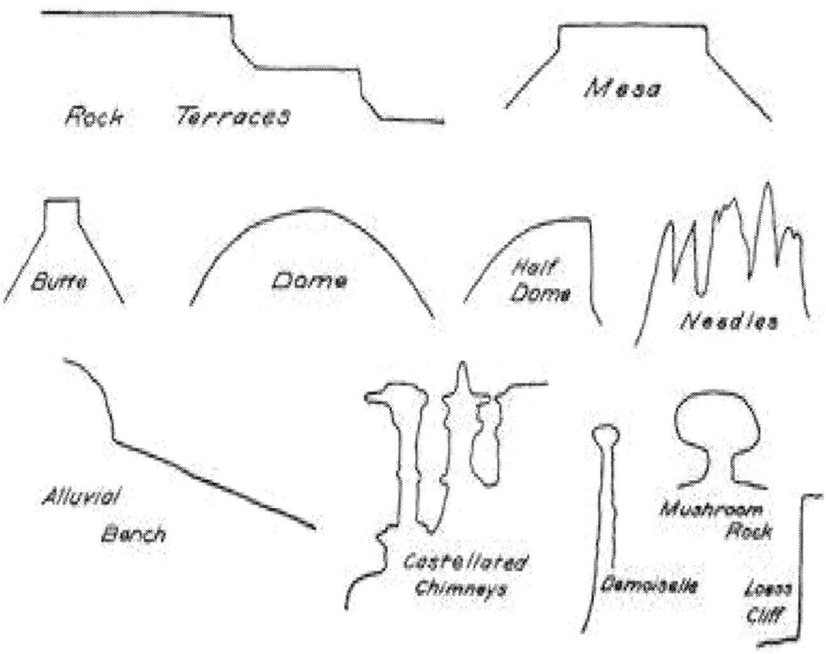

FIG. 237.—Character profiles in the landscapes of arid lands.

PLATE 9.

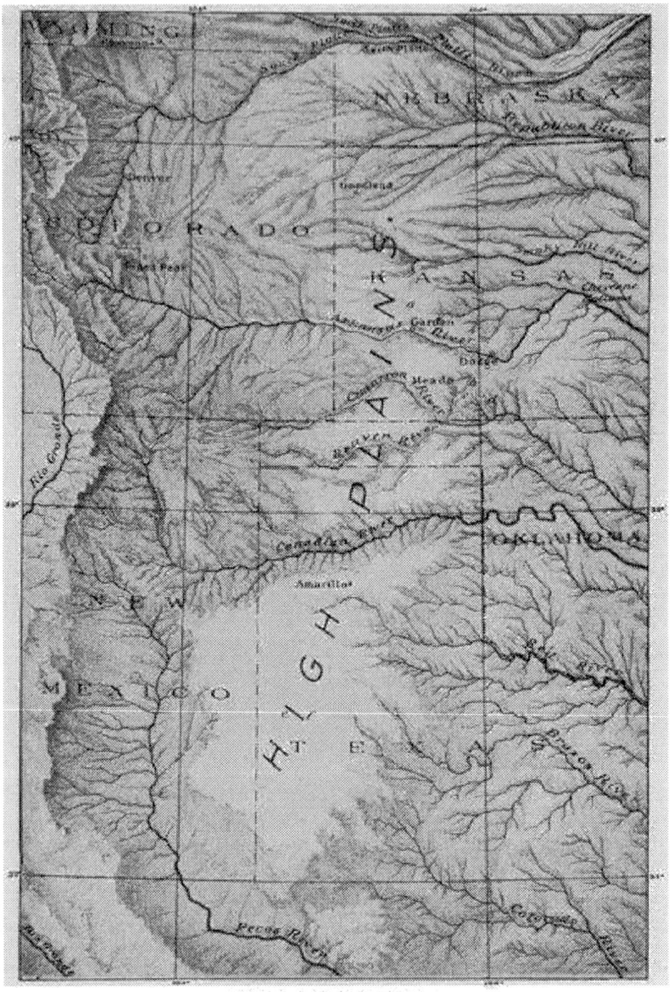

THE HIGH PLAINS

Scale of miles.

0 50 100 150 200

Reading References for Chapters XV and XVI

General:—

JOHANNES WALTHER. Das Gesetz der Wüstenbildung in Gegenwart und Vorzeit. Berlin, 1900, pp. 175, many plates. (This extremely valuable work is now out of print, but both a revised edition and an English translation are promised for 1912.)

SIEGFRIED PASSARGE. Die Kalihari. Berlin, 1904, pp. 662.

W. M. DAVIS. The Geographic Cycle in an Arid Climate, Jour. Geol., vol. 13, 1905, pp. 381-407.

ELLSWORTH HUNTINGTON. The Pulse of Asia. New York and Boston, 1907, pp. 415.

SVEN HEDIN. Scientific Results of a Journey through Central Asia, 1899-1900. Stockholm, 1904-1905, vols. 1 and 2, pp. 523 and 717, pls. 56 and 76.

JOSEPH BARRELL. Relative Geological Importance of Continental, Littoral and Marine Sedimentation, Jour. Geol., vol. 14, 1906, pp. 316-356, 429-457, 524-568.

E. F. GAUTIER. Études sahariennes, Ann. de Géogr., vol. 16, 1907, pp. 46-69, 117-138.

The self-registering gauge of past climates:—

G. K. GILBERT. Lake Bonneville, Mon. I, U. S. Geol. Surv., Chapter vi, pp. 214-318.

T. F. JAMIESON. The Inland Seas and Salt Lakes of the Glacial Period, Geol. Mag. decade III, vol. 2, 1885, pp. 193-200.

J. E. TALMAGE. The Great Salt Lake, Present and Past. Salt Lake City, 1900, pp. 116, plates.

E. HUNTINGTON. Some Characteristics of the Glacial Period in Non-glaciated Regions, Bull. Geol. Soc. Am., vol. 18, 1907, pp. 351-388, pls. 31-39.

T. C. CHAMBERLIN. The Future Habitability of the Earth, Rept. Smithson. Inst., 1910, pp. 371-389.

The red and brown desert varnish:—

I. C. RUSSELL. Subaërial Decay of Rocks and Origin of the Red Color of Certain Formations. Bull. 52, U. S. Geol. Surv., 1889, pp. 65, pls. 5.

Erosion in the desert:—

J. A. UDDEN. Erosion, Transportation, and Sedimentation performed by the Atmosphere, Jour. Geol., vol. 2, 1894, pp. 318-331.

S. PASSARGE. Die pfannenförmigen Hohlformen der südafrikanischen Steppen, Pet. Mitt., vol. 57, 1911, pp. 57-61, 130-135.

The dust carried out of the desert:—

F. VON RICHTOFEN. China, Ergebnisse eigene Reisen und darauf gegründeten Studien, Berlin, 1877, vol. 1, pp. 56-125.

E. HILGARD. The Loess of the Mississippi Valley, Am. Jour. Sci., (3), vol. 18, 1879, pp. 106-112.

T. C. CHAMBERLIN and R. D. SALISBURY. Preliminary Paper on the Driftless Area of the Upper Mississippi Valley, 6th Ann. Rept. U. S. Geol. Surv., 1885, pp. 278-307.

E. E. FREE. The movement of soil material by the wind, with a bibliography of eolian geology by S. C. Stuntz and E. E. Free, Bull. 68, U. S. Bureau of Soils, 1911, pp. 272, pls. 5.

M. NEUMAYR. Erdgeschichte, vol. 1, pp. 510-514.

E. DE MARTONNE. Géographie physique, pp. 663-668.

Dunes:—

VAUGHAN CORNISH. On the Formation of Sand-dunes, Geogr. Jour., vol. 9, 1897, pp. 278-309 (a most important paper).

F. SOLGER and Others. Dünenbuch. Enke, Stuttgart, 1910, pp. 373.

The zone of the dwindling river:—

E. HUNTINGTON. The Border Belts of the Tarim Basin, Bull. Am. Geogr. Soc., vol. 38, 1906, pp. 91-96; The Pulse of Asia, pp. 210-222, 262-279.

The war of dune and oasis:—

R. PUMPELLY. Explorations in Turkestan, Expedition of 1904, etc., Pub. 73, Carneg. Inst., Washington, vol. 1, pp. 1-13.

E. HUNTINGTON. The Oasis of Kharga, Bull. Am. Geogr. Soc., vol. 42. 1910, pp. 641-661.

TH. H. KEARNEY. The Country of the Ant Men, Nat. Geogr. Mag., vol. 22, 1911, pp. 367-382.

Features of the arid lands:—

C. E. DUTTON. Tertiary History of the Grand Cañon District, with Atlas, Mon. II, U. S. Geol. Surv., 1882, pp. 264, pls. 42, maps 23.

G. SWEINFURTH. Map Sheets of the Eastern Egyptian Desert. Berlin, 1901-1902, 8 sheets.

The origin of the high plains:—

W. D. JOHNSON. The High Plains and their Utilization, 21st Ann. Rept. U. S. Geol. Surv., Pt. iv, 1901, pp. 601-741.

CHAPTER XVII

REPEATING PATTERNS IN THE EARTH RELIEF

The weathering processes under control of the fracture system.— In an earlier chapter it was learned that the rocks which compose the earth's surface shell are intersected by a system of joint fractures which in little-disturbed areas divide the surface beds into nearly square perpendicular prisms (Fig. 36, p. 55), more or less modified by additional diagonal joints, and often also by more disorderly fractures. Throughout large areas these fractures may maintain nearly constant directions, though either one or more of the master series may be locally absent. This distinctive architecture of the surface shell of the lithosphere has exercised its influence upon the various weathering processes, as it has also upon the activities of running water and of other less common transporting agencies at the surface.

Within high latitudes, where frost action is the dominant weathering process, the water, by insinuating itself along the joints and through repeated freezings, has broken down the rock in the immediate neighborhood of these fractures, and so has impressed upon the surface an image of the underlying pattern of structure lines (plate 10 A).

In much lower latitudes and in regions of insufficient rainfall, the same structures are impressed upon the relief, but by other weathering processes. In the case of the less coherent deposits in these provinces, the initial forms of their erosional surface have sometimes been determined by the dash of rain from the sudden cloudburst. Thus the "bad lands" may have their initial gullies directed and spaced in conformity with the underlying joint structures (Fig. 238).

FIG. 238.—Rain sculpturing under control by joints. Coast of southern California (after a photograph by Fairbanks).

In such portions of the temperate regions as are favored by a humid climate, the mat of vegetation holds down a layer of soil, and mat and soil in coöperation are effective in preventing any such large measure of frostwork as is characteristic of the subpolar regions or of high levels in the arid lands. In humid regions the rocks become a prey especially to the processes of solution and accompanying chemical decomposition, and these processes, although guided by the course of the percolating ground water along the fracture planes, do not afford such striking examples of the control of surface relief.

Those limestones which slowly pass into solution in the percolating water do, however, quite generally indicate a localization of the solution along the joint channels (Fig. 239 and plate 6 B). Though in other rocks not so apparent, yet solutions generally take their courses along the same channels, and upon them is localized the development of the newly formed hydrated and carbonate minerals, as is well illustrated by the phenomenon of spheroidal weathering (Fig. 155, p. 150).

FIG. 239.—Outcrop of flaggy limestone which shows the effects of solution along neighboring joints in a sagging of the upper beds (after Gilbert, U. S. G. S.).

The fracture control of the drainage lines.—The etching out of the earth's architectural plan in the surface relief, which we have seen begun in the processes of weathering, is continued after the transporting agents have become effective. It is often easy to see that a river has taken its course in rectangular zigzags like the elbows of a jointed stove pipe, and that its walls are formed of joint planes from which an occasional squared buttress projects into the channel. This structure is rendered in the plan of the Abisko Cañon of northern Lapland (Fig. 240). We are later to learn that another great transporting agent, the water wave, makes a selective attack upon the lithosphere along the fractures of the joint system (Fig. 250, p. 233 and Fig. 254, p. 235).

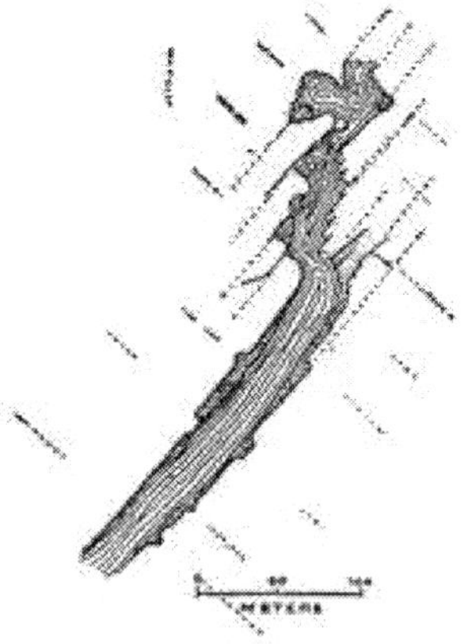

FIG. 240.—Map of the joint-controlled Abisko Cañon in northern Lapland (after Otto Sjögren).

Where the scale of the example is large, as in the cases which have been above cited, the actual position and directions of the joint wall are easily compared with the near-by elements of the river's course, so that the connection of the drainage lines with the underlying structure is at once apparent. In many examples where the scale is small, the evidence for the controlling influence of the rock structure in determining the courses of streams may be found in the peculiar character of the drainage plan. To illustrate: the course of the Zambesi River, within the gorge below the famous Victoria Falls, not only makes repeated turnings at a right angle, but its tributary streams, instead of making the usual sharp angle where they join the main stream, also affect the right angle in their junctions (Fig. 241).

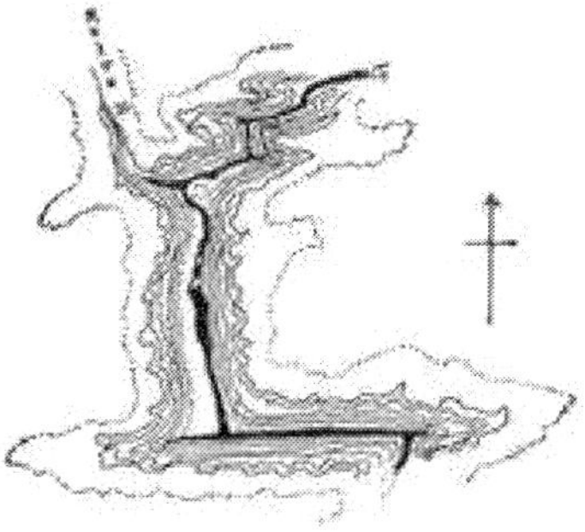

FIG. 241.—Map of the gorge of the Zambesi River below the Victoria Falls (after Lamplugh).

The repeating pattern in drainage networks.—It is a characteristic of the joint system that the fractures within each series are spaced with approximation to uniformity. If the plan of a drainage system has been regulated in conformity with the architecture of the underlying rock basement, the same repeating rectangles of the master joints may be expected to appear in the lines of drainage—the so-called drainage network.

Such rectangular patterns do very generally appear in the drainage network, though they are often masked upon modern maps by what, to the geologist, seems impertinent intrusion of the black lines of overprinting which indicate railways, lines of highway, and other culture elements. On river maps, which are printed without culture, the pattern is much more easily recognized (Figs. 242 and 243). Wherever the relief is strong, as is the case in the Adirondack Mountain province of the State of New York, individual hills may stand in relief between the bounding streams which compose the rectangular network, like the squared pedestals of monuments. Such a type of relief carved in repeating patterns has been described as "checkerboard topography."

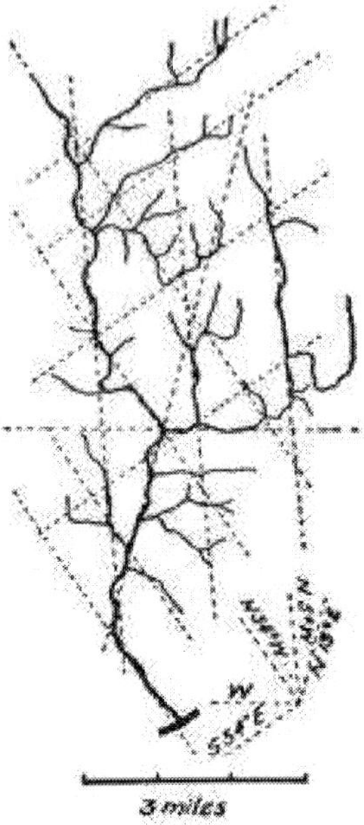

FIG. 242.—Controlled drainage network of the Shepaug River in Connecticut.

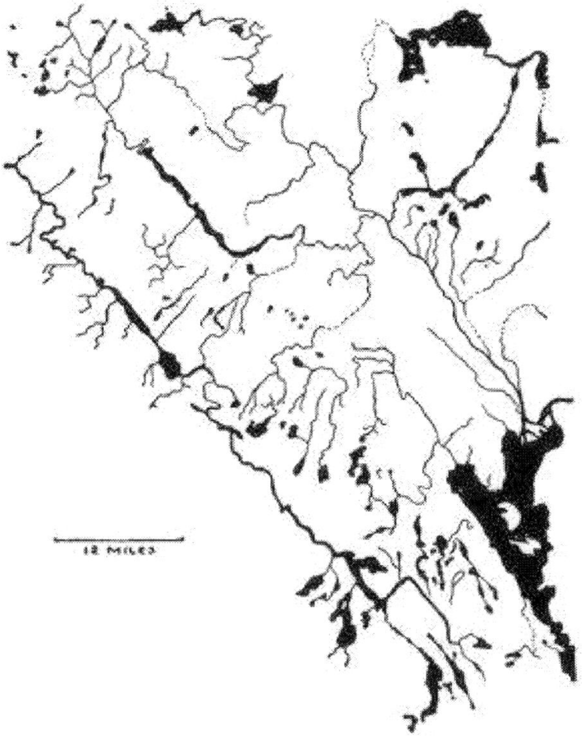

FIG. 243.—A river network of repeating rectangular pattern. Near Lake Temiskaming, Ontario (from the map by the Dominion Government).

The dividing lines of the relief patterns—lineaments.—The repeating design outlined in the river network of the Temiskaming district (Fig. 243) would appear in greater perfection if we could reproduce the relief without at the same time obscuring the lines of drainage; for where the pattern is not completely closed by the course of the stream, there is generally found either a dry valley or a ravine to complete the design. If these are not present, a bit of straight coast line, a visible line of fracture, a zone of fault breccia, or the boundary line separating different formations may one or more of them fill in the gaps of the parallel straight drainage lines which by their intersection bring out the pattern. These significant lines of landscapes which reveal the hidden architecture of the rock basement are described as *lineaments* (Fig. 82, p. 87). They are the character lines of the earth's physiognomy.

It is important to emphasize the essentially composite expression of the lineament. At one locality it appears as a drainage line, a little farther on it may be a line of coast; then, again, it is a series of aligned waterfalls, a visible fault trace, or a rectilinear boundary between formations; but in every case it is some surface expression of a buried fracture. Hidden as they so generally are, the fracture lines must be searched out by every means at our disposal, if we are not to be misled in accounting for the positions and the attitudes of disturbed rock masses.

As we have learned, during earthquake shocks, as at no other time, the surface of the earth is so sensitized as to betray the position of its buried fractures. As the boundaries of orographic blocks, certain of the fractures are at such times the seats of especially heavy vibrations; they are the seismotectonic lines of the earthquake province. Many lineaments are identical with seismotectonic lines, and they therefore afford a means of to some extent determining in advance the lines of greatest danger from earthquake shock.

The composite repeating patterns of the higher orders.—Not only do the larger joint blocks become impressed upon the earth relief as repeating diaper patterns, but larger and still larger composite units of the same type may, in favorable districts, be found to present the same characters. Attention has already been more than once directed to the fact that the more perfect and prominent fracture planes recur among the joints of any series at more or less regular intervals (Fig. 40, p. 57, and Fig. 41, p. 58). Nowhere, perhaps, is this larger order of the repeating pattern more perfectly exemplified than in some recent deposits in the Syrian desert (plate 10 B). It is usually, however, in the older sediments that such structures may be recognized; as, for example, in the squared towers and buttresses of the Tyrolean Dolomites (Fig. 244). Here the larger blocks appear in the thick bedded lower formation, the dolomite, divided into subordinate sections of large dimensions; but in the overlying formations in blocks of relatively small size, yet with similarly perfect subequal spacing.

FIG. 244.—Squared mountain masses which reveal a distribution of the joints in block patterns of different orders of magnitude. The Pordoi range of the Sella group of the Dolomites, seen from the Cima di Rossi (after Mojsisovics).

The observing traveler who is privileged to make the journey by steamer, threading its course in and out among the many islands and skerries of the Norwegian coast, will hardly fail to be struck by the remarkable profiles of most of the lower islands (Fig. 245). These profiles are generally convexly scalloped with a noteworthy regularity, and not in one unit only, but in at least two with one a multiple of the other (Fig. 246). As the steamer passes near to the islands, it is discovered that the smaller recognizable units in the island profiles are separated by widely gaping joints which do not, however, belong to the unit series, but to a larger composite group (Fig. 246 *b*). Frostwork, which depends for its action upon open spaces within the rocks, has here been the cause of the excessive weathering above the more widely gaping joints.

PLATE 10.

A. **View in Spitzbergen to illustrate the disintegration of rock under the control of joints. (*Photograph by O. Haldin.*)**

B. **Composite pattern of the joint structures within recent alluvial deposits. (*Photograph by Ellsworth Huntington.*)**

FIG. 245.—Island groups of the Lofoten archipelago off the northwest coast of Norway, which reveal repeating patterns of the relief in two orders of magnitude (after a photograph by Knudsen).

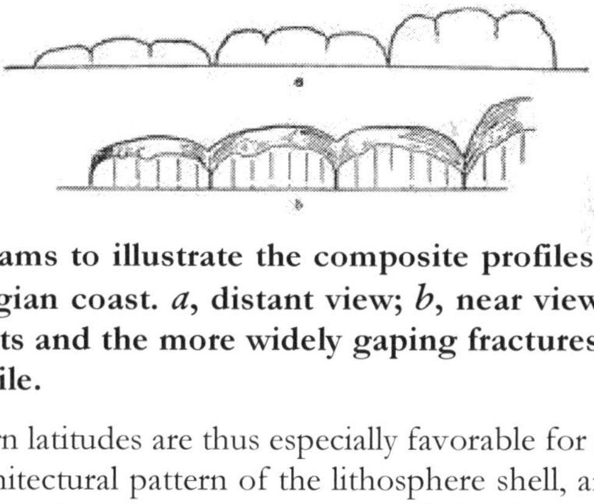

FIG. 246.—Diagrams to illustrate the composite profiles of the islands on the Norwegian coast. *a*, distant view; *b*, near view, showing the individual joints and the more widely gaping fractures beneath each sag in the profile.

High northern latitudes are thus especially favorable for revealing all the details in the architectural pattern of the lithosphere shell, and we need not be surprised that when the modern maps of the Norwegian coast are examined, still larger repeating patterns than any that may be seen in the field are to be made out. The Norwegian coast was long ago shown to be a complexly faulted region, and these larger divisions of the relief pattern, instead of being explained as a consequence solely of selective weathering, must be regarded as due largely to fault displacements of the type represented in our model (plate 4 C). Yet whether due to displacements or to the more numerous joints, all belong to the same composite system of fractures expressed in the relief.

Reading References for Chapter XVII

WILLIAM H. HOBBS. The River System of Connecticut, Jour. Geol., vol. 9, 1901, pp. 469-485, pl. 1; Lineaments of the Atlantic Border Region, Bull. Geol. Soc. Am., vol. 15, 1904, pp. 483-506, pls. 45-47; The Correlation of Fracture Systems and the Evidences for Planetary Dislocations within the Earth's Crust, Trans. Wis. Acad. Sci., etc., vol. 15, 1905, pp. 15-29; Repeating Patterns in the Relief and in the Structure of the Land, Bull. Geol. Soc. Am., vol. 22, 1911, pp. 123-176, pls. 7-13.

CHAPTER XVIII

THE FORMS CARVED AND MOLDED BY WAVES

The motion of a water wave.—The motions within a wave upon the surface of a body of water may be thought of in two different ways. First of all, there is the motion of each particle of water within an orbit of its own; and there is, further, the forward motion of propagation of the wave considered as a whole.

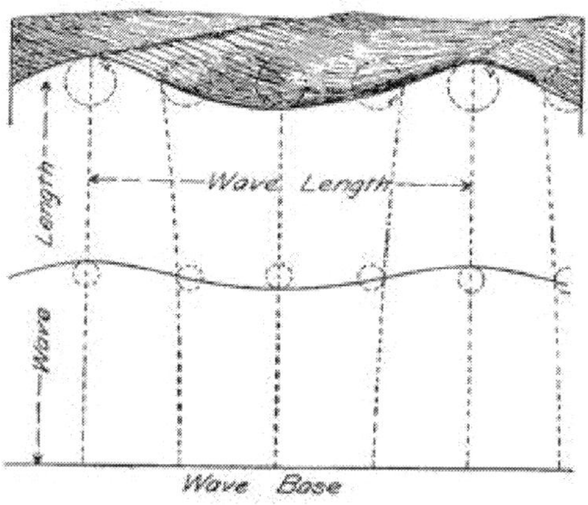

FIG. 247.—Diagram to show the nature of the motions within a free water wave.

The water particle in a wave has a continued motion round and round its orbit like that of a horse circling a race course, only that here the track is in a vertical plane, directed along the line of propagation of the wave (Fig. 247). Each particle of water, through its friction upon neighboring particles, is able to transmit its motion both along the surface and downward into the water below. The force which starts the water in motion and develops the wave, is the friction of wind blowing over the water surface, and the size of the orbit of the water particle at any point is proportional to the wind's force and to the stretch of water over which it has blown. The wind's effect is, therefore, cumulative—the wave is proportional to the wind's effect upon all water particles in its rear, added to the local wind friction.

The size or *height* of the wave is measured by the diameter of the orbit of motion of the surface particle, and this is the difference in height between trough and crest. The distance from crest to crest, or from trough to trough, is called the *wave length*. Though the wave motion is transmitted downward into the water there is a continued loss of energy which is here not compensated by added wind friction, and so the orbital motion grows smaller and smaller, until at the depth of about a wave length it has completely died out. This level of no motion is called the *wave base*. In quiet weather the level of no motion is practically at the water's surface, and inasmuch as the geological work of waves is in large part accomplished during the great storms, the term "wave base" refers to the lowest level of wave motion at the time of the heaviest storms. Upon the ocean the highest waves that have been measured have an amplitude of about fifty feet and a wave length of about six hundred feet.

Free waves and breakers.—So long as the depth of the water is below wave base, there is obviously no possibility of interference with the wave through friction upon the bottom. Under these conditions waves are described as *free waves*, and their forms are symmetrical except in so far as their crests are pulled over and more or less dissipated in the spray of the "white caps" at the time of high winds.

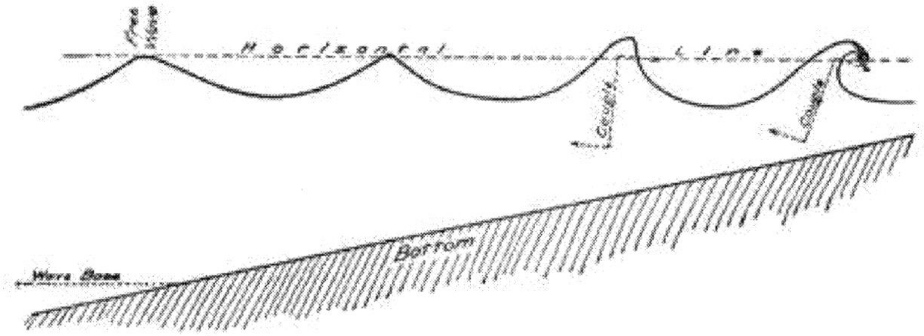

FIG. 248.—Diagram to illustrate the transformation of a free wave into a breaker as it approaches the shore.

As a wave approaches a shore, which generally has a gentle outward sloping surface, there is interposed in the way of a free forward movement the friction upon the bottom. This friction begins when the depth of water is less than wave base, and its effect is to hold back the wave at the bottom. Carried slowly upward in the water by the friction of particle upon particle, the effect of this holding back is a piling up of the water, which increases the wave height as it diminishes the wave length, and also interferes with wave symmetry (Fig. 248). Moving forward at the top under its inertia of motion and held back at the bottom by constantly increasing friction, a strong turning motion or couple is started about a horizontal axis, the immediate effect of which is to steepen the forward slope of the wave, and this continues until it overhangs, and, falling, "breaks" into surf. Such a breaking wave is called a "comber" or "breaker" (plate 11 B).

PLATE 11.

A. **Ripple markings within an ancient sandstone (courtesy of U. S. Grant).**

B. **A wave breaking as it approaches the shore. (*Photograph by Fairbanks.*)**

FIG. 249.—Notched rock cliff cut by waves and the fallen blocks derived from the cliff through undermining. Profile Rock at Farwell's Point near Madison, Wisconsin.

Effect of the breaking wave upon a steep rocky shore—the notched cliff.—If the shore rises abruptly from deeper water, the top of the breaking wave is hurled against the cliff with the force of a battering ram. During storms the water of shore waves is charged with sand, and each sand particle is driven like a stone cutter's tool under the stroke of his hammer. The effect is thus both to chip and to batter away the rock of the shore to the height reached by the wave, undermining it and notching the rock at its base (Fig. 249). When the notch has been cut in this manner to a sufficient depth, the overhanging rock falls by its own weight in blocks which are bounded by the ever present joints, leaving the upper cliff face essentially vertical.

FIG. 250.—A wave-cut chasm under control by joints, coast of Maine (after Tarr).

Coves, sea arches, and stacks.—It is the headland which is most exposed to the work of the waves, since with change of wind direction it is exposed upon more than a single face. The study of headlands which have been cut by waves shows that the joints within the rock play a large rôle in the shaping of shore features. The attack of the waves under the direction of these planes of ready separation opens out indentations of the shore (Fig. 250) or forms *sea caves* which, as they extend to the top of the cliff by the process of sapping, yield the *coves* which are so common a feature upon our rock-bound shores (Fig. 259, p. 238). With continuation of this process, the caves formed on opposite sides of the headland may be united to form a *sea arch* (Fig. 251).

FIG. 251.—The sea arch known as the Grand Arch upon one of the Apostle Islands in Lake Superior (after a photograph by the Detroit Photographic Company).

A later stage in this selective wave carving under the control of joints is reached when the bridge above the arch has fallen in, leaving a detached rock island with precipitous walls. Such an offshore island of rock with precipitous sides is known as a *stack* (Fig. 252), or sometimes as a "chimney", though this latter term is best restricted to other and similar forms which are the product of selective weathering (p. 300).

FIG. 252.—Stack near the shore of Lake Superior.

Whenever the rock is less firmly consolidated, and so does not stand upon such steep planes, the stack is apt to have a more conical form, and may not be preceded in its formation by the development of the sea arch (Fig. 260, p. 239). In the reverse case, or where the rock possesses an unusual tenacity, the stack may be largely undermined and stand supported like a table upon thick legs or pillars of rock (Fig. 253). In Fig. 254 is seen a group of stacks upon the coast of California, which show with clearness the control of the joints in their formation, but unlike the marble of the South American example the forms are not rounded, but retain their sharp angles.

FIG. 253.—The Marble Islands, stacks in Lake Buenos Aires, southern Andes (after F. P. Moreno).

The cut rock terrace.—When waves first begin their attack upon a steep, rocky shore, the lower limit of the action is near the wave base. The action at this depth is, however, less efficient, and as the recession of the cliff is one of the most rapid of erosional processes, the rock floor outside the receding cliff comes to slope gradually downward from the cliff to a maximum depth at the edge of the terrace, approximately equal to wave base (Fig. 255). This cut terrace is extended seaward or lakeward, as the case may be, in a *built terrace* constructed from a portion of the rock débris acquired from the cliff.

FIG. 254.—Squared stacks which reveal the position of the joint planes which have controlled in the process of carving by the waves. Pt. Buchon, California (after a photograph by Fairbanks).

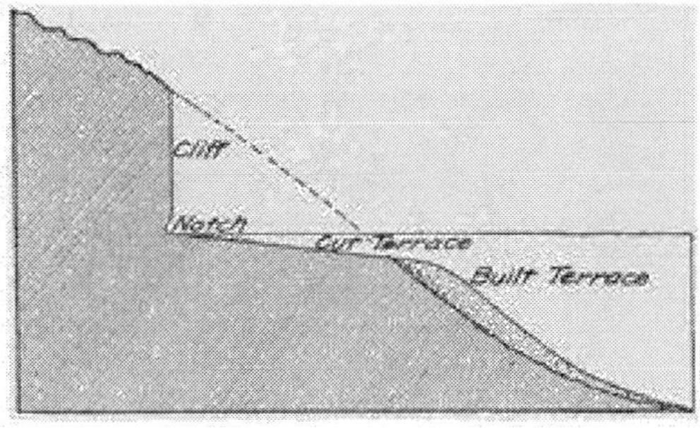

FIG. 255.—Ideal section of a steep rocky shore carved by waves into a notched cliff and cut terrace, and extended by a built terrace.

The broken wave, after rising upon the terrace under the inertia of its motion until all its energy has been dissipated, slides outward by gravity, and though checked and overridden by succeeding breakers, it continues its outward slide as the "undertow" until it reaches the end of the terrace. Here it suddenly enters deep water, and losing its velocity, drops its burden of rock, and builds the terrace seaward after the manner of construction of an embankment. As we are to see, the larger portion of the wave-quarried material is diverted to a different quarter.

FIG. 256.—Map showing the outlines of the Island of Heligoland at different stages in its recent history. The peripheries given are in miles.

To gain some conception of the importance of wave cutting as an eroding process, we may consider the late history of Heligoland, a sandstone island off the mouth of the Elbe in the North Sea (Fig. 256). From a periphery of 120 miles, which it possessed in the ninth century of the Christian era, the island has reduced its outline to 45 miles in the fourteenth century, 8 miles in the seventeenth, and to about 3 miles at the beginning of the twentieth century. The German government, which recently acquired this little remnant from England, has expended large sums of money in an effort to save this last relic.

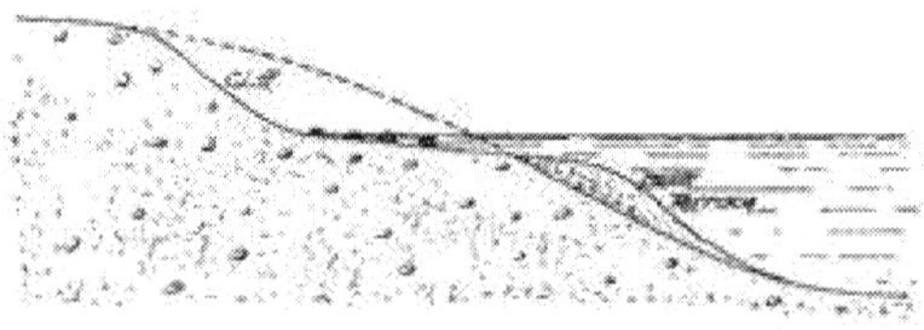

FIG. 257.—Cut and built terrace with bowlder pavement shaped by waves on a steep shore formed of loose materials.

The cut and built terrace on a steep shore of loose materials.—In materials which lack the coherence of firm rock, no vertical cliff can form; for as fast as undermined by the waves the loose materials slide down and assume a surface of practically constant slope—the "angle of repose" of the materials (Fig. 257). The terrace below this sloping cliff will not differ in shape from that cut upon a rocky shore; but whenever the materials of the shore include disseminated blocks too large for the waves to handle, they collect upon the terrace near where they have been exhumed, thus forming what has been called a "bowlder pavement" (Fig. 258).

FIG. 258.—Sloping cliff and terrace with bowlder pavement exposed at low tide upon the shore at Scituate, Massachusetts.

The edge of the cut and built terrace is, as already mentioned, maintained at the depth of wave base. If one will study the submerged contours of any of our inland lakes, it will be found that these basins are surrounded by a gently sloping marginal shelf,—the cut and built terrace,—and that the depth of this shelf at its outer edge is proportioned to the size of the lake. Upon Lake Mendota at Madison, Wisconsin, the large storm waves have a length of about twenty feet, which is the depth of the outer edge of the shore terraces (Fig. 267, p. 242). The shelf surrounding the continents has, with few local exceptions, a uniform depth of 100 fathoms, or about the wave base of the heaviest storm waves.

The work of the shore current.—In describing the formation of the built terrace, it was stated that the greater part of the rock material quarried upon headlands by the waves is diverted from the offshore terrace. This diversion is the work of the shore current produced by the wave.

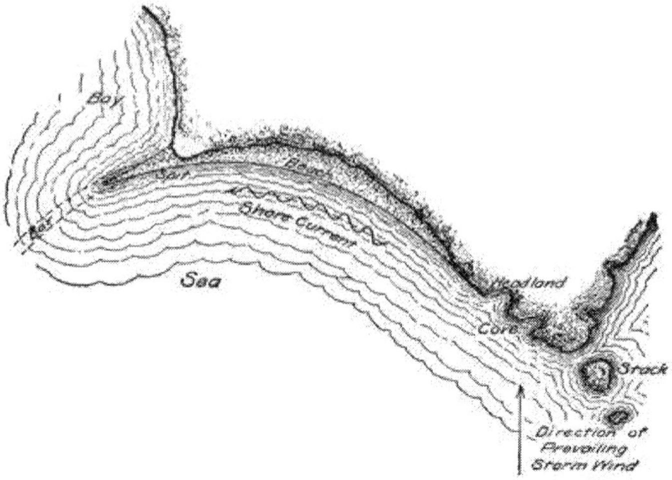

FIG. 259.—Map to show the nature of the shore current and the forms which are molded by it.

At but few places upon a shore will the storm waves beat perpendicularly, and then for but short periods only. The broken wave, as it crawls ever more slowly up the beach, carries the sand with it in a sweeping curve, and by the time gravity has put a stop to its forward movement, it is directed for a brief instant parallel to the shore. Soon, however, the pull of gravity upon it has started the backward journey in a more direct course down the slope of the terrace; and here encountering the next succeeding breaker, a portion of the water and the coarser sand particles with it are again carried forward for a repetition of the zigzag journey. This many times interrupted movement of the sand particles may be observed during a high wind upon any sandy lee shore. The "set" of the water along the shore as a result of its zigzag journeyings is described as the *shore current* (Fig. 259), and the effect upon sand distribution is the same as though a steady current moved parallel to the shore in the direction of the average trend of the moving particles.

The sand beach.—The first effect of the shore current is to deposit some portion of the sand within the first slight recess upon the shore in the lee of the cliff. The earlier deposits near the cliff gradually force the shore current farther from the shore and so lay down a sand selvage to the shore, which is shaped in the form of an arc or crescent and known as a *beach* (Fig. 259 and Fig. 260).

FIG. 260.—Crescent-shaped beach formed in the lee of a headland. Santa Catalina Island, California (after a photograph by Fairbanks).

FIG. 261.—Cross section of a beach pebble.

The shingle beach.—With heavy storms and an exceptional reach of the waves, the shore currents are competent to move, not the sand alone, but pebbles, the area of whose broader surface may be as great as the palm of one's hand. Such rock fragments are shaped by the continued wear against their neighbors under the restless breakers, until they have a lenticular or watch-shaped form (Fig. 261). Such beach pebbles are described as *shingle*, and they are usually built up into distinct ridges upon the shore, which, under the fury of the high breakers, may be piled several feet above the level of quiet water (Fig. 262). Such storm beaches have a gentle forward slope graded by the shore current, but a steep backward slope on the angle of repose. Most storm beaches have been largely shaped by the last great storm, such as comes only at intervals of a number of years.

FIG. 262.—Storm beach of coarse shingle about four feet in height at the base of Burnt Bluff on the northeast shore of Green Bay, Lake Michigan.

Bar, spit, and barrier.—Wherever the shore upon which a beach is building makes a sudden landward turn at the entrance to a bay, the shore currents, by virtue of their inertia of motion, are unable longer to follow the shore. The débris which they carry is thus transported into deeper water in a direction corresponding to a continuation of the shore just before the point of turning (see Fig. 259, p. 238). The result is the formation of a *bar*, which rises to near the water surface and is extended across the entrance to the bay through continued growth at its end, after the manner of constructing a railway embankment across a valley.

FIG. 263.—Spit of shingle on Au Train Island, Lake Superior (after Gilbert).

Over the deeper water near the bar the waves are at first not generally halted and broken, as they are upon the shore, and so the bar does not at once build itself to the surface, but remains an invisible bar to navigation. From its shoreward end, however, the waves of even moderate storms are broken, and the bar is there built above the water surface, where it appears as a narrow cape of sand or shingle which gradually thins in approaching its apex. This feature is the well-known *spit* (Fig. 263) which, as it grows across the entrance to the bay, becomes a *barrier* or *barrier beach* (Fig. 264).

The continuation of the visible in the usually invisible bar, is at the time of high winds made strikingly apparent, for the wave base is below the crest of the bar, and at such times its crescentic course beyond the spit can be followed by the eye in a white arc of broken water.

FIG. 264.—Barrier beach in front of a lagoon on Lake Mendota at Madison, Wisconsin. The shallow lagoon behind the barrier is filling up and is largely hidden in vegetation.

The construction of a barrier across the entrance to a bay transforms the latter into a separate body of water, a lagoon, within which silting up and peat formation usually lead to an early extinction (p. 429). The formation of barriers thus tends to straighten out the irregularities of coast lines, and opens the way to a natural enlargement of the land areas. While the coasts of the United Kingdom of Great Britain have been losing some four thousand acres through wave erosion, there has been a gain by growth in quiet lagoons which amounts to nearly seven times that amount. As evidence of the straightening of the shore line which results from this process, the coast of the Carolinas or of Nantucket (Fig. 459) may serve for illustration.

The land-tied island.—We have seen that wave erosion operates to separate small islands from the headlands, but the shore currents counteract this loss to the continents by throwing out barriers which join many separated islands to the mainland. Such land-tied islands are a common feature on many rocky coasts, and upon the New England coast they usually have been given the name of "neck." The long arc of Lynn Beach joins the former island of Nahant, through its smaller neighbor Little Nahant, to the coast of Massachusetts. A similar land-tied island is Marblehead Neck. The Rock of Gibraltar, formerly an island, is now joined to Spain by the low beach known as the "neutral ground." The Spanish name, *tombola*, has sometimes been employed to describe an island thus connected to the shore.

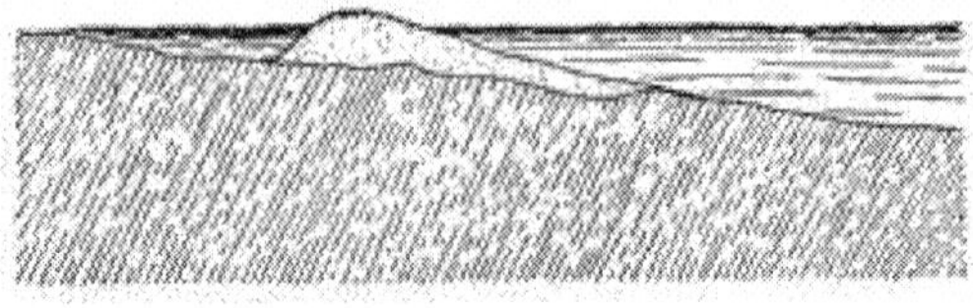

FIG. 265.—Cross section of a barrier beach with lagoon in its rear.

A barrier series.—The cross section of a barrier beach, like that of a storm beach upon the shore, slopes gently upon the forward side, and more steeply at the angle, of repose upon the rear or landward margin (Fig. 265). The thinning wedge of shore deposits which the barrier throws out to seaward raises the level of the lake bottom (Fig. 266), and when coast irregularities are favorable to it, new spits will develop upon the shore outside the earlier one, and a new bar, and in its turn a barrier, will be found outside the initial one, taking a course in a direction more or less parallel to it (Fig. 267).

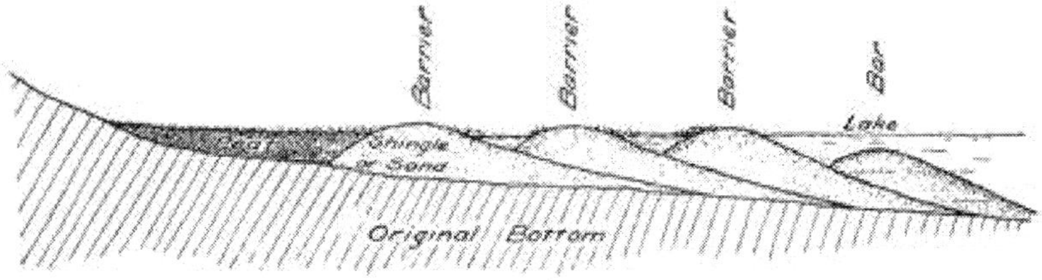

FIG. 266.—Cross section of a series of barriers and an outer bar.

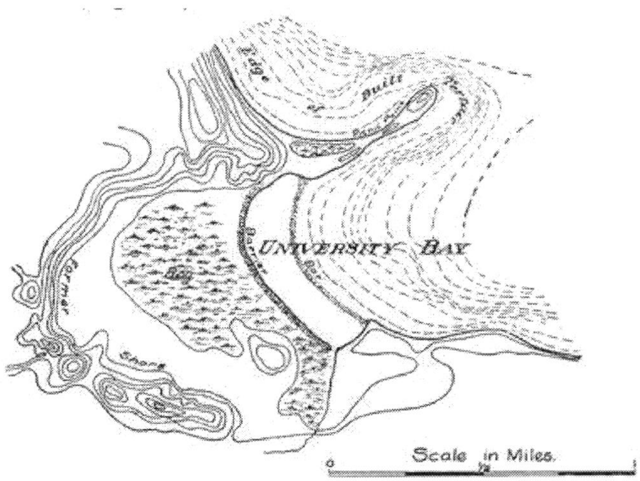

FIG. 267.—Formation of barrier series and an outer bar in University Bay of Lake Mendota, at Madison, Wisconsin. The water contour interval is five feet, and the land contour interval ten feet (based on a map by the Wisconsin Geological Survey).

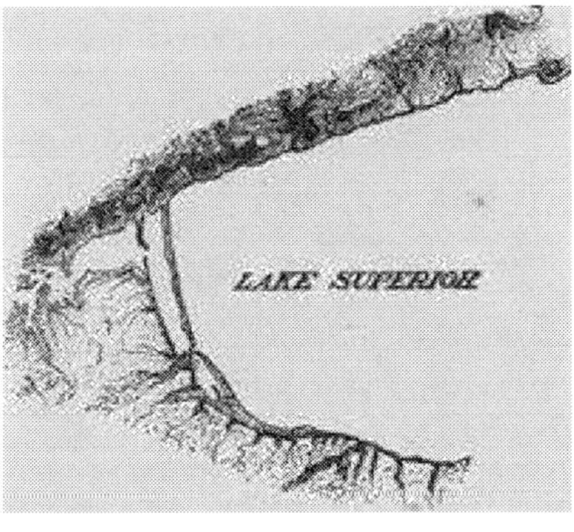

FIG. 268.—Series of barriers at the western end of Lake Superior (after Gilbert).

So soon as the first barrier is formed, processes are set in operation which tend to transform the newly formed lagoon into land, and so with a series of barriers, a zone of water lilies between the outer barrier and the bar, a bog, and a land platform may represent the successive stages in this acquisition of territory by the lands. A noteworthy example of barrier series and extension of the land behind them, is afforded by the bay at the western end of Lake Superior (Fig. 268).

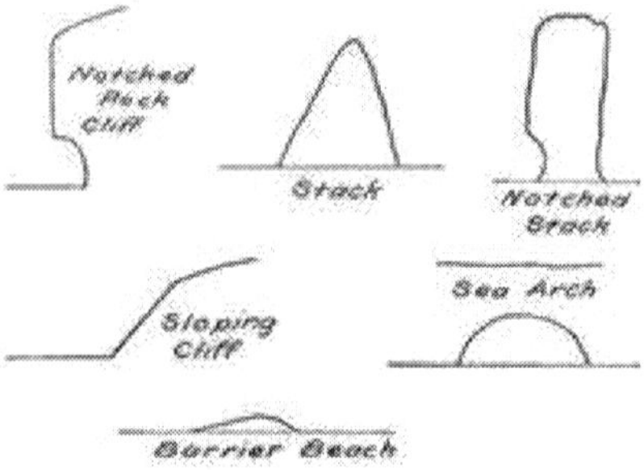

FIG. 269.—Character profiles resulting from wave action upon shores.

Character profiles.—The character profiles yielded by the work of waves are easy of recognition (Fig. 269). The vertical cliff with notch at its base is varied by the stack of sugar-loaf form carved in softer rocks, or the steeper notched variety cut from harder masses. Sea caves and sea arches yield variations of a curve common to the undercut forms. Wherever the materials of the shore are loosely consolidated only, the sloping cliff is formed at the angle of repose of the materials. The barrier beach, though projecting but a short distance above the waves, shows an unsymmetrical curve of cross section with the steeper slope toward the land.

READING REFERENCES FOR CHAPTER XVIII

G. K. GILBERT. The Topographic Features of Lake Shores, 5th Ann. Rept. U. S. Geol. Surv., 1885, pp. 69-123, pls. 3-20; Lake Bonneville, Mon. I, U. S. Geol. Surv., 1890, Chapters ii-iv, pp. 23-187.

VAUGHAN CORNISH. On Sea Beaches and Sand Banks, Geogr. Jour., vol. 11, 1898, pp. 528-543, 628-658.

F. P. GULLIVER. Shore Line Topography, Proc. Am. Acad. Arts and Sci., vol. 34, 1899, pp. 149-258.

N. S. SHALER. The Geological History of Harbors, 13th Ann. Rept. U. S. Geol. Surv., 1893, pp. 93-209.

SIR A. GEIKIE. The Scenery of Scotland, 1901, pp. 46-89.

W. H. WHEELER. The Sea Coast. Longmans, London, 1902, pp. 1-78.

G. W. VON ZAHN. Die zerstörende Arbeit des Meeres an Steilküsten nach Beobachtungen in der Bretagne und Normandie in den Jahren 1907 und 1908, Mitt. d. Geogr. Ges. Hamb., vol. 24, 1910, pp. 193-284, pls. 12-27.

CHAPTER XIX

COAST RECORDS OF THE RISE OR FALL OF THE LAND

The characters in which the record has been preserved.—The peculiar forms into which the sea has etched and molded its shores have been considered in the last chapter. Of these the more significant are the notched rock cliff, the cut rock terrace, the sea cave, the sea arch, the stack, and the sloping cliff and terrace, among the carved features; and the barrier beach and built terrace, among the molded forms. It is important to remember that the molded forms, by the very manner of their formation, stand in a definite relationship to the carved features; so that when either one has been in part effaced and made difficult of determination, the discovery of the other in its correct natural position may remove all doubt as to the origin of the relic.

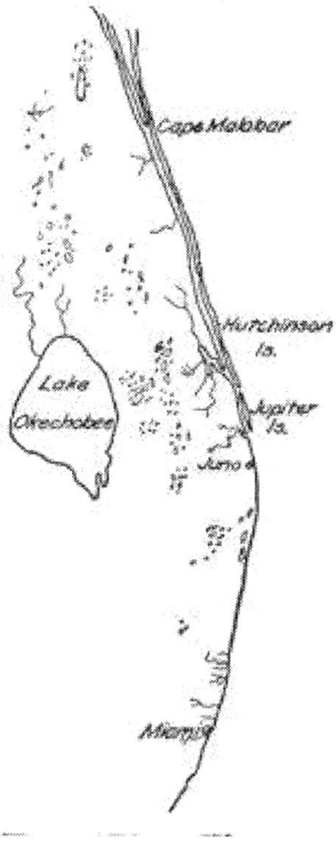

FIG. 270.—The east coast of Florida, with shore line characteristic of a raised coast.

In studies of the change of level of the land, it is customary to refer all variations to the sea level as a zero plane of reference. It is not on this account necessary to assume that the changes measured from this arbitrary datum plane are the absolute upward or downward oscillations which would be measured from the earth's center; for the sea, like the land, has been subject to its changes of level. There need, however, be no apology for the use of the sea surface as a plane of reference; for it is all that we have available for the purpose, and the changes in level, even if relative only, are of the greatest significance. It is probable that in most cases where the coast line is rising from uplift, some portion of the sea basin not far distant is becoming deepened, so that the visible change of level is the algebraic sum of the two effects.

Even coast line the mark of uplift.—It was early pointed out in this volume (p. 158) that the floor of the sea in the neighborhood of the land presents a relatively even surface. The carving by waves, combined with the process of deposition of sediments, tends to fill up the minor irregularities of surface and preserve only the features of larger scale, and these in much softened outlines. Upon the continents, on the contrary, the running water, taking advantage of every slight difference in elevation and searching out the hidden structure planes within the rock, soon etches out a surface of the most intricate detail. The effect of elevation of the sea floor into the light of day will therefore be to produce an even shore line devoid of harbors (Fig. 270). If the coast has risen along visible planes of faulting near the sea margin, the coast line, in addition to being even, will usually be made up of notably straight elements joined to one another.

FIG. 271.—Ragged coast line of Alaska, the effect of subsidence.

A ragged coast line the mark of subsidence.—When in place of uplift a subsidence occurs upon the coast, the intricately etched surface, resulting from erosion beneath the sky, comes to be invaded by the sea along each trench and furrow, so that a most ragged outline is the result (Fig. 271). Such a coast has many harbors, while the uplifted coast is as remarkable for its lack of them.

Slow uplift of the coast—the coastal plain and cuesta.—A gradual uplift of the coast is made apparent in a progressive retirement of the sea across a portion of its floor, thus exposing this even surface of recent sediments. The former shore land will be easily recognized by it's etched surface, which will now come into sharp contrast with the new plain. It is therefore referred to as the *oldland* and the newly exposed *coastal plain* as the *newland* (Fig. 272).

FIG. 272.—Portion of Atlantic coastal plain and neighboring oldland of the Appalachian Mountains.

But the near-shore deposits upon the sea floor had an initial dip or slope to seaward, and this inclination has been increased in the process of uplift. The streams from the oldland have trenched their way across these deposits while the shore was rising. But the process being a slow one, deposits have formed upon the seaward side of the plain after the landward portion was above tide, and the coastal plain may come to have a "belted" or zoned character. The streams tributary to those larger ones which have trenched the plain may encounter in turn harder and softer layers of the plain deposits, and at each hard layer will be deflected along its margin so as to enter the main streams more nearly at right angles. They will also, as time goes on, migrate laterally seaward through undermining of the harder layers, and thus will be shaped alternating belts of lowland separated by escarpments in the harder rock from the residual higher slopes. Belts of upland of this character upon a coastal plain are called *cuestas* (Fig. 273).

FIG. 273.—Ideal form of cuestas and intermediate lowlands carved from a coastal plain (after Davis).

The sudden uplifts of the coasts.—Elevations of the coast which yield the coastal plain must be accounted among the slower earth movements that result in changes of level. Such movements, instead of being accompanied by disastrous earthquakes, were probably marked by frequent slight shocks only, by subterranean rumblings, or, it may be, the land rose gradually without manifestations of a sensible character.

Upon those coasts which are often in the throes of seismic disturbance, a quite different effect is to be observed. Here within the rocks we will probably find the marks of recent faulting with large displacements, and the movements have been upon such a scale that shore features, little modified by subsequent weathering, stand well above the present level of the seas. Above such coasts, then, we recognize the characteristic marks of wave action, and the evidence that they have been suddenly uplifted is beyond question.

FIG. 274.—Uplifted sea cave, ten feet above the water upon the coast of California; the monument to a former earthquake (after a photograph by Fairbanks).

FIG. 275.—Double-notched cliff near Cape Tiro, Celebes (after a photograph by Sarasin).

The upraised cliff.—Upon the coast of southern California may be found all the features of wave-cut shores now in perfect preservation, and in some cases as much as fifteen hundred feet above the level of the sea. These features are monuments to the grandest of earthquake disturbances which in recent time have visited the region (Fig. 274). Quite as striking an example of similar movements is afforded by notched cliffs in hard limestone upon the shore of the Island of Celebes (Fig. 275). But the coast of California furnishes the other characteristic coast features in the high sea arch and the stack as additional monuments to the recent uplift. Let one but imagine the stacks which now form the Seal Rocks off the Cliff House at San Francisco to be suddenly raised high above the sea, and the forms which they would then present would differ but little from those which are shown in Fig. 276.

FIG. 276.—Jasper rock stacks uplifted on the coast of California (after a photograph by Fairbanks).

The uplifted barrier beach.—Within the reëntrants of the shore, the wave-cut cliff is, as we know, replaced by the barrier beach, which takes its course across the entrance to a bay. After an uplift, such a barrier composed of sand or shingle should be connected with the headlands, often with a partially filled lagoon behind it. Its cross section should be steep in the direction of the lagoon, but quite gradual in front (Fig. 277).

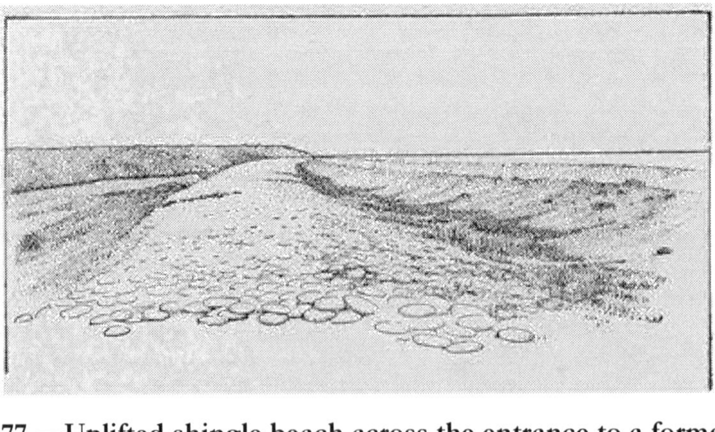

FIG. 277.—Uplifted shingle beach across the entrance to a former bay upon the coast of southern California (after a photograph by Fairbanks).

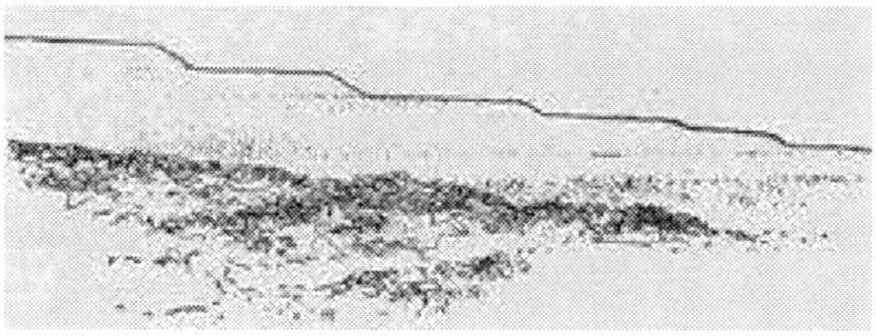

FIG. 278.—Raised beach terraces near Elie, Fife, Scotland.

Coast terraces.—Upon those shores where to-day high mountains front the sea, the coast may generally be seen to rise in a series of terraces (Fig. 278). This is notably true of those coasts which are to-day racked by earthquakes, such as is the eastern margin of the Pacific from Alaska to Patagonia. The traveler by steamer along the coast from San Francisco to Chili has for weeks almost constantly in sight these giant steps on which the mountains have been uplifted from the sea. In Alaska we are fortunate in having the history of the later stages in this uplift (Fig. 279). As described in a former chapter, portions of this shore rose in the month of September of the year 1899 in some places as high as forty-seven feet, to the accompaniment of a terrific earthquake and sea wave. Above the terrace which marks the beach line of 1899 there is a higher terrace of similar form now overgrown with trees, but none the less clearly to be recognized as a shore line of the past century which preceded in the long sequence the uplift of 1899.

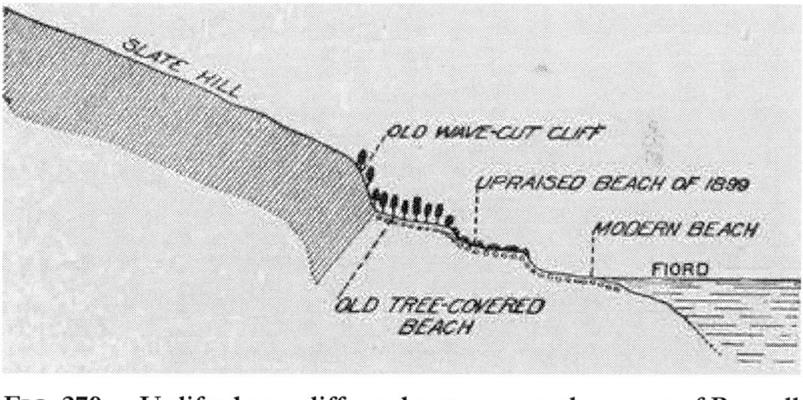

FIG. 279.—Uplifted sea cliffs and terraces on the coast of Russell Fjord, Alaska (after Tarr and Martin).

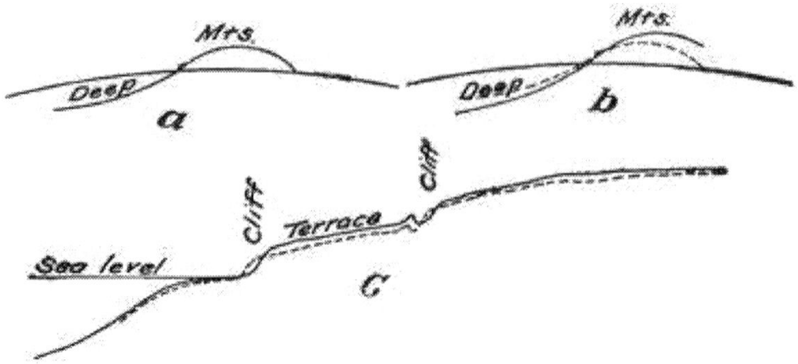

FIG. 280.—Diagrams to show how excessive sinking upon the sea floor will cause the shore to migrate landward as it is uplifted.

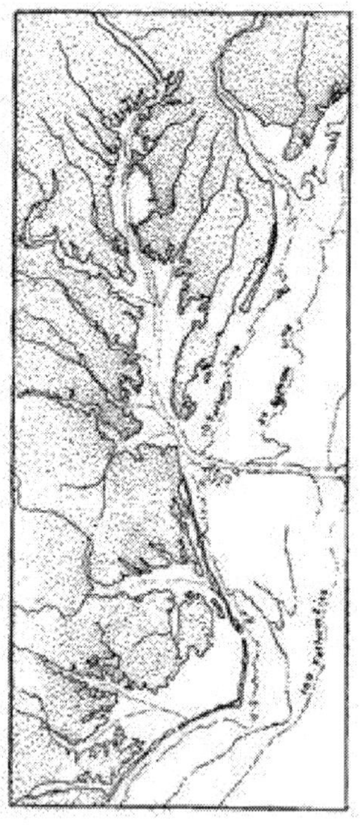

FIG. 281.—A drowned river mouth, or estuary upon a coastal plain.

As was noted in our study of earthquakes, the recent instrumental records of distant earthquakes tell us that the movements upon the sea floor are many times larger than those upon the continents, and that while the mountainous coasts are generally rising, the deeps of the sea are sinking. The effect of this over-balance of sinking, or resultant shrinking of the earth's shell, may be to compress the mountain district and so cause the shore line to move landward at the same time that it moves upward (Fig. 280).

The sunk or embayed coast.—When now, upon the other hand, a section of the coast line sinks with reference to the sea, the water invades all the near-shore valleys, thus "drowning" them and yielding the drowned river mouth or *estuary*. If the relief of the shore was slight, as it generally is upon a coastal plain, slight depression only will produce broad estuaries, such as Chesapeake Bay at the drowned mouth of the Susquehanna (Fig. 281).

If, on the other hand, the relief of the shore is strong and the subsidence is large, the entire coast line will be transformed into an archipelago of steep-walled rocky islets which rise abruptly from the sea (Figs. 282 and 284). A plateau which is intersected by deep and steep-walled valleys of U-section (p. 341) under large submergence yields the *fjords* so characteristic of Scandinavia or Alaska. A ragged coast line, fringed with islands as a result of submergence, is described as an *embayed coast.*

FIG. 282.—Archipelago of steep rocky islets due to large submergence of a coast having strong relief. Entrance to Esquimalt Harbor, Vancouver Island (after a photograph by Fairbanks).

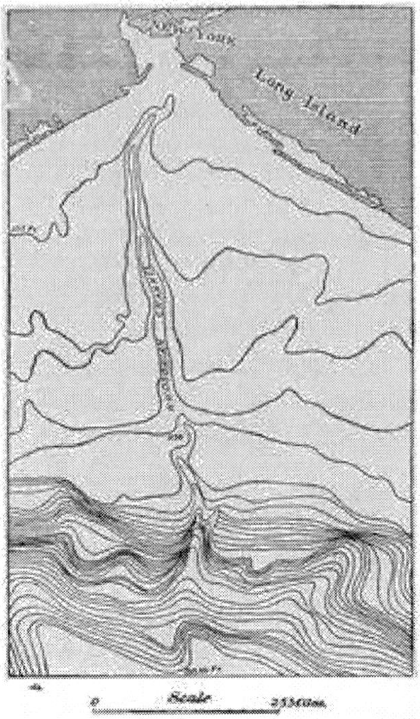

FIG. 283.—The submerged Hudsonian channel which continues the Hudson River across the continental shelf.

Submerged river channels.—The sinking of a coast of small relief be sufficient to completely submerge river valleys, whose channels then begin to fill with sediment and whose courses can only be followed in soundings. One of the most interesting of such channels is that which continues the Hudson River across the continental shelf into the deeper sea (Fig. 283).

Records of an oscillation of movement.—Because a coast is deeply embayed is no ground for assuming that a subsidence is now in progress, or is, in fact, the latest movement recorded upon the coast. In many cases it is easy to see that such is not the case. The coast of Maine is perhaps as typical of an embayed shore line as any that might be cited, but a study of the river valleys in the neighborhood shows clearly that the present submergence of their mouths is a fraction only of an earlier one which has left a record of its existence in beds of marine clay which outline the earlier and far deeper indentations (Fig. 284).

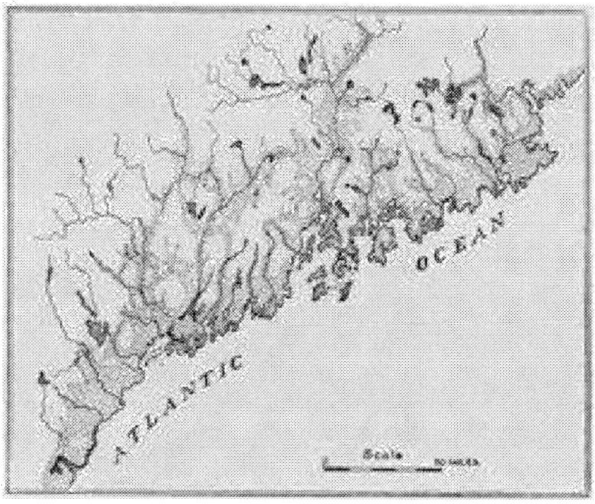

FIG. 284.—Marine clay deposits near the mouths of the rivers of Maine which preserve a record of earlier subsidence (after Stone).

If now we give a closer examination to the coast, it is found that there are marks of recent uplift in an abandoned shore line now far above the reach of the waves. There is here, then, the record, first of subsidence and consequent embayment, and, later, of an uplift which has reduced the raggedness of the coast outline exposed the clay deposits, and raised the strands of the period of deep subsidence to their present position.

In countries which possess a more ancient civilization than our own, the record of such oscillations in the level of the ground has sometimes been entered upon human monuments, so that it is possible to date more or less definitely the periods of subsidence or elevation. At the little town of Pozzuoli, upon the shore of the Bay of Naples, is found one of the mos instructive of these records.

FIG. 285.—View of the three standing columns of the temple of Jupiter Serapis at Pozzuoli, showing the dark and rough band nine feet in

width affected by the rock-boring mollusks which now live in the Bay of Naples.

In the ruins of the ancient temple of Jupiter Serapis are three marble monoliths 40 feet in height, curiously marked by a roughened surface between the heights of 12 and 21 feet above their pedestals (Fig. 285). Closer inspection shows that this roughened surface has been produced by a marine, rock-boring mollusk, the *lithodomus*, which lives in the waters of the Bay of Naples, and the shells of this animal are still to be found within the cavities upon the surface of the columns. Without recounting details which have been many times recited since these interesting monuments were first geologically explored by Babbage and Lyell, it may be stated that a record is here preserved, first of subsidence amounting to some 40 feet, and of subsequent elevation, of the low coast land on which stood the temple in the old Roman city of Puteoli (Fig. 286).

At the time of deepest submergence the top of the lithodomus zone upon the column stood at the level of the water in the Bay of Naples, the smoother lower zone being buried at the time in the sand at the bottom, and thus made inaccessible for the lithodomi. It is to be added that studies made in the environs of Pozzuoli have fully confirmed the changes of level revealed by the columns, through the discovery of now elevated shore lines which are referable to the period of deep submergence.

Simultaneous contrary movements upon a coast.—In our study of the changes in the level of the ground that take place during earthquakes, it was learned that neighboring sections of the earth's crust may be moved at different rates or even in opposite directions, notwithstanding the fact that the general movement of the province is one of uplift. Thus during the Alaskan earthquake of 1899, although portions of the coast line were elevated by as much as forty-seven feet, neighboring sections were raised by smaller amounts, and some small sections were sunk and so far submerged that the salt water and the beach sand were washed about the roots of forest trees.

FIG. 286.—Pozzuoli in the 3rd, 9th, and 20th Centuries.

A region racked by heavy earthquakes, where the present configuration of the ground speaks strongly for a movement of somewhat similar nature, but with average movement of elevation much greater to the northward than in the opposite direction, is the extended coast line of Chili. This country is characterized by a great central north and south valley which separates the coast range from the high chain of the Cordilleras to the eastward. To the southward the floor of this valley descends, and has its continuance in the Gulf of Corcovado behind the island of Chiloe and the Chonos archipelago. The known recent uplift of the coast of Chili, particularly in the northern sections and during the earthquakes of the eighteenth, nineteenth, and twentieth centuries, lends great interest to this topographic peculiarity. Indications are not lacking that, during the earthquake of Concepcion in 1835, and of Valparaiso in 1907, the measure of uplift was greater to the north than it was to the south.

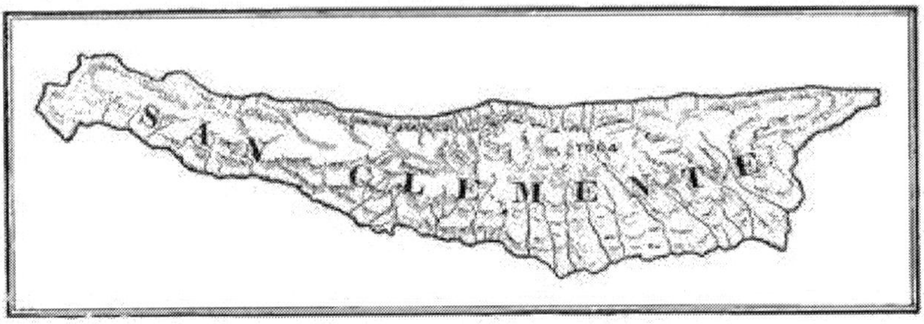

FIG. 287.—Map of San Clemente Island, California, showing the characteristic topography of recent uplift (after U. S. Coast and Geodetic Survey).

The contrasted islands of San Clemente and Santa Catalina.—Perhaps the most striking example of simultaneous opposite movements observable in neighboring portions of the earth's crust is furnished by the coast of southern California. The coast itself at San Pedro and the island of San Clemente, some fifty miles off this point, in common with most portions of the neighboring coast land, have been rising in interrupted movements from the sea, and offer in rare perfection the characteristic coast terraces (Fig. 287 and Fig. 278, p. 250). Midway between these two rising sections of the crust, and less than twenty-five miles distant from either, is the island of Santa Catalina, which has been sinking beneath the waves, and apparently at a similarly rapid rate (Fig. 288). The topography of the island shows the intricate detail of a maturely eroded surface, while that of the neighboring San Clemente shows only the widely spaced, deep cañons of the infantile stage of erosion (Fig. 165 and pl. 12 A). While Santa Catalina has been sinking, San Pedro Hill has risen 1240 feet, and San Clemente, 1500 feet. It is characteristic of a sinking coast line that the cliff recession is abnormally rapid, and evidence for this is furnished by the shores of Santa Catalina, upon which the waves are cutting the cliffs back into the beds of cañons, and so causing small falls to develop at the cañon mouths.

PLATE 12.

A. **V-shaped cañon cut in an upland recently elevated from the sea, San Clemente Island, California (after W. S. Tangier-Smith).**

B. **A "hogback" at the base of the Bighorn Mountains, Wyoming (after Darton).**

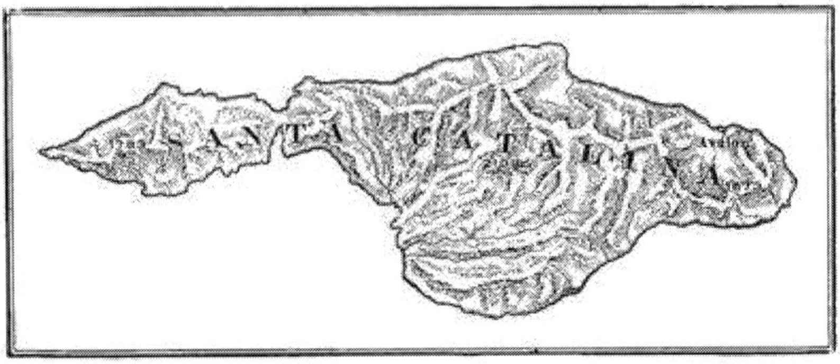

FIG. 288.—Map of Santa Catalina Island, California, showing the characteristic surface of an area which has long been above the waves, and the entire absence of coast terraces (after U. S. C. and G. S.).

The Blue Grotto of Capri.—We may now return to the Bay of Naples for additional evidence that oscillations of level in neighboring portions of the same coast are not necessarily synchronous, and that near-lying sections may even move in opposite directions at the same time, as has already been shown for the islands off the California coast. For the Pozzuoli shore of the bay it was learned that within the Christian Era a complete cycle of downward, followed by later upward, movement has been largely accomplished. Across the bay, and less than 20 miles distant, is the Blue Grotto of Capri, a sea cave cut in limestone above an earlier cave of the same nature which is now deep below the water surface. It is the refracted sunlight which enters the cave through this lower submerged opening and has been robbed on the way of all but its blue rays which gives to the famous grotto its special charm (Fig. 289).

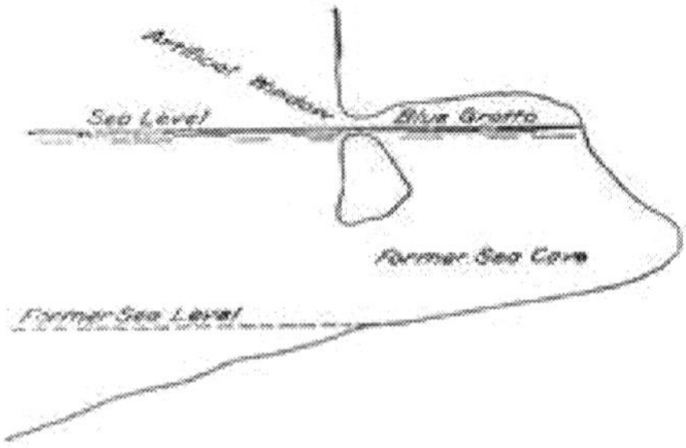

FIG. 289.—Cross section of the Blue Grotto on the Island of Capri, showing the submerged sea cave through which most of the light

enters the grotto, and the higher artificial window now widened by wave action (after von Knebel).

It is known that the former, and now submerged, sea cave was in use by Roman patricians as a cool retreat from the oppressive hot wind known as the sirocco, and that an artificial entrance or window was cut where is now the only accessible entrance to the grotto. In the ancient writings, no mention is made, however, of the remarkable blue illumination for which it is now famous, and the conditions at the time, as we may see, were not such as to make this possible. Later subsidence of the coast has brought the ancient window to the sea level, where it has been considerably enlarged by the waves. The earlier grotto, abandoned as its entrance was closed, was rediscovered in 1826 by the painter and poet, August Kopisch.

A grotto with green illumination (the Grotto Verde) is situated upon the opposite side of the island, and a blue grotto, having its origin in similar conditions to those of the famous Blue Grotto, is found upon the island of Busi off the Dalmatian coast.

Character profiles.—In the landscape of a coast which has been slowly uplifted the characteristic line is the profile of the cuesta, with short perpendicular element joined to a gently sloping and longer section and continued in the horizontal portion corresponding to the lowland (Fig. 290). Rapidly uplifted coasts offer in contrast the lines characteristic of wave erosion and deposition, but at higher levels and in repeated sections. Most prominent of all is the staircase constructed of coast terraces, with either vertical or sloping risers and with outwardly inclining and gently graded treads. Near the steep riser in the staircase may sometimes be seen the sugar-loaf outline of the stack cut in softer material, or the obelisk-like pillar undercut at its base, which is carved in firmer rock masses. With excessively rapid uplift, the double-notched cliff or the double sea arch may appear in the landscape. Upon a submerged coast the most significant lines in the view are those of the rock islet and the steep-walled fjord.

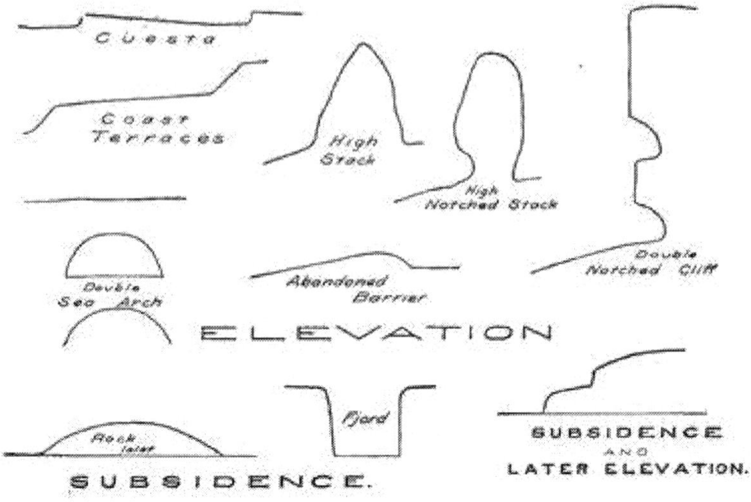

FIG. 290.—Character profiles in coast landscapes where there has been either elevation or depression.

READING REFERENCES FOR CHAPTER XIX

General:—

SIR CH. LYELL. Principles of Geology, vol. 2, pp. 180-197.

ED. SUESS. The Face of the Earth, Clarendon Press, Oxford, 1906, vol. 2, Chapters i and xiv, pp. 1-29, 535-556.

ROBERT SIEGER. Seenschwankungen und Strandverschiebungen in Scandinavien, Zeit. d. Gesell. f. Erdk., Berlin, vol. 28, 1893, pp. 1-106, 393-688, pl. 7.

Elevated shore lines:—

F. B. TAYLOR. The Highest Old Shore Line of Mackinac Island, Am. Jour. Sci., vol. 43, 1892, pp. 210-218.

THOMAS L. WATSON. Evidences of Recent Elevation of the Southern Coast of Baffins Land, Jour. Geol., vol. 5, 1897, pp. 17-33.

J. W. GOLDTHWAIT. The Abandoned Shore Lines of Eastern Wisconsin. Bull. 17, Wis. Geol. and Nat. Hist. Surv., 1907, pp. 134, pls. 1-37.

Evidences of depression:—

W. B. SCOTT. Introduction to Geology, New York, 1907, pp. 33-36.

W. J. McGee. The Gulf of Mexico as a Measure of Isostacy, Am. Jour. Sci. (3), vol. 44, 1892, pp. 177-192.

A. Lindenkohl. Notes on the Submarine Channel of the Hudson River, etc., Am. Jour. Sci. (3), vol. 41, 1891, pp. 489-499, pl. 18.

J. W. Spencer. The Submarine Great Cañon of the Hudson River, *ibid.* (4), vol. 19, 1905, pp. 1-15; Submarine Valleys off the American Coast and in the North Atlantic, Bull. Geol. Soc. Am., vol. 14, 1903, pp. 207-226, pls. 19-20.

F. Nansen. The Bathymetrical Features of the North Polar Sea, with a Discussion of the Continental Shelves and Previous Oscillations of Shore Line, Norwegian North Polar Expedition, vol. 4, 1904, pp. 99-231, pl. 1.

W. v. Knebel. Höhlenkunde, etc., Braunschweig, 1906, pp. 175-177 (the blue grotto of Capri).

Oscillation of movement:—

C. Lyell. Principles of Geology, vol. 2, pp. 164-176 (Temple of Jupiter Serapis).

E. Ray Lankester. Extinct Animals, New York, 1905, pp. 31-42.

H. W. Fairbanks. Oscillations of the Coast of California during the Pliocene and Pleistocene, Am. Geol., vol. 20, 1897, pp. 213-245.

G. H. Stone. Mon. 34, U. S. Geol. Surv., 1899, pp. 56-58, pl. 2.

Bailey Willis. Ames Knob, North Haven, Maine. Bull. Geol. Soc. Am., vol. 14, 1903, pp. 201-206, pls. 17-18.

Simultaneous contrary movements on a coast:—

A. C. Lawson. The Post-Pliocene Diastrophism of the Coast of Southern California, Bull. Univ. Calif. Dept. Geol., vol. 1, 1893, pp. 115-160, pls. 8-9.

W. S. Tangier-Smith. A Geological Sketch of San Clemente Island, 18th Ann. Rept. U. S. Geol. Surv., Pt. ii, 1898, pp. 459-496, pls. 84-96.

R. S. Tarr and L. Martin. Recent Changes of Level in the Yakutat Bay Region, Alaska, Bull. Geol. Soc. Am., vol. 17, 1906, pp. 29-64, pls. 12-23.

CHAPTER XX

THE GLACIERS OF MOUNTAIN AND CONTINENT

Conditions essential to glaciation.—Wherever for a sufficiently protracted period the annual snowfall of a district is in excess of the snow that is melted, a residue must remain from each season to be added to that of succeeding ones. Eventually so much snow will have accumulated that under its own weight and in obedience to its peculiar properties, a movement will begin within the mass tending to spread it and so to reduce the slope of its upper surface (Frontispiece plate). The conditions favorable to glaciation are, therefore, heavy precipitation and low annual temperature. If the precipitation is scanty, the small snowfall is soon melted; and if the temperature be too high, the moisture is precipitated not in the form of snow but as rain. It is important here to keep in mind that snow is a poor heat conductor and itself protects its deeper layers from melting.

The snow-line.—Because of the low temperatures glaciers should be most abundant or most extensive in high latitudes and in high altitudes. The largest are found in polar and subpolar regions, and they are elsewhere encountered only at considerable elevations. The largest glaciers are the vast sheets of ice which inwrap the continents of Greenland and Antarctica, but glaciers of large size are to be found upon other large land masses of the Arctic, as well as in Alaska, in the southern Andes, and in New Zealand. Much smaller glaciers are characteristic of certain highlands within temperate and tropical regions, but because of specially favorable conditions both of altitude and precipitation the Himalayas, although in relatively low latitudes, nourish glaciers of large proportions. In general, it may be said that the nourishing grounds of glaciers are largely restricted to those areas where snow covers the ground throughout the year. The lower margin of such areas is designated the *snow line*, and varies but little from the line on which the average summer temperature is at the freezing point of water—the so-called *summer isotherm of 32° Fahrenheit.* Within the tropics this line may rise as high as 18,000 feet above the sea, whereas in polar latitudes it descends to sea level.

Importance of mountain barriers in initiating glaciers.—The precipitation within any district depends, however, not alone upon the amount of moisture which is brought to it in the clouds, but upon the amount which is abstracted before the clouds have passed over it. The capacity of space to hold moisture increases with its temperature, and hence any lowering of this temperature will reduce the capacity. If lowered sufficiently, the point of complete saturation will be reached and further cooling must result in precipitation. Hence, anything which forces an air current to rise into more rarefied zones above, will lower the pressure upon it and so bring about a cooling effect in which no heat is abstracted. This so-called *adiabatic refrigeration* of a gas may be illustrated by the cool current which issues in a jet from a warm expanded rubber tire after the cock has been opened; or even better, by the instant solidification at extreme low temperatures of such normal gases as carbonic acid when they are allowed to issue under heavy pressure from a small orifice.

As applied to moisture-laden and near-surface winds, the effective agents of adiabatic cooling are the upland areas upon the continents, and especially the ranges of mountains. These barriers force the moving clouds to rise, cool, and deposit their moisture. It is, therefore, the highland barriers which face the oncoming, moisture-laden winds that receive the heaviest precipitation. Within temperate regions, because of the prevalence of westerly winds, those barriers which face the western shores receive the heaviest fall. Within the tropics, on the other hand, it is the barriers facing the eastern shores which, because of the easterly "trades", are most favorable to precipitation.

Thus it is in the Sierra Nevadas of California, and not in the Rockies or the Appalachians, that the glaciers of the United States are found. The highland of the Swiss Alps lying likewise athwart the "westerlies" of the temperate zone acquires the moisture for nourishment of its glaciers from the western ocean—here the Atlantic (Fig. 291). Within the tropics the conditions are reversed, and it is in general the ranges which lie nearer the eastern coasts that are the more favored. If no barrier is found upon this coast, the clouds may travel over vast stretches of country before being arrested by mountains and robbed of their moisture. Thus in tropical Brazil the glaciers are found in the Andes upon the Pacific coast though nourished by clouds from the Atlantic.

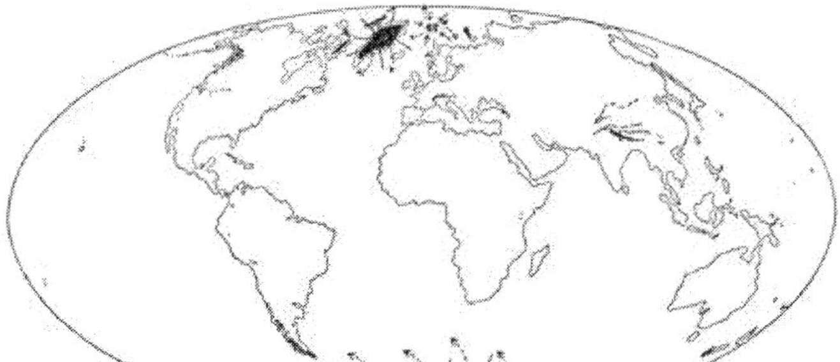

FIG. 291.—Map showing the distribution of existing glaciers, and the two important wind poles of the earth.

Sensitiveness of glaciers to temperature changes.—How sensitive is the adjustment between snow precipitation and temperature may be strikingly illustrated by the statement on excellent authority that if the average annual temperature of the air within the Scottish Highlands should be lowered by only three degrees Fahrenheit, small glaciers would be the result; and a moderate temperature fall within the region surrounding the Laurentian lakes of North America would bring on glaciation, otherwise expressed as a depression of the snow line of the region.

The cycle of glaciation.—Though to-day buried beneath its ice mantle, it is known that Greenland had more than once in earlier geological ages a notably mild climate, and in some future age it may revert to this condition. In other regions, also, we have evidence that such a rotation of climatic changes has been successively accomplished, the climate having steadily increased in severity towards a culminating point, and been followed by a reverse series of changes. Such a complete period may be called a *cycle of glaciation*. While the climate is steadily becoming more rigorous, we have to do with an *advancing hemicycle* of glaciation, but after the culminating point has been reached, the period of amelioration of climate is the *receding hemicycle*.

The advancing hemicycle.—There is little reason to doubt that whatever be the cause of the climatic changes which bring on glacial conditions, these changes come on by insensible gradations. The first visible evidence of the increased severity of the climate is the longer persistence of the winter snows, at first within the more elevated districts. In such positions drifts must eventually continue throughout the warm season and so contribute to the snow accumulations of the succeeding winter. This point once reached, small glaciers are inevitable, even should the average temperature fall no further, for the snow left over in each season must steadily increase the depth of the deposits until the weight brings about an internal motion of the mass from higher to lower levels.

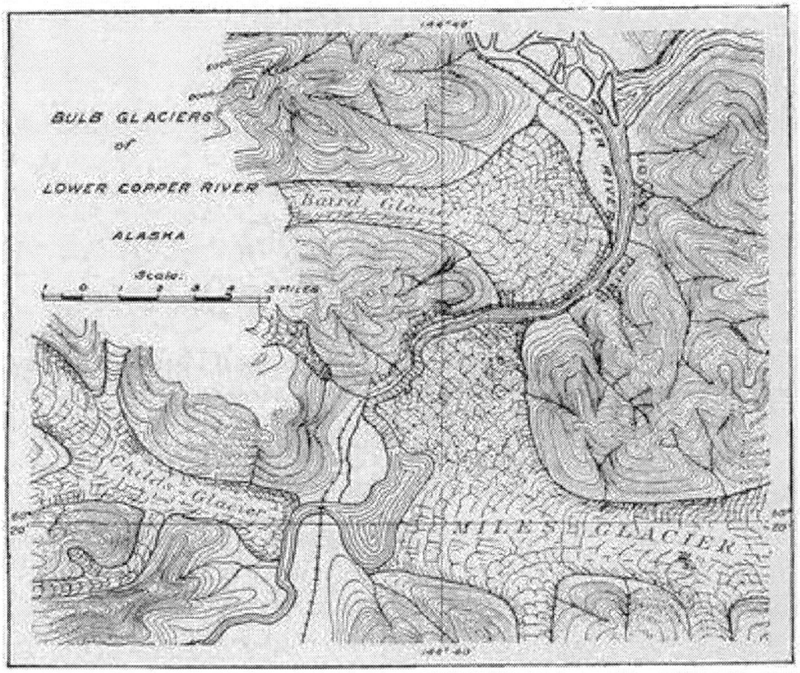

FIG. 292.—An Alaskan glacier spreading out at the foot of the range which nourishes it.

The inherited depressions of the upland—the gentle hollows at the heads of rivers—will first be filled, and so the valleys below become the natural channels for the outflow of the early glaciers. With a continued lowering of the annual temperature and consequent increased snowfall, the early glaciers become more and more amply nourished. Snow and ice will, therefore, cover larger areas of the upland, and the glaciers will push their fronts farther down the valleys before they are wasted in the warm air of the lower levels. As the valleys become thus more completely invested by the glacier they are likewise filled to greater and greater depths, and they may thus submerge portions of the walls that separate adjacent valleys. Reaching at last the front of the upland area, the glaciers may now be so well nourished at their heads that they push out upon the flatter foreland and without restraint from retaining walls spread broadly upon it (Fig. 292).

FIG. 293.—Surface of a glacier whose upper layers spread with slight restraint from retaining walls. Surface of the Folgefond, an ice cap of southern Norway.

The culmination of the progressive climatic change may ere this have been reached and milder conditions have ensued. If, however, the severity of the climate should be still further increased, the expanded fronts of neighboring glaciers will coalesce to form a common ice fan or apron along the foot of the upland (Plate 18 B). This could hardly take place without a still further deepening of the ice within the valleys above, and, probably, a progressive submergence of the lower crests in the valley walls. This may even continue until all parts of the upland area have been buried. The snow and ice now take the form of a covering cap or carapace, and the upper portions being no longer restrained at the sides, now spread into a broad dome, as would a viscous liquid like thick molasses when poured out upon the floor (Fig. 293). The lower zones of the mass and the thinner marginal portions still have their motion to a greater or less extent controlled by the irregularity of the rock floor against which they rest.

The reverse series of changes in the glacier is inaugurated by an amelioration of the climate, and here, therefore, the advancing hemicycle becomes merged in the receding hemicycle of glaciation.

Continental and mountain glaciers contrasted.—The time when the rock surface becomes submerged beneath the glacier is, as regards both the surface forms and the erosive work, a critical point of much significance; for the ice cap and larger continental glacier obviously protect the rock surface from the action of those chemical and mechanical processes in which the atmosphere enters as chief agent, and which are collectively known as weathering processes. Until submergence is accomplished, larger or smaller portions of the rock surface project either through or between the ice masses and are, therefore, exposed to direct attack by the weather (see below, p. 370).

FIG. 294.—Section through a mountain glacier (in solid black), showing how its surface is determined by the irregularities in the rock basement (after Hess).

Snow which falls in the mountains is not allowed to remain long where it falls. By the first high wind it is swept off the more elevated and exposed surfaces and collected under eddies in any existing hollows, but especially those upon the lee slopes of the range. We are to learn that glaciers carve the mountains by enlarging the hollows which they find and producing great basins for the collection of their snows; but with the initiation of glaciation the inherited hollows are in most cases the unimportant depressions at the heads of streams. Whatever they may be and however formed, the snow first fills those hollows which are sheltered from the wind, and as it accumulates and becomes distributed as ice, assumes a surface of its own that is dependent upon the form and the position of the basin which it occupies (see Fig. 294).

FIG. 295.—Profile across the largest of the Icelandic ice caps, with the vertical scale greatly exaggerated (after Thoroddsen and Spethmann).

When the quantity of accumulated snow is so great that all hollows of the rock surface are filled, its own surface is no longer controlled by retaining rock walls, and it now assumes a form largely independent of the irregularities in the upland. Experience shows that this surface is approximately that of a flat dome or shield, and as it covers all the upland, save where the ice thins upon its margins, this type of glacier is called an *ice cap* (Fig. 295). All types of glacier in which rock projects above the highest levels of the ice and snow are known as *mountain glaciers*.

FIG. 296.—Ideal section across a continental glacier, with the vertical scale and the projecting rock masses of the marginal zone greatly magnified.

The flat domes of ice which mantle the continents of Greenland and Antarctica, though resembling in form the smaller ice cap, are yet because of their vast size so distinct from them, particularly in the manner of their nourishment, that they belong in a separate class described as *inland ice* or *continental glaciers*. Though they have some affinities with ice caps, they are most sharply differentiated from all types of mountain glaciers. Of them it is true that the lithosphere projects through them only in the neighborhood of their margins (Fig. 296), whereas in the case of mountain glaciers rock may project at any level but *always above the highest snow surface*. Ice caps may be regarded as intermediate between the two main classes of mountain and continental glaciers (Fig. 297). Because of the large rôle which continental glaciers have played in geological history, it is thought best to consider them first, leaving for later discussion the no less interesting but less important mountain glaciers.

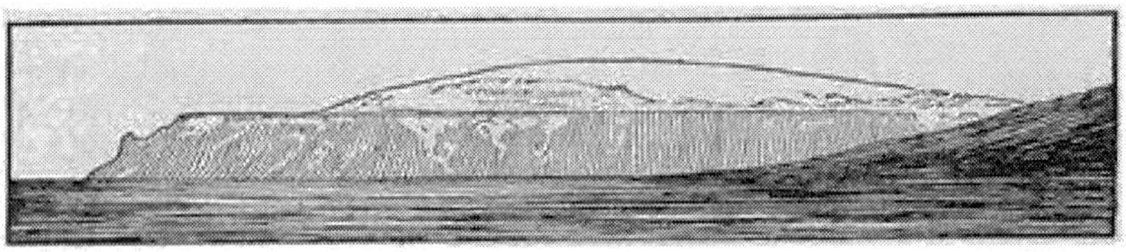

FIG. 297.—View of the Eyriks-Jökull, an ice-cap of Iceland (after Grossman).

The nourishment of glaciers.—The life of a glacier is dependent upon the continued deposition of snow in aggregate amount in excess of that which is lost by melting or by other depleting processes. Whenever, on the other hand, the waste exceeds the precipitation, the glacier is in a receding condition and must eventually disappear, if such conditions are sufficiently long continued. The source of the snow is the water of the ocean evaporated into the atmosphere and transported over the land in the form of clouds. We are to learn that the changes which this moisture undergoes before its delivery to the glacier are notably different for the classes of continental and mountain glacier.

The upper and lower cloud zones of the atmosphere.—Before we can comprehend the nature of the processes by which glaciers are nourished, it will be necessary to review the results of recent studies made upon the earth's atmospheric envelope. It must be kept in mind that the sun's rays are chiefly effective in warming the atmosphere through being first absorbed by some solid body such as rock or water and their heat then communicated by contact to the immediately adjacent air layers. The layers thus warmed being now lighter than before, they rise and are replaced by colder air, which in its turn is warmed and likewise set in upward motion. Such currents developed in the air by contact with warmer solid bodies constitute the process known as convection.

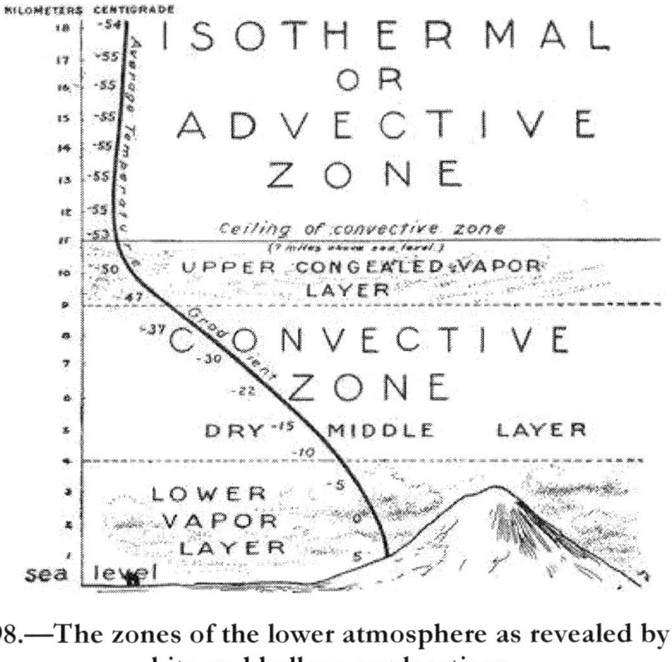

FIG. 298.—The zones of the lower atmosphere as revealed by recent kite and balloon explorations.

To a relatively small degree the atmosphere is heated by the direct absorption of the sun's rays which pass through it. Since air has weight, it compresses the lower layers near the earth, and hence as we ascend from the earth's surface the air becomes continually lighter. Convection currents must, therefore, adjust themselves by the air expanding as it rises. But expansion of a gas always results in its cooling, as every one must have observed who has placed his finger in the air current which escapes from the open valve of a warm rubber tire. Dry air is cooled a degree Fahrenheit for every six hundred feet of ascent in the atmosphere. At a height of about seven miles above the earth's surface all rising air currents have cooled to about 68° below the zero of the Fahrenheit scale, and exploration with balloons has shown that the currents rise no farther. At this level they move horizontally, just as rising vapor spreads out in a room beneath the ceiling. Above this level, as far as exploration has gone, or to a height of more than twelve miles, the temperature remains nearly constant, and this upper zone is, therefore, called the *isothermal* or the *advective zone*—the uniform temperature zone of the lower atmosphere. Beneath the convective ceiling the process of convection is characteristic, and this zone is therefore described as the *convective zone* (Fig. 298).

A large part of the moisture which rises from the ocean's surface is condensed to vapor before it has ascended three miles, and in this form it makes its transit over land as fleecy or stratiform clouds—the so-called cumulus and stratus clouds and their many intermediate varieties (see Frontispiece). This lower layer within the convective zone is, therefore, a moist one overlaid by a relatively drier middle layer of the convective zone. That moisture which rises above the lower cloud layer is congealed by adiabatic cooling to fine ice needles visible as the so-called cirrus clouds which float as feathery fronds beneath the convective ceiling (see frontispiece at right upper corner of picture). Thus we have within the convective zone an upper layer more or less charged with water in the form of ice needles. It is the clouds of the lower zone whose moisture in the form of vapor supplies the nourishment of mountain glaciers, and the high cirrus clouds whose congealed moisture, after interesting transformations, is responsible for the continued existence of continental glaciers.

As we are to see, there are other noteworthy differences between continental and mountain glaciers, in the manner of their sculpture of the lithosphere, so that long after they have disappeared the characters of each are easily identified in their handiwork. How the lower clouds are forced upward and so compelled to give up their moisture to feed the mountain glaciers, and how the upper clouds are pulled downward to nourish the glaciers of continents, can be best understood after the characteristics of each glacier class have been studied.

CHAPTER XXI

THE CONTINENTAL GLACIERS OF POLAR REGIONS

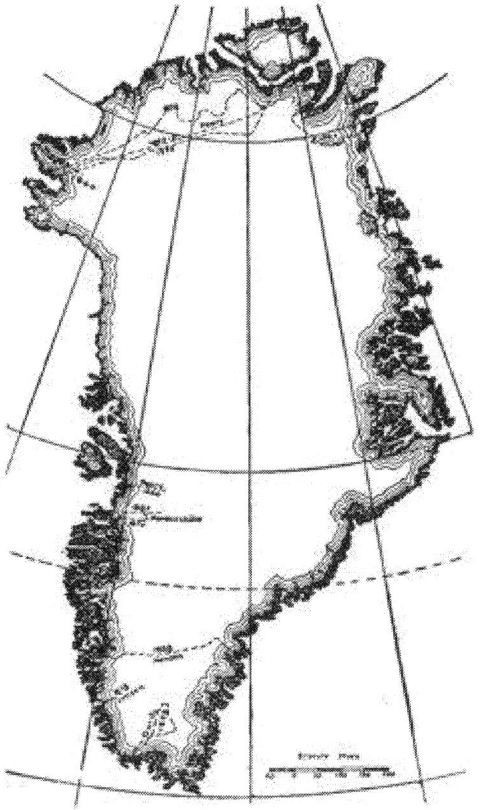

FIG. 299.—Map of Greenland showing the area of inland-ice and the routes of different explorers.

The inland ice of Greenland.—In Greenland and in Antarctica the land is almost or quite buried under a cover of snow and ice—the so-called "inland ice"—which always assumes the surface of a very flat dome or shield. In Greenland there is found a marginal ribbon of land generally from five to twenty-five miles in width (Fig. 299), but in Antarctica all the land, with the exception of a few mountain peaks, is inwrapped in a mantle of ice which is also extended upon the sea in a broad shelf of snow and ice. Neither of these vast glaciers has been explored except in its marginal portion, yet such is the symmetry of the profiles along the routes traversed, and such the flatness and monotony of the snow surface within the margins, that there is little reason to doubt that the profile made along Nansen's route in southern

Greenland would, save only for magnitude, fairly represent a section across the middle of the continent (Fig. 300).

The mountain rampart and its portals.—As soon as we examine the coastal belt we observe that the "Great Ice" of Greenland is held in by a wall of mountains and so prevented from spreading out to its natural surface in the marginal portions. Through portals of the inclosing mountain ranges—the *outlets*—it sends out *tongues* of ice which in many respects resemble certain types of mountain glaciers.

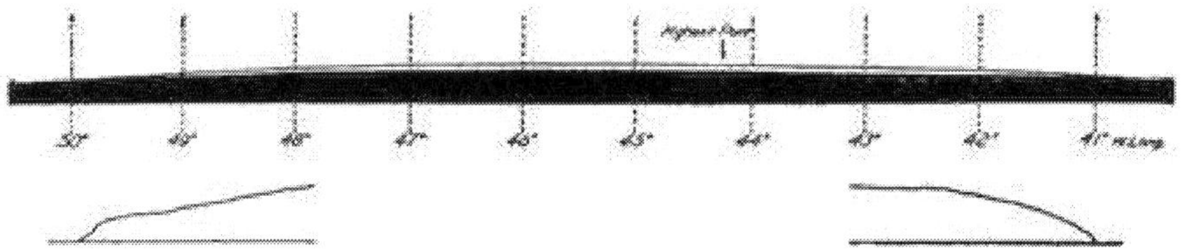

FIG. 300.—Profile in natural proportions across the southern end of the continental glacier of Greenland, constructed upon an arc of the earth's surface and based upon Nansen's profile corrected by Hess. The marginal portions of the profile are represented below upon a magnified scale in order to bring out the characters of the marginal slopes.

Such measurements as have been made upon the inland ice of Greenland at points back from, but yet comparatively near to, the outlets, show that it has here a surface rate of motion amounting to less than an inch per day, and it is highly probable that at moderate distances from the margin this amount diminishes to zero. Upon the outlets, on the contrary, surface rates as high as 59 feet per day have been measured, and even 100 feet per day has been reported. We are thus justified in saying that glacier flow within the outlets is from 700 to 1000 times as great as it is upon the near-by inland ice, and that the glacier is in a measure drained through the portals of the inclosing ranges. Back from these outlet streams of ice, or tongues, the surface of the inland ice is depressed to form a dimple or "basin of exudation" as is the surface of a reservoir above the raceway when the water is being rapidly drawn away (Fig. 301).

PLATE 13.

A. **Precipitous front of the Bryant glacier outlet of the Greenland inland-ice (after Chamberlin).**

B. **Lateral stream beside the Benedict glacier outlet, Greenland (after R. E. Peary).**

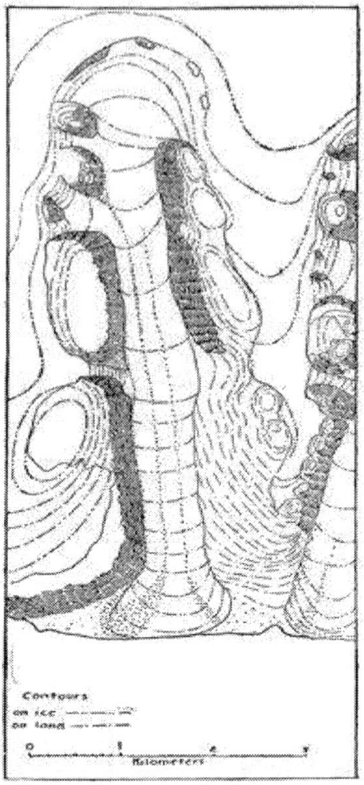

FIG. 301.—Map of a glacier tongue, with dimple showing above and due to indraught of the ice. Umanakfjord, Greenland (after von Drygalski).

Fissures in the ice, the so-called crevasses, are the recognized marks of ice movement, and these are always concentrated at the steep slopes of the ice surface in the neighborhood of its margins. Upon the Greenland ice, crevasses are restricted in their distribution to a zone which extends from seven to twenty-five miles within the ice border.

The marginal rock islands.—From its margin the ice surface rises so steeply as to be climbed only with difficulty, but this gradient steadily diminishes until at a distance of between seventy-five and a hundred miles its slope is less than two degrees. Where crossed by Nansen near latitude 64° N. the broad central area of ice was so nearly level as to appear to be a plain.

As we pass across the irregular ice margin in the direction of the interior, larger and larger proportions of the land's surface are submerged, until only projecting peaks rise above the ice as islands which are described as *nunataks* (Fig. 302).

Though not a universal observation, it has been often noted that the absorption of the sun's rays by rock masses projecting through the snow results in a radiation of the heat and a lowering by melting of the surrounding snow and ice. For this reason nunataks are often surrounded by a deep trench due to a melting of the snow. Such a depression in the ice surface about the margin of a nunatak, from its resemblance to a trench about an ancient castle, has been designated a *moat* (Fig. 303). For the same reason, the outlet tongues of ice which descend in deep fjords between walls of rock are melted away from the walls and a lateral stream of water is sometimes found to flow between ice and rock (pl. 13 B).

FIG. 302.—Edge of the Greenland inland ice, showing the nunataks diminishing in size toward the interior. The lines upon the ice are medial moraines starting from nunataks (after Libbey).

Rock fragments which travel with the ice.—Rock surfaces which are exposed to the atmosphere are in high latitudes broken down through the freezing of water within their crevices. The fragments resulting from this rending process fall upon the glacier surface and are carried forward as passengers in the direction of the ice margin. They are either visible as long and narrow ridges or trains following the directions of the steepest slope (Fig. 302), or they become buried under fresh falls of snow and only again become visible where summer melting has lowered the glacier surface in the vicinity of its margin. These longitudinal trains of rock fragments upon the glacier surface always have their starting point at the lower margin of one of the nunataks, and are known as *medial moraines* (Fig. 301, p. 273, and Fig. 302). Inside the zone of nunataks the glacier surface is, however, clear of rock débris except where dust has been blown on by the wind, and this extends for a few miles only. The material of the medial moraines is a collection of angular blocks whose surfaces are the result of frost rending, for in their travel above the ice they are subjected to no abrading processes.

FIG. 303.—Moat surrounding a nunatak in Victoria Land (after Scott).

A contrasted type of surface moraines upon the Greenland glacier, instead of being parallel to the direction of ice movement, is directed transversely or parallel to the margins. The materials of these moraines are more rounded fragments of rock which have come up from the bottom layers, and we shall again refer to the origin of such moraines after the subglacial conditions have been considered.

The grinding mill beneath the ice.—If, now, we examine the front of a glacier tongue which goes out from the inland ice, we find that while the upper portion is white and mainly free from rock débris (plate 13 A), the lower zone is of a dark color and crowded with layers of pebbles and bowlders which have been planed, polished, and scratched in a quite remarkable manner. The ice front is itself subject to forward and retrograde migrations of short period, but it is easily seen that in the main its larger movement has been a retrograde one. The ground from which it has lately withdrawn is generally a hard rock floor unweathered, but smooth, polished, and scratched in the same manner as the bowlders which are imbedded within the ice. It is perfectly apparent that the latter have been derived from some portion of the rock basement upon which the glacier still rests, and that floor and bowlders have alike been ground smooth by mutual contact under pressure.

This erosion beneath the ice is accomplished by two processes; namely, *plucking* and *abrasion.* Wherever the rock over which the glacier moves has stood up in projecting masses and is riven by fissure planes of any kind, the ice has found it easy to remove it in larger or smaller fragments by a quarrying process described as plucking. The rock may be said to be torn away in blocks which are largely bounded by the preëxisting fissure planes. Over relatively even surfaces plucking has little importance, but where there are noteworthy inequalities of surface upon the glacier bed, those sides which are away from the oncoming ice (*lee* side) are degraded by plucking in such a manner as sometimes to leave steep and ragged fracture surfaces. The tools of the ice thus acquired in the process of plucking are quickly frozen into the lowest ice layers, and being now dragged along the floor they abrade in the same manner as does a rasp or file. These tools of the ice are themselves worn away in the process and are thus given their characteristic shapes. Just as the lapidary grinds the surface of a jewel into facets by imbedding the gem in a matrix, first in one and then in another position, each time wearing down the projecting irregularities through contact with the abrading surface; so in like manner the rock fragment is held fast at the bottom of the glacier until "soled" or "shod", first upon one side and then upon another. Accidental contact with some obstruction upon the floor may suffice to turn the fragment and so expose a new surface to wear upon the abrading floor. Minor obstructions coming in contact with one side of the fragment only, may turn it in its own plane without overturning. Evidence of such interruptions can be later read in the different directions of striæ upon the same facet (plate 17 A).

FIG. 304.—A glacier pavement of Permo-Carboniferous age in South Africa. The striæ running in the direction of the observer are prominent and a noteworthy gouging of the surface is to be noted to the right in the middle distance (after Davis).

The floor beneath the glacier is reduced by the abrading process to a more or less smooth and generally flattened or rounded surface—the so-called *glacier pavement* (Fig. 304). To accomplish this all former mantle rock due to weathering processes must first be cleared away, and the firm unaltered rock beneath is wherever susceptible of it given a smooth polish although locally scored and scratched by the grinding bowlders. The earlier projections of the surface of the floor, if not entirely planed away, are at least transformed into rounded shoulders of rock, which from their resemblance to closely crowded backs in a flock of sheep have been called "sheep backs" or "*roches moutonnées.*" Thus the effect of the combined action of the processes of plucking and abrasion is to reduce the accent of the relief and to mold the contours of the rock in smoothly flowing curves, generally of large radius.

The lifting of the grinding tools and their incorporation within the ice.—Wherever the ice is locally held in check by the projecting nunataks, relief is found between such obstructions, and there the flow of the ice has a correspondingly increased velocity (Fig. 305 *b*). If the obstructions are not of large dimensions, the ice which flows around the outer edges is soon joined to that which passes between the obstructions and so normal conditions of flow are restored below the nunataks. The locally rapid flow of the ice is, therefore, restricted to a relatively short distance, the passageway between the nunataks, and the conditions are thus to be likened to the fall of water at a raceway due to the sudden descent of its surface from the level of the reservoir to the level of the stream in the outlet. As is well known, there is under these conditions a prodigious scour upon the bottom which tends to dig a pit just above and below the dam—a *scape colk*—and carry the materials up to the surface below the pit. Such a tendency was well illustrated by the behavior of the water at the opening of the Neu Haufen dam below the city of Vienna (Fig. 305 *a*). In the case of ice, material from the bottom may by the upward current be brought up to the surface of the glacier at the lower edge of the colk and thus produce a type of local surface moraine of horseshoe form with its direction generally transverse to the direction of ice movement (Fig. 305 *b*).

FIG. 305.—*a*, Map showing pit excavated by the current below the opening in a dam. *b*, Nunataks and surface moraines on the Greenland ice. Dalager's Nunataks (after Suess).

Any obstruction upon the pavement of the glacier apparently exerts a larger or smaller tendency to elevate the bowlders and pebbles and incorporate them within the ice. Rock débris thus incorporated is described as *englacial* drift. In the case of Greenland glaciers this material seems at the ice front to be largely restricted to the lower 100 feet (plate 13 A).

Near the front of the inland ice the increased slope of the upper surface greatly increases the flow of the upper ice layers in comparison with those nearer the bottom, so that the upper layers override the lower as they would an obstruction. The englacial drift is either for this reason or because of rock obstructions brought to the surface, where it yields parallel ridges corresponding in direction to the glacier margin. Such transverse surface moraines are thus in many respects analogous to those which appear about the lower margins of scape colks. In contrast to the longitudinal or medial surface moraines the materials of the transverse moraines are more faceted and rounded—they have been abraded upon the glacier pavement.

Melting upon the glacier margins in Greenland.—During the short but warm summer season, the margins of the Greenland ice are subject to considerable losses through surface melting. When the uppermost ice layer has attained a temperature of 32° Fahrenheit, melting begins and moves rapidly inward from the glacier margin. In late spring the surface of the outer marginal zone is saturated with water, and this zone of slush advances inward with the season, but apparently never transgresses the inner border of what we have generally referred to as the marginal zone of the ice characterized by relatively steep slopes, crevasses, and nunataks. Upon the ice within this zone are found streams large enough to be designated as rivers and these are connected with pools, lakes, and morasses. The dirt and rock fragments imbedded in the ice are melted out in the lowering of the surface, so that late in the season the ice presents a most dirty aspect. At the front of the great mountain glaciers of Alaska, a more vigorous operation of the same process has yielded a surface soil in which grow such rank forests as entirely to mask the presence of the ice beneath.

In addition to the visible streams upon the surface of the Greenland ice, there are others which flow beneath and can be heard by putting the ear to the surface. All surface streams eventually encounter the marginal crevasses and plunge down in foaming cascades, producing the well known "glacier wells" or "glacier mills." The progress of the water is now throughout in tunnels within the ice until it again makes its appearance at the glacier margin.

FIG. 306.—Marginal moraine now forming at the edge of Greenland inland ice, showing a smooth rock pavement outside it. A small lake with a partial covering of lake ice occupies a hollow of this pavement (after von Drygalski).

The marginal moraines.—Study of both the Greenland and Antarctic glaciers has shown that if we disregard the smaller and short-period migrations of the ice front, the general later movement has been a retrograde one—we live in a receding hemicycle of glaciation. The earlier Greenland glacier has now receded so as to expose large areas of the former glacier pavement. In places this pavement is largely bare, indicating a relatively rapid retirement of the ice front, but at all points at which the ice margin was halted there is now found a ridge of unassorted rock materials which were dropped by the ice as it melted (Fig. 306). Such ridges, composed of the unassorted materials described as *till*, come to have a festooned arrangement largely concentric to the ice margin, and are the *marginal* or *terminal moraines* (see Fig. 336, p. 312). Marginal moraines, if of large dimensions, usually have a hummocky surface, and are apt to be composed of rock fragments of a wide range of size from rock flour (clay) to large bowlders (plate 17 A), which may represent many types since they have been plucked by the glacier or gathered in at its surface from many widely separated localities.

FIG. 307.—Small lake impounded between the ice front and a moraine which it has recently built. Greenland (after von Drygalski).

As the glacier front retires from the moraine which it has built up, the water which emerges from beneath the ice is impounded behind the new dam so as to form a lake of crescentic outline (Fig. 307). Such lakes are particularly short-lived, for the reason that the water finds an outlet over the lowest point in the crest of the moraine and easily cuts a gorge through the loose materials, thus draining the lake (Fig. 308). Thereafter, the escaping water flows in a braided stream across the late lake bottom and thence at the bottom of the gorge through the moraine.

FIG. 308.—View of a drained lake bottom between the moraine-covered ice front in the foreground and an abandoned marginal moraine in the middle distance. The water flows from the ice front in a braided stream and passes out through the moraine in a narrow gorge. Variegated glacier, Alaska (after Lawrence Martin).

The outwash plain or apron.—The water which descends from the glacier surface in the glacier wells or mills, eventually arrives at the bottom, where it follows a sinuous course within a tunnel melted out in the ice. Much of this water may issue at the ice front beneath the coarse rock materials which are found there, and so be discovered with the ear rather than by the eye. The water within the tunnels not flowing with a free surface but being confined as though it were in a pipe, may, however, reach the glacier margin under a hydrostatic pressure sufficient to carry it up rising grades. Inasmuch as it is heavily charged with rock débris and is suddenly checked upon arriving at the front it deposits its burden about the ice margin so as to build up plains of assorted sands and gravels, and over this surface it flows in ever shifting serpentine channels of braided type (Fig. 308). Such plains of glacier outwash are described as *outwash plains* or *outwash aprons.*

Rising as it does under hydrostatic pressure the water issuing at the glacier front may find its way upward in some of the crevasses and so emerge at a level considerably above the glacial floor. It may thus come about that the outwash plain is built up about the nose of the glacier so as partially to bury it from sight. When now the ice front begins a rapid retirement, a depression or *fosse* (Fig. 309 and Fig. 339, p. 314) is left behind the outwash plain and in front of the moraine which is built up at the next halting place.

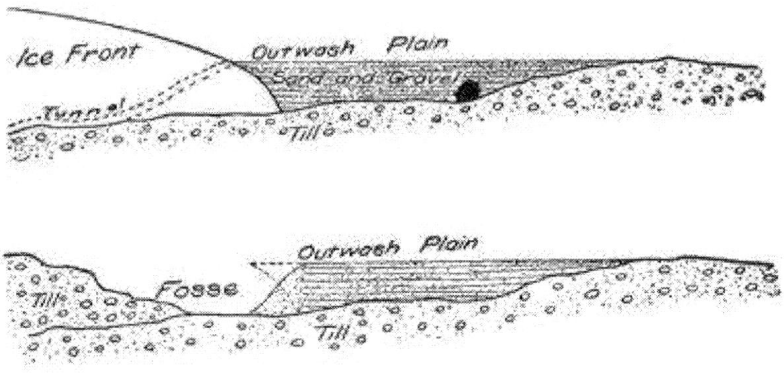

FIG. 309.—Diagrams to show the manner of formation and the structure of an outwash plain, and the position of the fosse between this and the moraine.

The continental glacier of Antarctica.—In Victoria Land, upon the continent of Antarctica, so far as exploration has yet gone, the continental glacier is held back by a rampart of mountains, as has been shown to be true of the inland ice of Greenland. The same flat dome or shield has likewise been found to characterize its upper surface (Fig. 310).

The most noteworthy differences between the inland ice masses of Greenland and Antarctica are to be ascribed to the greater severity of the Antarctic climate and to the more ample nourishment of the southern glacier measured by the land area which it has submerged. There is here no marginal land ribbon as in Greenland, but the glacier covers all the land and is, moreover, extended upon the sea as a broad floating terrace—the shelf ice (Fig. 311). This barrier at its margin puts a bar to all further navigation, rising as it does in some cases 280 feet above the sea and descending to even greater depths below (plate 15 B).

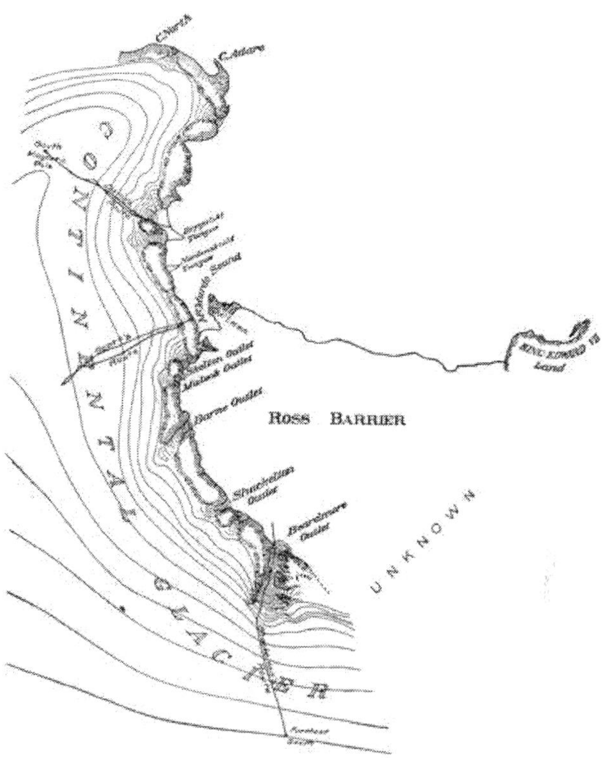

FIG. 310.—Map showing the inland ice of Victoria Land bordered by the shelf ice of the Great Ross Barrier. The arrows show the direction of the prevailing winds (based on maps by Scott and Shackleton).

In that portion of Antarctica which was explored by the German expedition, the inland ice is not as in Victoria Land restrained within walls of rock, but is spread out upon the continent so as to assume its natural ice slopes, which are therefore much flatter than those examined in Greenland and Victoria Land. Here in Kaiser Wilhelm Land the ice rises at its sea margin in a cliff which is from 130 to 165 feet in height, then upon a fairly steeply curving slope to an elevation of perhaps a thousand feet. Here the grades have become relatively level, and on ever flatter slopes the surface appears to continue into the distant interior (plate 14). Near the ice margin numerous fissures betray a motion within the mass which exact measurements indicate to be but one foot per day, and at a distance of a mile and a quarter from the margin even this slight value has diminished by fully one eighth. It can hardly be doubted that at moderate distances only within the ice margin, the glacier is practically without motion.

PLATE 14.

View of the margin of the Antarctic continental glacier in Kaiser Wilhelm Land (after E. v. Drygalski).

Rain or general melting conditions being unknown in Antarctica, a striking contrast is offered to the marginal zone of the Greenland continent. This is to a large extent explained by the existence upon the northern land mass of a coastland ribbon which becomes quickly heated in the sun's rays, and both by warming the air and by radiating heat to the ice it causes melting and produces local air temperatures which in summer may even be described as hot. About Independence Bay in latitude 82° N. and near the northernmost extremity of Greenland, Peary descended from the inland ice into a little valley within which musk oxen were lazily grazing and where bees buzzed from blossom to blossom over a gorgeous carpet of flowers.

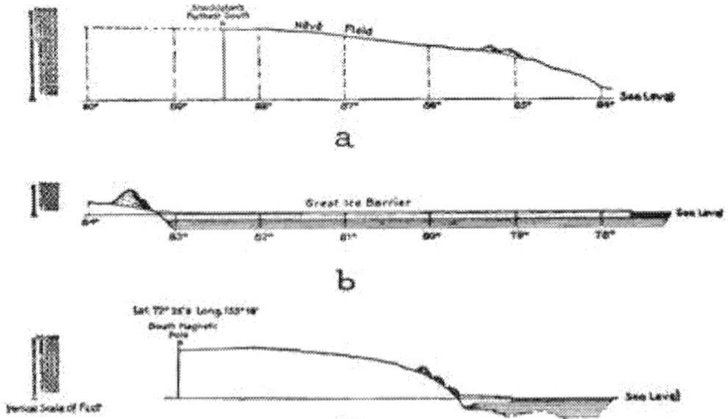

FIG. 311.—Sections across the inland ice of Victoria Land, Antarctica, with the shelf ice in front (after Shackleton).

Nourishment of continental glaciers.—Explorations upon and about the glaciers of Greenland and Antarctica have shown that the circulation of air above these vast ice shields conforms to a quite simple and symmetrical model subject to spasmodic pulsations of a very pronounced type. Each great ice mass with its atmospheric cover constitutes a sort of refrigerating air engine and plays an important part in the wind system of the globe. (See Fig. 291, p. 263). Both the domed surface and the low temperature of the glacier are essential to the continuation of this pulsating movement within the atmosphere (Fig. 312). The air layer in contact with the ice is during a period of calm cooled, contracted, and rendered heavier, so that it begins to slide downward and outward upon the domed surface in all directions. The extreme flatness of the greater portion of the glacier surface—a fraction only of one degree—makes the engine extremely slow in starting, but like all bodies which slide upon inclined planes, the velocity of its movement is rapidly accelerated, until a blizzard is developed whose vigor is unsurpassed by any elsewhere experienced.

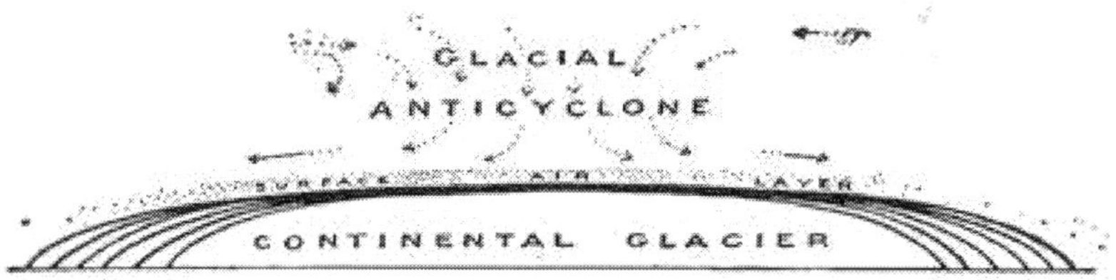

FIG. 312.—Diagram to show the nature of the fixed glacial anticyclone above continental glaciers and the process by which their surface is shaped.

The effect of such centrifugal air currents above the glacier is to suck down the air of the upper currents in order to supply the void which soon tends to develop over the central portion of the glacier dome. This downward vortex, fed as it is by inward-blowing, high-level currents, and drained by outwardly directed surface currents, is what is known as an *anticyclone*, here fixed in position by the central embossment of the dome.

The air which descends in the central column is warmed by compression, or adiabatically, just as air is warmed which is forced into a rubber tire by the use of a pump. The moisture congealed in the cirrus clouds floating in the uppermost layer of the convective zone, is carried down in this vortex and first melted and in turn evaporated, due to the adiabatic effect. This fusion and evaporation of the ice by its transformation of latent, to sensible, heat, in a measure counteracts, and so retards, the adiabatic elevation of temperature within the column. Eventually the warm air now charged with water vapor reaches the ice surface, is at once chilled, and its burden of moisture precipitated in the form of fine snow needles, the so-called "frost snow", which in accompaniment to the sudden elevation of temperature is precipitated at the termination of a blizzard.

The warming of the air has, however, had the effect of damping as it were, the engine stroke, and, as the process is continued, to start a reverse or upward current within the chimney of the anticyclone. The blizzard is thus suddenly ended in a precipitation of the snow, which by changing the latent heat of condensation to sensible heat tends to increase this counter current.

FIG. 313.—Snow deltas about the margins of the Fan glacier outlet of Greenland (after Chamberlin).

The glacier broom.—During the calm which succeeds to the blizzard, heat is once more abstracted from the surface air layer, and a new outwardly directed engine stroke is begun. The tempest which later develops acts as a gigantic centrifugal broom which sweeps out to the margins of the glacier all portions of the latest snowfall which have not become firmly attached to the ice surface. The sweepings piled up about the margin of continental glaciers have been described as fringing glaciers, or the glacial fringe. The northern coast of Greenland and Grant Land are bordered by a fringe of this nature (plate 14, and Fig. 315, p. 288). It is by the operation of the glacier broom that the inland ice is given its characteristic shield-like shape (Fig. 312). The granular nature of the snow carried by the wind is well brought out by the little snow deltas about the margins of Greenland ice tongues (Fig. 313). Obviously because of the presence of the vigorous anticyclone, no snows such as nourish mountain glaciers can be precipitated upon continental glaciers except within a narrow marginal zone, and, as shown by Nansen rock dust from the coastland ribbon and from the nunataks of Greenland, is carried by a few miles inside the western margin, and not at all within the eastern.

FIG. 314.—Sea ice of the Arctic region in lat. 80° 5′ N. and long. 2° 52′ E. (after Duc d'Orleans).

Field and pack ice.—Within polar regions the surface of the sea freezes during the long winter season, the product being known as *sea-ice* or *field-ice* (Fig. 314). This ice cover may reach a thickness by direct freezing of eight or more feet, and by breaking up and being crowded above and below neighboring fragments may increase to a considerably greater thickness. Ice thus crowded together and more or less crushed is described as *pack ice* or *the pack.*

The pack does not remain stationary but is continually drifting with the wind and tide, first in one direction and then in another, but with a general drift in the direction of the prevailing winds. Because of the vast dimensions of the pack, the winds over widely separated parts may be contrary in direction, and hence when currents blow toward each other or when the ice is forced against a land area, it is locally crushed under mighty pressures and forced up into lines of *hummocks*—the so-called *pressure ridges.* At other times, when the winds of widely separated areas blow away from each other, the pack is parted, with the formation of lanes or *leads* of open water.

If seen in bird's-eye view the lines of hummocks would according to Nansen be arranged like the meshes of a net having roughly squared angles and reaching to heights of 15 to 25, rarely 30, feet above the general surface of the pack. The ice within each mesh of the network is a *floe,* which at the times of pressure is ground against its neighbors and variously shifted in position. At the margin of the pack these floes become separated and float toward lower latitudes until they are melted.

The drift of the pack.—The discovery of the drift in the Arctic pack is a romantic chapter in the history of polar exploration, and has furnished an example of faith in scientific reasoning and judgment which may well be compared with that of Columbus. The great figure in this later discovery is the Norwegian explorer Fridtjof Nansen, and to the final achievement the ill-fated *Jeannette* expedition contributed an important part.

The *Jeannette* carrying the American exploring expedition was in 1879 caught in the pack to the northward of Wrangel Island (Fig. 315), and two years later was crushed by the ice and sunk to the northward of the New Siberian Islands. In 1884 various articles, including a list of stores in the handwriting of the commander of the *Jeannette*, were picked up at Julianehaab near the southern extremity of Greenland but upon the western side of Cape Farewell. Nansen, having carefully verified the facts, concluded that the recovered articles could have found their way to Julianehaab only by drifting in the pack across the polar sea, and that at the longest only five years had been consumed in the transit. After being separated from the pack the articles must have floated in the current which makes southward along the east coast of Greenland and after doubling Cape Farewell flows northward upon the west coast. It was clear that if they had come through Smith Sound they would inevitably have been found upon the other shore of Baffin Bay. In confirmation of this view there was found at Godthaab, a short distance to the northward of Julianehaab (Fig. 315), an ornamented Alaskan "throwing stick" which probably came by the same route. Moreover, large quantities of driftwood reach the shores of Greenland which have clearly come from the Siberian coast, since the Siberian larch has furnished the larger quantity.

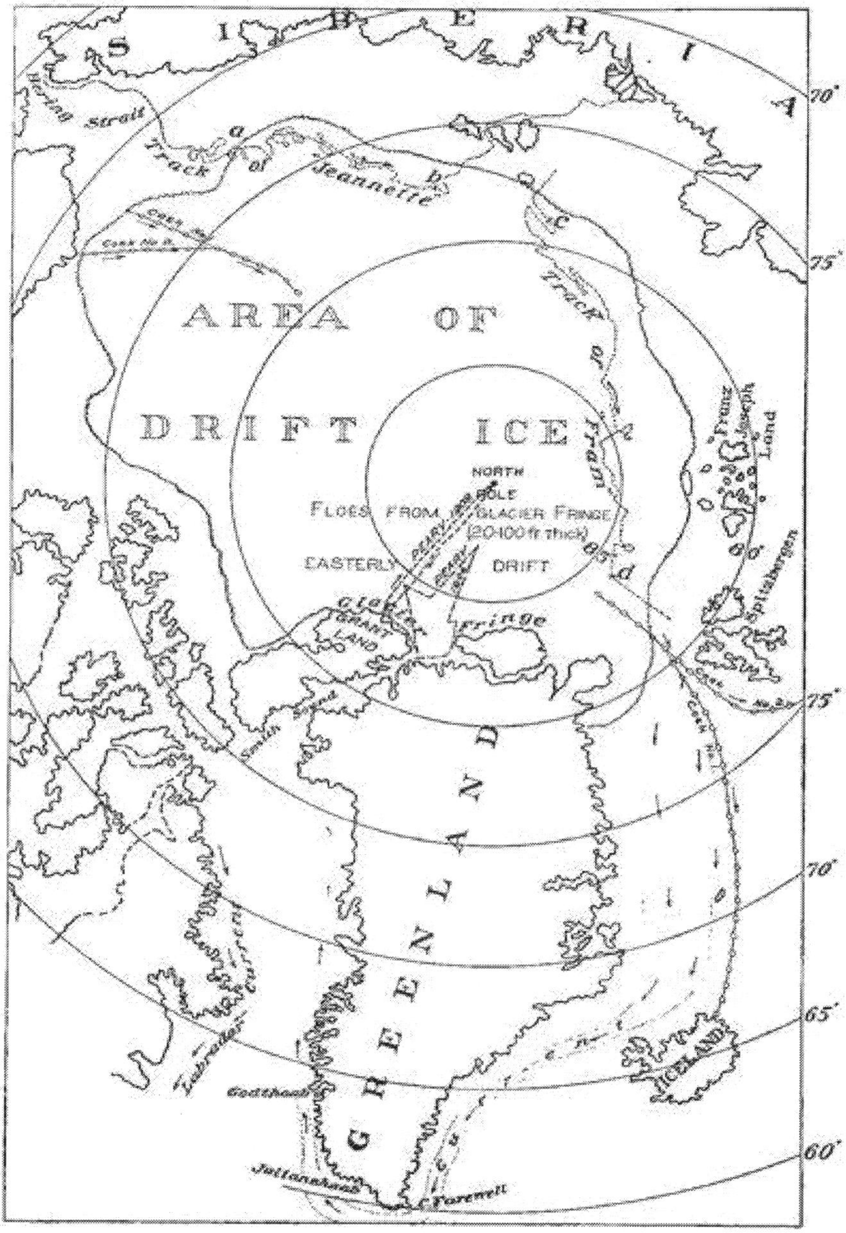

FIG. 315.—Map of the north polar regions, showing the area of drift ice and the tracks of the *Jeannette* and the *Fram* (compiled from various maps).

Pinning his faith to these indubitable facts, Nansen built the *Fram* in such a manner as to resist and elude the enormous pressures of the ice pack, stocked her with provisions sufficient for five years, and by allowing the vessel to be frozen into the pack north of the New Siberian Islands, he consigned himself and his companions to the mercy of the elements. The world knows the result as one of the most remarkable achievements in the long history of polar exploration. The track of the *Fram*, charted in Fig. 315, considered in connection with that of the *Jeannette*, shows that the Arctic pack drifts from Bering Sea westward until near the northeastern coast of Greenland.

Special casks were for experimental purposes fastened in the ice to the north of Behring Strait by Melville and Bryant, and two of these were afterwards recovered, the one near the North Cape in northern Norway, and the other in northeastern Iceland (see map, Fig. 315). Peary's trips northward in 1906 and 1909 from the vicinity of Smith Sound have indicated that between the Pole and the shores of Greenland and Grant Land the drift is throughout to the eastward, corresponding to the westerly wind. Upon this border the great area of Arctic drift ice is in contact with great continental glaciers bordered by a glacier fringe. Admiral Peary has shown that instead of consisting of frozen sea ice, the pack is here made up of great floes from 20 to 100 feet in thickness and that these have been derived from the glacier fringe.

Whenever the blizzards blow off the inland ice from the south, leads are opened at the margin of the fringe and may carry strips from the latter northward across the lead. With favorable conditions these leads may be closed by thick sea ice so that with the occurrence of counter winds from the north they do not entirely return to their original position. A continuance of this process may have resulted in the heavy floe ice to the northward of Greenland, which, acting as an obstruction, may have forced the thinner drift ice to keep on the European side of the Arctic pack.

FIG. 316.—The shelf ice of Coats Land with the surrounding pack ice showing in the foreground (after Bruce).

About the Antarctic continent there is a broad girdle of pack ice which, while more indolent in its movements than the Arctic pack, has been shown by the expeditions of the *Belgica* and the *Pourquoi-Pas* to possess the same kind of shifting movements. In the southern spring this pack floats northward and is to a large extent broken up and melted on reaching lower latitudes.

FIG. 317.—Tidewater cliff at the front of a glacier tongue from which icebergs are born.

The Antarctic shelf ice.—It has been already pointed out that the inland ice of Antarctica is in part at least surrounded by a thick snow and ice terrace floating upon the sea and rising to heights of more than 150 feet above it (plate 15 B and Fig. 316). The visible portions of this shelf-ice are of stratified compact snow, and the areas which have thus far been studied are found in bays from which dislodgment is less easily effected. The origin of the shelf ice is believed to be a sea-ice which because not easily detached at the time of the spring "break-up" is thickened in succeeding seasons chiefly by the deposition of precipitated and drifted snow upon its surface, so that it is bowed down under the weight and sunk to greater and greater depths in the water. To some extent, also, it is fed upon its inner margin by overflow of glacier ice from the inland ice masses.

Icebergs and snowbergs and the manner of their birth.—Greenland reveals in the character of its valleys the marks of a large subsidence of the continent—the serpentine inlets or fjords by which its coast is so deeply indented. Into the heads of these fjords the tongues from the inland ice descend generally to the sea level and below. The glacier ice is thus directly attacked by the waves as well as melted in the water, so that it terminates in the fjords in great cliffs of ice (Fig. 317). It is also believed to extend beneath the water surface as a long toe resting upon the bottom (Fig. 319).

PLATE 15.

A. **An Antarctic ice foot with boat party landing (after R. F. Scott).**

B. **A near view of the front of the Great Ross Barrier, Antarctica (after R. F. Scott).**

FIG. 318.—A Greenlandic iceberg after a long journey in warm latitudes.

The exposed cliff is notched and undercut by the waves in the same manner as a rock cliff, and the upper portions override the lower so that at frequent intervals small masses of ice from this front separate on crevasses, and toppling over, fall into the water with picturesque splashes. Such small bergs, whose birth may be often seen at the cliff front of both the Greenland and Alaskan glaciers, have little in common with those great floating islands of ice that are drifted by the winds until, wasted to a fraction only of their former proportions, they reach the lanes of transatlantic travel and become a serious menace to navigation (Fig. 318).

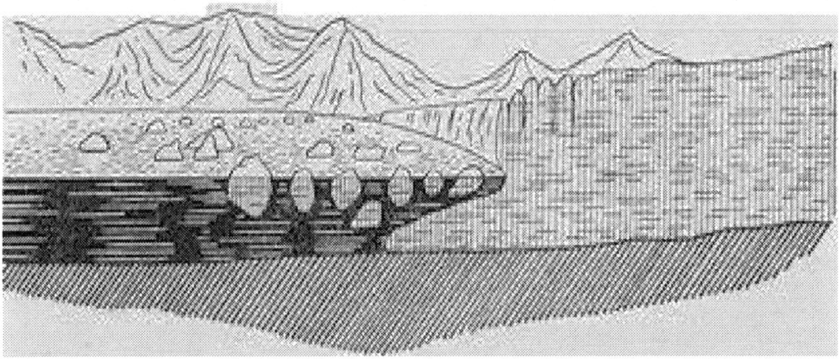

FIG. 319.—Diagram showing one way in which northern icebergs may be born from the glacier tongue (after Russell).

Northern icebergs of large dimensions are born either by the lifting of a separated portion of the extended glacier toe lying upon the bottom of the fjord, or else they separate bodily from the cliff itself, apparently where it reaches water sufficiently deep to float it. In either case the buoyancy of the sea water plays a large rôle in its separation.

If derived from the submerged glacier toe (Fig. 319), a loud noise is heard before any change is visible, and an instant later the great mass of ice rises out of the water some distance away from the cliff, lifting as it does so a great volume of water which pours off on all sides in thundering cascades and exposes at last a berg of the deepest sapphire blue. The commotion produced in the fjord is prodigious, and a vessel in close proximity is placed in jeopardy.

Even larger bergs are sometimes seen to separate from the ice cliff, in this case an instant before or simultaneously, with a loud report, but such bergs float away with comparatively little commotion in the water.

FIG. 320.—A northern iceberg surrounded by sea ice.

The icebergs of the south polar region are usually built upon a far grander scale than those of the Arctic regions, and are, further, both distinctly tabular in form and bounded by rectangular outlines (Fig. 321). Whereas the large bergs of Greenlandic origin are of ice and blue in color, the tabular bergs of Antarctica might better be described as *snowbergs*, since they are of a blinding whiteness and their visible portions are either compacted snow or alternating thick layers of compact snow and thin ribbons of blue ice, the latter thicker and more abundant toward the base. All such bergs have been derived from the shelf ice and not from the inland ice itself. Blue icebergs which have been derived from the inland ice have been described from the one Antarctic land that has been explored in which that ice descends directly to the sea.

FIG. 321.—Tabular Antarctic iceberg separating from the shelf ice (after Shackleton).

In both the northern and southern hemispheres those bergs which have floated into lower latitudes have suffered profound transformations. Their exposed surfaces have been melted in the sun, washed by the rain, and battered by the waves, so that they lose their relatively simple forms but acquire rounded surfaces in place of the early angular ones (Fig. 318, p.291). Sir John Murray, who had such extended opportunities of studying the southern icebergs from the deck of the *Challenger*, has thus described their beauties:

"Waves dash, against the vertical faces of the floating ice island as against a rocky shore, so that at the sea level they are first cut into ledges and gullies, and then into caves and caverns of the most heavenly blue, from out of which there comes the resounding roar of the ocean, and into which the snow-white and other petrels may be seen to wing their way through guards of soldier-like penguins stationed at the entrances. As these ice islands are slowly drifted by wind and current to the north, they tilt, turn and sometimes capsize, and then submerged prongs and spits are thrown high into the air, producing irregular pinnacled bergs higher, possibly, than the original table-shaped mass."

READING REFERENCES FOR CHAPTERS XX AND XXI

General:—

I. C. RUSSELL. Glaciers of North America. Ginn, Boston, 1897, pp. 210, pls. 22.

CHAMBERLIN and SALISBURY. Geology, vol. 1, pp. 232-308.

H. HESS. Die Gletscher, Braunschweig, 1904, pp. 426 (illustrated).

WILLIAM H. HOBBS. Characteristics of Existing Glaciers. Macmillan, 1911, pp. 301, pls. 34.

Special districts of mountain glaciers:—

JAMES D. FORBES. Travels Through the Alps of Savoy and other Parts of the Pennine Chain with Observations on the Phenomena of Glaciers. Edinburgh, 1845, pp. 456, pls. 9, maps 2.

A. PENCK, E. BRÜCKNER, et L. DU PASQUIER. Le système glaciare des alpes, etc., Bull. Soc. Sc. Nat. Neuchâtel, vol. 22, 1894, pp. 86.

E. RICHTER. Die Gletscher der Ostalpen. Stuttgart, 1888, pp. 306, 7 maps.

JAMES D. FORBES. Norway and Its Glaciers, etc. Edinburgh, 1853, pp. 349, pls. 10, map.

I. C. RUSSELL. Existing Glaciers of the United States, 5th Ann. Rept. U. S. Geol. Surv., 1885, pp. 307-355, pls. 32-55; Glaciers of Mt. Ranier, 18th Ann. Rept. U. S. Geol. Surv., 1898, pp. 349-423, pls. 65-82.

W. H. SHERZER. Glaciers of the Canadian Rockies and Selkirks, Smith. Cont. to Knowl. No. 1692, Washington, 1907, pp. 135, pls. 42.

H. F. REID. Studies of Muir Glacier, Alaska, Nat. Geogr. Mag., vol. 4, 1892, pp. 19-84, pls. 1-16.

I. C. RUSSELL. Malaspina Glacier, Jour. Geol., vol. 1, 1893, pp. 219-245.

G. K. GILBERT. Harriman Alaska Expedition, vol. 3, Glaciers, 1904, pp. 231, pls. 37.

W. M. CONWAY. Climbing and Exploration in the Karakoram Himalayas, Maps and Scientific Reports, 1894, map sheets I-II.

FANNY BULLOCK WORKMAN and WILLIAM HUNTER WORKMAN. The Hispar Glacier, Geogr. Jour., vol. 35, 1910, pp. 105-132, 7 pls. and map.

The cycle of glaciation:—

WILLIAM H. HOBBS. The Cycle of Mountain Glaciation, Geogr. Jour., vol. 36, 1910, pp. 146-163, 268-284.

Upper and lower cloud zones of the atmosphere:—

R. ASSMANN, A. BERSON, and H. GROSS. Wissenschaftliche Luftfahrten ausgeführt vom deutschen Verein zur Förderung der Luftschiffahrt in Berlin, 1899-1900, 3 vols.

E. GOLD and W. A. HARWOOD. The Present State of our Knowledge of the Upper Atmosphere as Obtained by the Use of Kites, Balloons, and Pilot-balloons, Rept. Brit. Assoc. Adv. Sci., 1909, pp. 1-55.

W. H. MOORE. Descriptive Meteorology, Appleton, New York, 1910, pp. 95-136.

WILLIAM H. HOBBS. The Pleistocene Glaciation of North America Viewed in the Light of our Knowledge of Existing Continental Glaciers, Bull. Am. Geogr. Soc., vol. 42, 1911, pp. 647-650.

The continental glacier of Greenland:—

F. NANSEN. The First Crossing of Greenland, 2 vols, Longmans, London, 1890 (the scientific results are contained in an appendix to volume 2, pp. 443-497).

R. E. PEARY. A Reconnaissance of the Greenland Inland Ice, Jour. Am. Geogr. Soc., vol. 19, 1887, pp. 261-289; Journeys in North Greenland, Geogr. Jour., vol. 11, 1898, pp. 213-240.

T. C. CHAMBERLIN. Glacier Studies in Greenland, Jour. Geol., vol. 2, 1894, pp. 649-668, 768-788, vol. 3, pp. 61-69, 198-218, 469-480, 565-582, 668-681, 833-843, vol. 4, pp. 582-592, 769-810, vol. 5, pp. 229-245; Recent glacial studies in Greenland (Presidential address), Bull. Geol. Soc. Am., vol. 6, 1895, pp. 199-220, pls. 3-10.

R. S. TARR. The Margin of the Cornell Glacier, Am. Geol., vol. 20, 1897, pp. 139-156, pls. 6-12.

R. D. SALISBURY. The Greenland Expedition of 1895, Jour. Geol., vol. 3, 1895, pp. 875-902.

E. V. DRYGALSKI. Grönland Expedition der Gesellschaft für Erdkunde zu Berlin 1891-1893, Berlin, 1897, 2 vols., pp. 551 and 571, pls. 53, maps 10.

WILLIAM H. HOBBS. Characteristics of the Inland Ice of the Arctic Regions, Proc. Am. Phil. Soc., vol. 49, 1910, pp. 57-129, pls. 26-30.

The Antarctic continental glacier:—

R. F. SCOTT. The Voyage of the *Discovery*. London, 2 vols., 1905.

E. H. SHACKLETON. The Heart of the Antarctic. London, 2 vols., 1910.

E. VON DRYGALSKI. Zum Kontinent des eisigen Südens, Deutsche Südpolar-Expedition, Fahrten und Forschungen des "Gauss", 1901-1903, Berlin, 1904, pp. 668, pls. 21.

OTTO NORDENSKIÖLD and J. S. ANDERSSON. Antarctica or Two Years Amongst the Ice of the South Pole. London, 1905, pp. 608, illustrated.

E. PHILIPPI. Ueber die fünf Landeis-Expeditionen, etc., Zeit. f. Gletscherk., vol. 2, 1907, pp. 1-21.

Nourishment of continental glaciers:—

WILLIAM H. HOBBS. Characteristics of the Inland Ice of the Arctic Regions, Proc. Am. Phil. Soc., vol. 49, 1910, pp. 96-110; The Ice Masses on and about the Antarctic Continent, Zeit. f. Gletscherk., vol. 5, 1910, pp. 107-120; Characteristics of Existing Glaciers. New York, 1911, pp. 143-161, 261-289. Pleistocene Glaciation of North America Viewed in the Light of our Knowledge of Existing Continental Glaciers, Bull. Am. Geogr. Soc., vol. 43, 1911, pp. 641-659.

Field and pack ice:—

EMMA DE LONG. The Voyage of the *Jeannette*, the ship and ice journals of George W. de Long, etc. Berlin, 1884, 2 vols., chart in back of vol. 1.

ROBERT E. PEARY. The Discovery of the North Pole (for further references on both sea and pack ice and Antarctic shelf ice, consult Hobbs's Characteristics of Existing Glaciers, pp. 210-213, 242-244).

Icebergs:—

WYVILLE THOMSON. Challenger Report, Narrative, vol. 1, 1865, Pt. i, pp. 431-432, pls. B-D.

I. C. RUSSELL. An Expedition to Mt. St. Elias, Nat. Geogr. Mag., vol. 3, 1891, pp. 101-102, fig. 1.

H. F. REID. Studies of Muir Glacier, Alaska, *ibid.*, vol. 4, 1892, pp. 47-48.

E. VON DRYGALSKI. Grönland-Expedition, etc., vol. 1, pp. 367-404.

M. C. ENGELL. Ueber die Entstehung der Eisberge, Zeit. f. Gletscherk., vol. 5, 1910, pp. 112-132.

CHAPTER XXII

THE CONTINENTAL GLACIERS OF THE "ICE AGE"

Earlier cycles of glaciation.—Our study of the rocks composing the outermost shell of the lithosphere tells us that in at least three widely separated periods of its history the earth has passed through cycles of glaciation during which considerable portions of its surface have been submerged beneath continental glaciers. The latest of these occurred in the yesterday of geology and has often been referred to as the "ice age", because until quite recently it was supposed to be the only one of which a record was preserved.

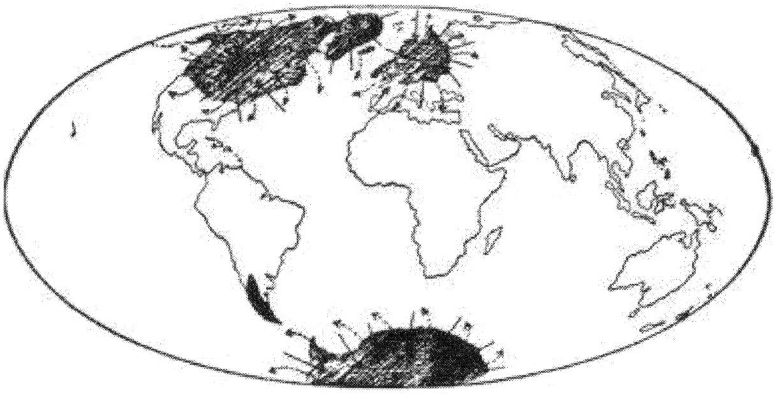

FIG. 322.—Map of the globe showing the areas which were covered by the continental glaciers of the so-called "ice-age" of the Pleistocene period. The arrows show the directions of the centrifugal air currents in the fixed anticyclones above the glaciers.

FIG. 323.—Glaciated granite bowlder which has weathered out of a moraine of Permo-Carboniferous age upon which it rests. South Australia (after Howchin).

This latest ice age represents four complete cycles of glaciation, for it is believed that the continental ice developed and then completely disappeared during a period of mild climate before the next glacier had formed in its place, and that this alternation of climates was no less than three times repeated, making four cycles in all. At nearly or quite the same time ice masses developed in northern North America and in northern Europe, the embossments of the ice domes being located in Canada and in Scandinavia respectively (Fig. 322). There appears to have been at this time no extensive glaciation of the southern hemisphere, though in the next earlier of the known great periods of glaciation—the so-called Permo-Carboniferous—it was the southern hemisphere, and not the northern, that was affected (Fig. 323 and Fig. 304, p.276). From the still earlier glacial period our data are naturally much more meager, but it seems probable that it was characterized by glaciated areas within both the northern and the southern hemispheres.

FIG. 324.—Map to show the glaciated and nonglaciated regions of North America (after Salisbury and Atwood).

Contrast of the glaciated and nonglaciated regions.—Since we have now studied in brief outline the characteristics of the existing continental glaciers, we are in a position to review the evidences of former glaciers, the records of which exist in their carvings, their gravings, and their deposits.

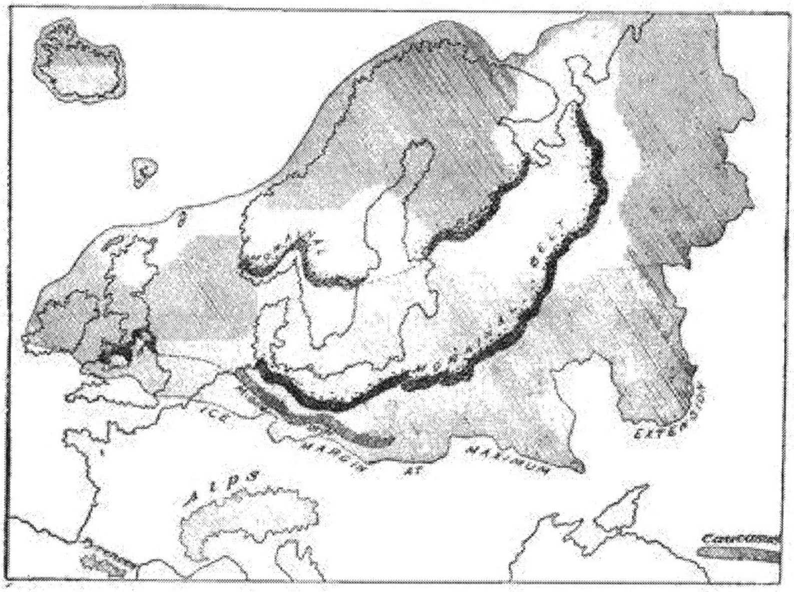

FIG. 325.—Map of the glaciated and nonglaciated areas of northern Europe. The strongly marked morainal belts respectively south and north of the Baltic depression represent halting places in the retreat of the latest continental glacier (compiled from maps by Penck and Leverett).

An observant person familiar with the aspects of Nature in both the northern and southern portions of the central and eastern United States must have noticed that the general courses of the Ohio and Missouri rivers define a somewhat marked common border of areas which in most respects are sharply contrasted (Fig. 324). Hardly less striking is the contrast between the glaciated and the nonglaciated regions upon the continent of Europe (Fig. 325).

It is the northern of the two areas which in each case reveals the characteristic evidences of glaciation, while there is entire absence of such marks to the southward of the common border. Within the American glaciated region there is, however, an area surrounded like an island, and within this district (Fig. 324) none of the marks characteristic of glaciation are to be found. This area is usually referred to as the "driftless area", and occupies portions of the states of Wisconsin, Illinois, Minnesota, and Iowa. Even better than the area to the southward of the Ohio and Missouri rivers, it permits of a comparison of the nonglaciated with the drift-covered region.

FIG. 326.—"Stand Rock" near the "Dells" of the Wisconsin river, an unstable erosion remnant characteristic of the driftless area of North America (after Salisbury and Atwood).

The "driftless area."—Within this district, then, we have preserved for our study a landscape which remains largely as it was before the several ice invasions had so profoundly transformed the general surface of the surrounding country. Speaking broadly, we may say that it represents an uplifted and in part dissected plain, which to the south and east particularly reveals the character of nearly mature river erosion (Fig. 177, p. 170). The rock surface is here everywhere mantled by decomposed and disintegrated rock residues of local origin. The soluble constituents of the rock, such as the carbonates, have been removed by the process of leaching, so that the clays no longer effervesce when treated with dilute mineral acid.

Wherever favored by joints and by an alternation of harder and softer rock layers, picturesque unstable erosion remnants or "chimneys" may stand out in relief (Fig. 326). Furthermore, the driftless area is throughout perfectly drained—it is without lakes or swamps—since all valleys are characterized throughout by forward grades. The side valleys enter the main valleys as do the branches a tree trunk; in other words, the drainage is described as arborescent. In so far as any portions of a plane surface now remain in the landscape, they are found at the highest levels (plate 16 A). The topography is thus the result of a partial removal by erosion of an upland and may be described as *incised topography*. Nowhere within the area are there found rock masses foreign to the region, but all mantle rock is the weathered product of the underlying ledges.

PLATE 16.

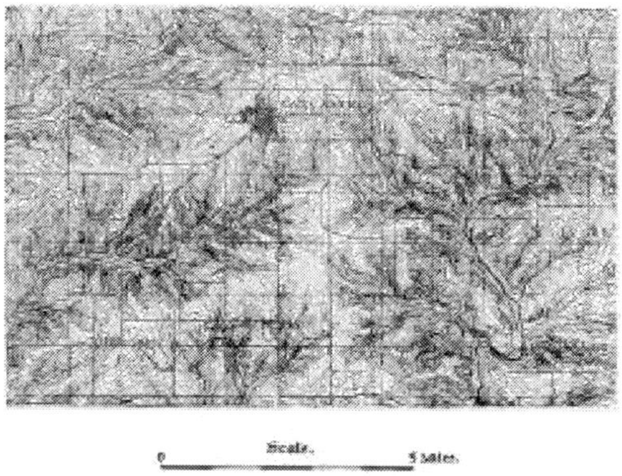

A. **Incised topography within the "driftless area" (U. S. Geol. Survey).**

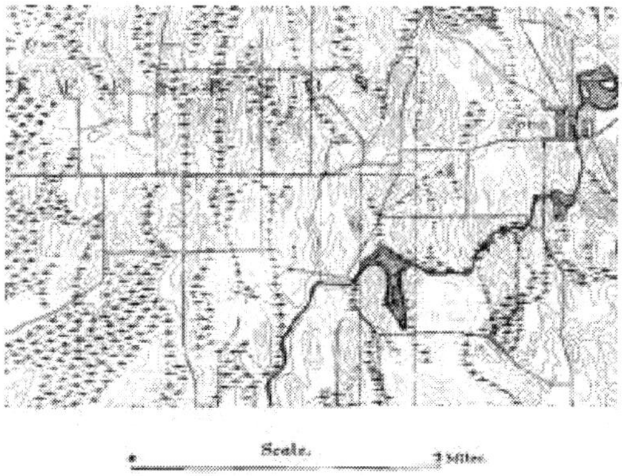

B. **Built-up topography within glaciated region (U. S. Geol. Survey).**

Characteristics of the glaciated regions.—The topography of the driftless area has been described as *incised*, because due to the partial destruction of an uplifted plain; and this surface is, moreover, perfectly drained. The characteristic topography of the "drift" areas is by contrast *built up*; that is to say, the features of the region instead of being *carved* out of a plain are the result of *molding* by the process of deposition (plate 16 B). In so far as a plane is recognizable, it is to be found not at the highest, but at the lowest level—a surface represented largely by swamps and lakes—and above this plain rise the characteristic rounded hills of various types which have been *built up* through deposition. The process by which this has been accomplished is one easy to comprehend. As it invaded the region, the glacier planed away beneath its marginal zone all weathered mantle rock and deposited the planings within the hollows of the surface (Fig. 327). The effect has been to flatten out the preëxisting irregularities of the surface, and to yield at first a gently undulating plain upon which are many undrained areas and a haphazard system of drainage (Fig. 328). All unstable erosion remnants, such as now are to be found within the driftless area, were the first to be toppled over by the invading glacier, and in their place there is left at best only rounded and polished "shoulders" of hard and unweathered rock—the well-known *roches moutonnées*.

FIG. 327.—Diagram showing the manner in which a continental glacier obliterates existing valleys (after Tarr).

The glacier gravings.—The tools with which the glacier works are never quite evenly edged, and instead of an in all respects perfect polish upon the rock pavement, there are left furrowings, gougings, and scratches. Of whatever sort, these scorings indicate the lines of ice movement and are thus indubitable records graven upon the rock floor. When mapped over wide areas, a most interesting picture is presented to our view, and one which supplements in an important way the studies of existing continental glaciers (Fig. 334, p.308, and Fig. 336, p. 312).

FIG. 328.—Lake and marsh district in northern Wisconsin, the effect of glacial deposition in former valleys (after Fairbanks).

It has been customary to think of the glacier as everywhere eroding its bed, although the only warrant for assuming degradation by flow of the ice is restricted to the marginal zone, since here only is there an appreciable surface grade likely to induce flow. Both upon the advance and again during the retreat of a glacier, all parts of the area overridden must be subjected to this action. Heretofore pictured in the imagination as enlarged models of Alpine glaciers, the vast ice mantles were conceived to have spread out over the country as the result of a kind of viscous flow like that of molasses poured upon a flat surface in cold weather. The maximum thickness of the latest American glacier of the ice age has been assumed to have been perhaps 10,000 feet near the summit of its dome in central Labrador. From this point it was assumed that the ice traveled southward up the northern slope of the Laurentian divide in Canada, and thence to the Ohio river, a distance of over 1300 miles. If such a mantle of ice be represented in its natural proportions in vertical section, to cover the distance from center to margin we may use a line six inches in length, and only $^{1}/_{100}$ of an inch thick. Upon a reduced scale these proportions are given in Fig. 329. Obviously the force of gravity acting within a viscous mass of such proportions would be incompetent to effect a transfer of material from the center to the periphery, even though the thickness should be doubled or trebled. Yet until the fixed glacial anticyclone above the glacier had been proven and its efficiency as a broom recognized, no other hypothesis than that of viscous flow had been offered in explanation. The inherited conception of a universal plucking and abrasion on the bed of the glacier is thus made untenable and can be accepted for the marginal portion only.

FIG. 329.—Cross section in approximate natural proportions of the latest North American continental glacier of Pleistocene age from its center to its margin.

Not only do the rock scorings show the lines of ice movement, but the directions as well may often be read upon the rock. Wherever there are pronounced irregularities of surface still existing on the pavement, these are generally found to have gradual slopes upon the side from which the ice came, and relatively steep falls upon the lee or "pluck" side. If, however, we consider the irregularities of smaller size, the unsymmetrical slopes of these protruding portions of the floor are found to be reversed—it is the steep slope which faces the oncoming ice and the flatter slope which is upon the lee side. Such minor projections upon the floor usually have their origin in some harder nodule which deflects the abrading tools and causes them to pass, some on the one side and some upon the other. By this process a staple-shaped groove comes to surround the nodule, leaving an unsymmetrical elevated ridge within, which is steep upon the stoss side and slopes gently away to leeward.

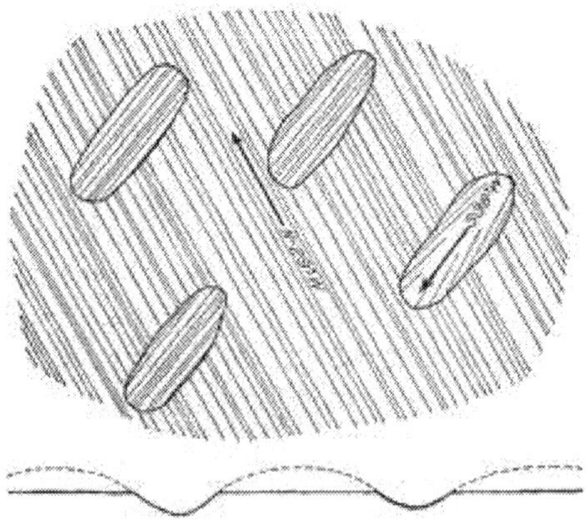

FIG. 330.—Limestone surface at Sibley, Michigan.

Younger records over older—the glacier palimpsest.—Many important historical facts have been recovered from the largely effaced writing upon ancient palimpsests, or parchments upon which an earlier record has been intentionally erased to make room for another. In the gravings upon the glacier pavement, earlier records have been likewise in large part effaced by later, though in favorable localities the two may be read together. Thus, as an example, at the great limestone quarries of Sibley, in southeastern Michigan, the glaciated rock surface wherever stripped of its drift cover is a smoothly polished and relatively level floor with striæ which are directed west-northwest. Beneath this general surface there are, however, a number of elliptical depressions which have their longer axes directed south-southwest, one being from twenty-five to thirty feet long and some ten feet in depth (Fig. 330). These boat-shaped depressions are clearly the remnants of an earlier more undulating surface which the latest glacier has in large part planed away, since the bottoms of the depressions are no less perfectly glaciated but have their striæ directed in general near the longer axis of the troughs. Palimpsest-like there are here also the records of more than one graving.

FIG. 331.—Map to show the outcroppings of peculiar rock types in the region of the Great Lakes, and some of the localities where "float copper" has been collected (float copper localities after Salisbury).

The dispersion of the drift.—Long before the "ice age" had been conceived in the minds of Agassiz and his contemporaries, it had been remarked that scattered over the North German plain were rounded fragments of rock which could not possibly have been derived from their own neighborhood but which could be matched with the great masses of red granite in Sweden well known as the "Swedish granite." Buckland, an English geologist, had in 1815 accounted for such "erratic" blocks of his own country, here of Scotch granite, by calling in the deluge of Noah; but in the late thirties of the nineteenth century, Sir Charles Lyell, with the results of English Arctic explorers in mind, claimed that such traveled blocks had been transported by icebergs emanating from the polar regions. A relic of

Buckland's earlier view we have in the word "diluvium" still occasionally used in Germany for glacier transported materials; while the term "drift" still remains in common use to recall Lyell's iceberg hypothesis, even though the original meaning of the term has been abandoned. Drift is now a generic term and refers to all deposits directly or indirectly referable to the continental glaciers.

In general the place of derivation of the glacial drift may be said to be some point more distant from and within the former ice margin at the time when it was deposited; in other words, the dispersion of the drift was centrifugal with reference to the glacier.

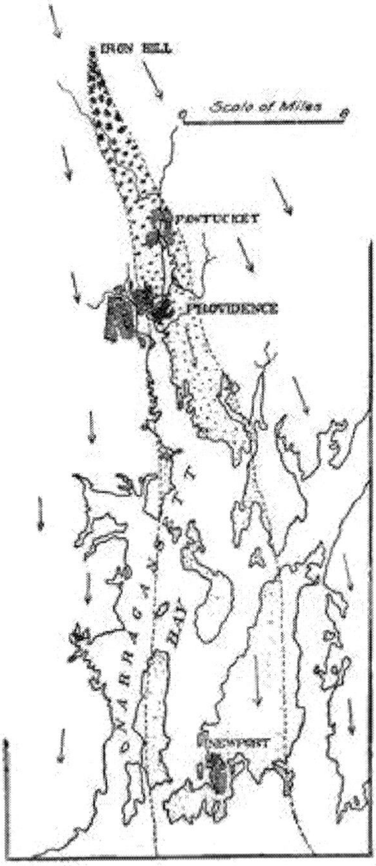

FIG. 332.—Map of the "bowlder train" from Iron Hill, R. I. (based upon Shaler's map, but with the directions of glacial striæ added).

Wherever rocks of unusual and therefore easily recognizable character can be shown to occur in place and with but limited areas, the dispersion of such material is easy to trace. The areas of red Swedish and Scotch granite have been used to follow out in a broad way the dispersion of drift over northern Europe. Within the region of the Great Lakes of North America are areas of limited size which are occupied by well marked rock types, so that the journeyings of their fragments with the continental glacier can be mapped with some care. Upon the northern shore of Georgian Bay occurs the beautiful jasper conglomerate, whose bright red pebbles in their white quartz field attract such general notice. At Ishpeming in the northern peninsula of Michigan is found the equally beautiful jaspilite composed of puckered alternating layers of black hematite and red jasper. On Keweenaw Peninsula, which protrudes into Lake Superior from its southern shore, is found that remarkable occurrence of native copper within a series of igneous rocks of varied types and colors. Fragments of this copper, some weighing several hundreds of pounds each and masked in a coat of green malachite, have under the name of "drift" or "float" copper been collected at many localities within a broad "fan" of dispersal extending almost to the very limits of glaciation (Fig. 331).

Some miles to the north of Providence in Rhode Island there is a hill known as Iron Hill composed in large part of black magnetite rock, the so-called Cumberlandite. From this hill as an apex there has been dispersed a great quantity of the rock distributed as a well marked "bowlder train" within which the size and the frequency of the dispersed bowlders is in inverse ratio to the distance from the parent ledge (Fig. 332). Similar though less perfect trains of bowlders are found on the lee side of most projecting masses of resistant rocks within the area of the drift.

Large bowlders when left upon a ledge of notably different appearance easily attract attention, and have been described as "perched bowlders." Resting as they sometimes do upon a relatively small area, they may be nicely balanced and thus easily given a pendular or rocking motion. Such "rocking stones" are common enough, especially among the New England hills (plate 17 B). Many such bowlders have made somewhat remarkable peregrinations with many interruptions, having been carried first in one direction by an earlier glacier to be later transported in wholly different directions at the time of new ice invasions.

PLATE 17.

A. **Soled glacial bowlders which show differently directed striæ upon the same facet.**

B. **Perched bowlder upon a striated ledge of different rock type, Bronx Park, New York (after Lungstedt).**

C. **Characteristic knob and basin surface of a moraine.**

The diamonds of the drift.—Of considerable popular, even if not economic, interest are the diamonds which have been sown in the drift after long and interrupted journeyings with the ice from some unknown home far to the northward in the wilderness of Canada. The first stone to be discovered was taken by workmen from a well opening near the little town of Eagle in Wisconsin in the year 1876. Its nature not being known, it remained where it was found as a curiosity only, and it was not until 1883 that it was taken to Milwaukee and sold to a jeweler equally ignorant of its value, and for the merely nominal sum of one dollar. Later recognized as a diamond of the unusual weight of sixteen carats, it was sold to the Tiffanys and became the cause of a long litigation which did not end until the Supreme Court of Wisconsin had decided that the Milwaukee jeweler, and not the finder, was entitled to the price of the stone, since he had been ignorant of its value at the time of purchase.

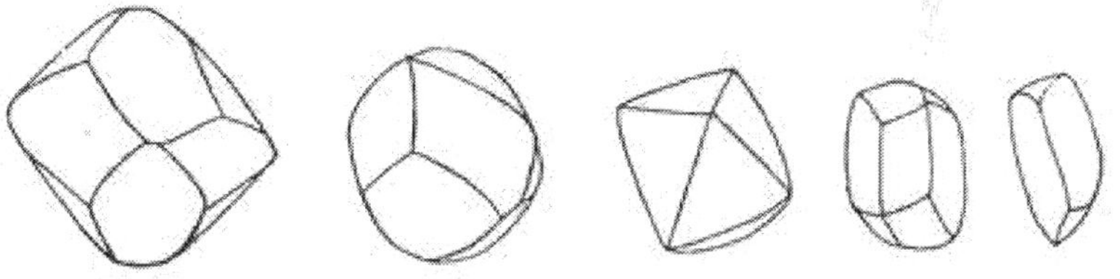

FIG. 333.—Shapes and approximate natural sizes of some of the more important diamonds from the Great Lakes region of the United States. In order from left to right these figures represent the Eagle diamond of sixteen carats, the Saukville diamond of six and one half carats, the Milford diamond of six carats, the Oregon diamond of four carats, and the Burlington diamond of a little over two carats.

An even larger diamond, of twenty-one carats weight, was found at Kohlsville, and smaller ones at Oregon, Saukville, Burlington, and Plum Creek in the state of Wisconsin; at Dowagiac in Michigan; at Milford in Ohio, and in Morgan and Brown counties in Indiana. The appearance of some of the larger stones in their natural size and shape may be seen in Fig. 333.

While the number of the diamonds sown in the drift is undoubtedly large, their dispersion is such that it is little likely they can be profitably recovered. The distribution of the localities at which stones have thus far been found is set forth upon Fig. 334. Obviously those that have been found are the ones of larger size, since these only attract attention. In 1893, when the finding of the Oregon stone drew attention to these denizens of the drift, the writer prophesied that other stones would occasionally be discovered under essentially the same conditions, and such discoveries are certain to continue in the future.

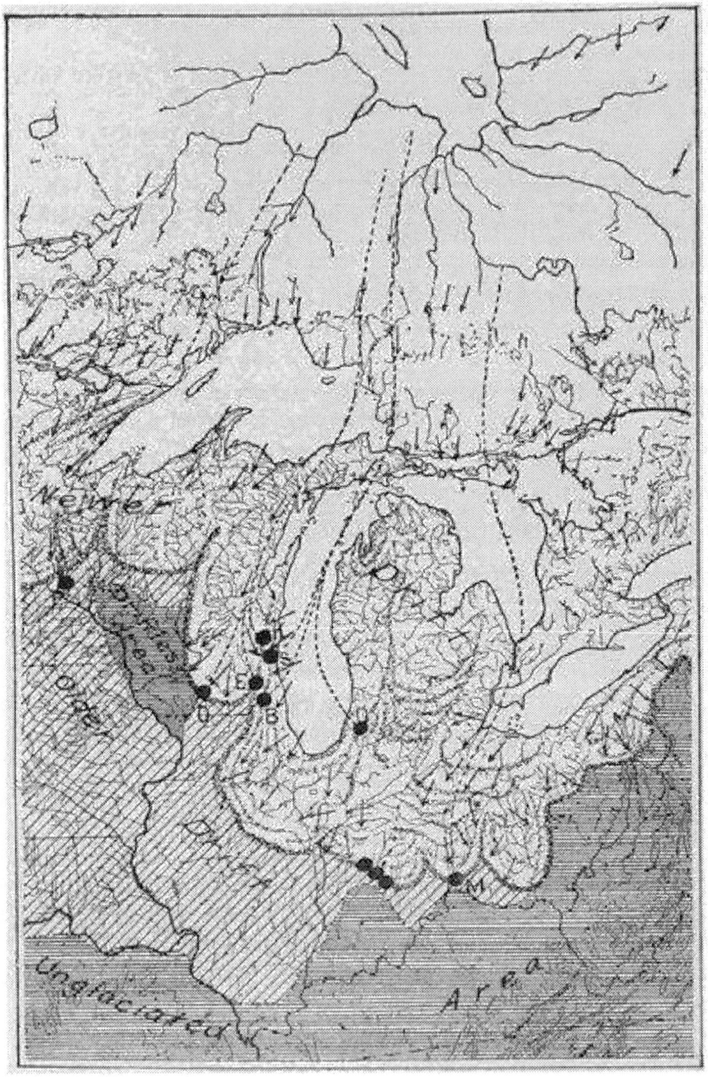

FIG. 334.—Glacial map of a portion of the Great Lakes region, showing the unglaciated area and the areas of older and newer drift. The driftless area, the moraines of the later ice invasion, and the distribution of diamond localities upon the latter are also shown. With

the aid of the directions of striæ some attempt has been made to indicate the probable tracks of more important diamonds, which tracks converge in the direction of the Labrador peninsula.

Tabulated comparison of the glaciated and nonglaciated regions.—It will now be profitable to sum up in parallel columns the contrasted peculiarities of the glaciated and the unglaciated regions.

Unglaciated Region	Glaciated Region
TOPOGRAPHY	
The topography is *destructional*; the remnants of a plain are found at the highest levels or upon the hill tops; hills are *carved* of a high plain; unstable erosion remnants are characteristic.	The topography is *constructional*; the remnants of a plain are found at the lowest levels in lakes and swamps; hills are *molded* above a plain in characteristic forms; no unstable erosion remnants, but only rounded shoulders of rock.
DRAINAGE	
The area is completely drained, and the drainage network is *arborescent.*	The area includes undrained areas,—lakes and swamps,—and the drainage system is *haphazard.*
ROCK MANTLE	
The exposed rock is decomposed and disintegrated to a considerable depth; it is all of local derivation and hence of few types—*homogeneous*; the fragments are angular; soils are leached and hence do not contain carbonates.	No decomposed or disintegrated rock is "in place", but only hard, fresh surface; loose rock material is all foreign and of many izes and types—*heterogeneous*; rock bowlders and pebbles are faceted and polished as well as striated, usually in several directions upon each facet; soils are rock flour—the grist of the glacial mill.
ROCK SURFACE	
Rock surface is rough and irregular.	Rock surface is planed or grooved, and polished. Shows glacial striæ.

Unassorted and assorted drift.—The drift is of two distinct types; namely, that deposited directly by the glacier, which is without stratification, or unassorted; and that deposited by water flowing either beneath or from the ice, and this like most fluid deposited material is assorted or stratified. The unassorted material is described as *till*, or sometimes as "bowlder clay"; the assorted is sand or gravel, sometimes with small included bowlders, and is described as *kame gravel*. To recall the parts which both the glacier and the streams have played in its deposition, all water-deposited materials in connection with glaciers are called *fluvio-glacial*.

FIG. 335.—Section in coarse till. Note the range in size of the materials, the lack of stratification, and the "soled" form of the bowlders.

Till is, then, characterized by a noteworthy lack of homogeneity, both as regards the size and the composition of its constituent parts. As many as twenty different rock types of varied textures and colors may sometimes be found in a single exposure of this material, and the entire gamut is run from the finest rock flour upon the one hand to bowlders whose diameter may be measured in feet (Fig. 335).

In contrast with those derived by ordinary stream action, the pebbles and bowlders of the till are faceted or "soled", and usually show striations upon their faces. If a number of pebbles are examined, some at least are sure to be found with striations in more than one direction upon a single facet. As a criterion for the discrimination of the material this may be an important mark to be made use of to distinguish in special cases from rock fragments derived by brecciation and slickensiding and distributed by the torrents of arid and semiarid regions.

Inasmuch as the capacity of ice for handling large masses is greater than that of water, assorted drift is in general less coarse, and, as its name implies, it is also stratified. From ordinary stream gravels, the kame gravels are distinguished by the form of their pebbles, which are generally faceted and in some cases striated. In proportion, however, as the materials are much worked over by the water, the angles between pebble faces become rounded and the original shapes considerably masked.

Features into which the drift is molded.—Though the preëxisting valleys were first filled in by drift materials, thus reducing the accent of the relief, a continuation of the same process resulted in the superimposition of features of characteristic shapes upon the imperfectly evened surface of the earlier stages. These features belong to several different types, according as they were built up outside of, at and upon, or within the glacier margin. The extra-marginal deposits are described as *outwash plains* or *aprons*, or sometimes as *valley trains*; the marginal are either *moraines* or *kames*; while within the border were formed the *till plain* or *ground moraine*, and, locally also, the *drumlin* and the *esker* or *os*. These characteristic features are with few exceptions to be found only within the area covered by the latest of the ice invasions. For the earlier ones, so much time has now elapsed that the effect of weathering, wash, and stream erosion has been such that few of the features are recognizable.

Marginal and extra-marginal features are extended in the direction of the margin or, in other words, perpendicular to the local ice movement; while the intra-marginal deposits are as noteworthy for being perpendicular to the margin, or in correspondence with the direction of local ice movement. Each of these features possesses characteristic marks in its form, its size, proportions, surface molding and orientation, as well as in its constituent materials. It should perhaps be pointed out that the existing continental glaciers, being in high latitudes, work upon rock materials which have been subjected to different weathering processes from those characteristic of temperate latitudes. Moreover, the melting of the Pleistocene glaciers having taken place in relatively low latitudes, larger quantities of rock débris were probably released from the ice during the time of definite climatic changes, and hence heavier drift accumulations have for both of these reasons resulted.

Marginal or "kettle" moraines.—Wherever for a protracted period the margin of the glacier was halted, considerable deposits of drift were built up at the ice margin. These accumulations form, however, not only about the margin, but upon the ice surface as well; in part due to materials collected from melting down of the surface, and in part by the upturning of ice layers near the margin (see *ante*, p. 277).

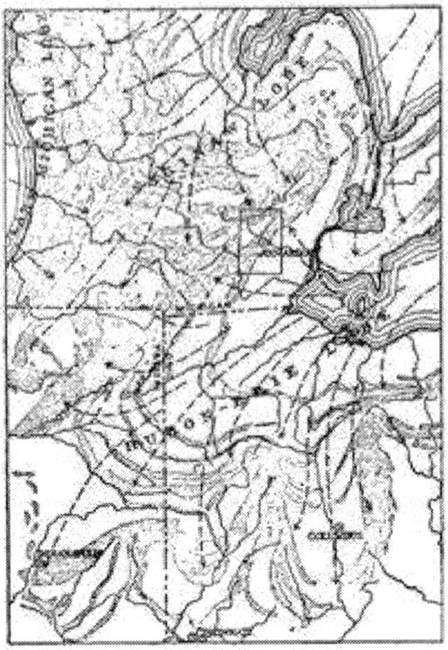

FIG. 336.—Sketch map of portions of Michigan, Ohio, and Indiana, showing the festooned outlines of the moraines about the former ice lobes, and the directions of ice movement as determined by the striæ upon the rock pavement (after Leverett).

An important rôle is played by the thaw water which emerges at the ice margin, especially within the reëntrants or recesses of the outline. The materials of moraines are, therefore, till with large local deposits of kame gravel, and these form in a series of ridges corresponding to the temporary positions of the ice front. Their width may range from a few rods to a few miles, their height may reach a hundred feet or more, and they stretch across the country for distances of hundreds or even thousands of miles, looped in arcs or scallops which are always convex outward and which meet in sharp cusps that in a general way point toward the embossment of the former glacier (Fig. 334, p. 308, and Fig. 336). These festoons of the moraines outline the ice lobes of the latest ice invasion, which in North America were centered over the depressions now occupied by the Laurentian lakes. There was, thus, a Lake Superior lobe, a Lake Michigan lobe, etc. With the aid of these moraine maps we may thus in imagination picture in broad lines the frontal contours of the earlier glaciers. At specially favorable localities where the ice front has crossed a deep valley at the edge of the Driftless Area, we may, even in a rough way, measure the slope of the ice face. Thus near Devils Lake in southern Wisconsin the terminal moraine crosses the former valley of the Wisconsin River, and in so doing has dropped a distance of about four hundred feet within the distance of a half mile or thereabouts (Fig. 337).

The characteristic surface of the marginal moraine is responsible for the name "kettle" moraine so generally applied to it. The "kettles" are roughly circular, undrained basins which lie among hummocks or knobs, so that the surface has often been referred to as "knob and basin" topography (plate 17 C).

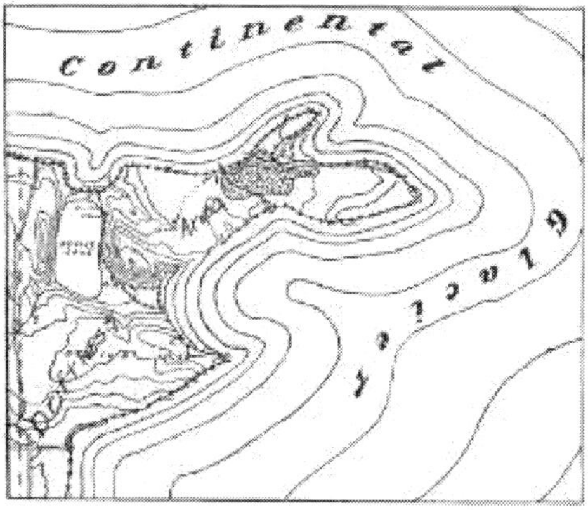

FIG. 337.—Map of the vicinity of Devils Lake, Wisconsin, located within a reëntrant of the "kettle" moraine upon the margin of the Driftless Area. The lake lies within an earlier channel of the Wisconsin River which has been blocked at both ends, first by the glacier and later by its moraine. The stippled area upon the heights and next the moraine represents the clay deposits of a former lake (based on map by Salisbury and Atwood).

FIG. 338.—Moraine with outwash apron in front, the latter in part eroded by a river. Westergötland, Sweden (after H. Munthe).

FIG. 339.—Fosse between an outwash plain (in the foreground) and the moraine, which rises to the left in the middle distance. Ann Arbor, Michigan.

Kames.—Within reëntrants or recesses of the ice margin the drift deposits were especially heavy, so that high hills of hummocky surface have been built up, which are described as *kames*. Most of the higher drift hills have this origin. They rarely have any principal extension along a single direction, but are composed in large part of assorted materials. In contrast with other portions of the morainal ridges they lack the prominent basins known as kettles. Other *kames* are high hills of assorted materials not in direct association with moraines and believed to have been built up beneath glacier wells or mills (p. 278).

Outwash plains.—Upon the outer margin of the moraine is generally to be found a plain of glacial "outwash" composed of sand or gravel deposited by the braided streams (Fig. 308, p. 280) flowing from the glacier margin. Such plains, while notably flat (Fig. 338), slope gently away from the moraine. Between the outwash plain and the moraine there is sometimes found a pit, or *fosse* (Fig. 309, p. 281), where a part of the ice front was in part buried in its own outwash (Fig. 339).

FIG. 340.—View looking along an esker in southern Maine (after Stone).

Pitted plains and interlobate moraines.—Where glacial outwash is concentrated within a long and narrow reëntrant, separating glacial lobes, strips of high plain are sometimes built up which overtop the other glacial deposits of the district. The sand and gravel which compose such plains have a surface which is pitted by numerous deep and more or less circular lakes, so that the term "pitted plain" has been applied to them. The surface of such a plain steadily rises toward its highest point in the angle between the ice lobes. Though consisting almost entirely of assorted materials, and built up largely without the ice margins, such gently sloping pitted platforms are described as *interlobate moraines*. Upon a topographic map the course of such an interlobate moraine may often be followed by the belts of small pit lakes (see Fig. 336).

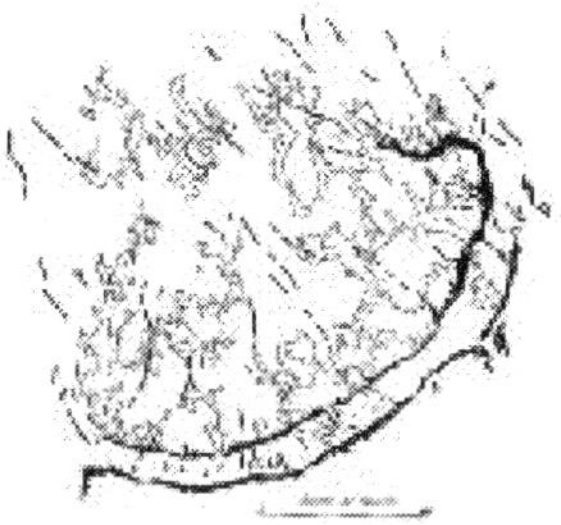

FIG. 341.—Outline map showing the eskers of Finland trending southeasterly toward the festooned moraines at the margin of the ice. The characteristic lakes of a glaciated region appear behind the moraines (after J. J. Sederholm).

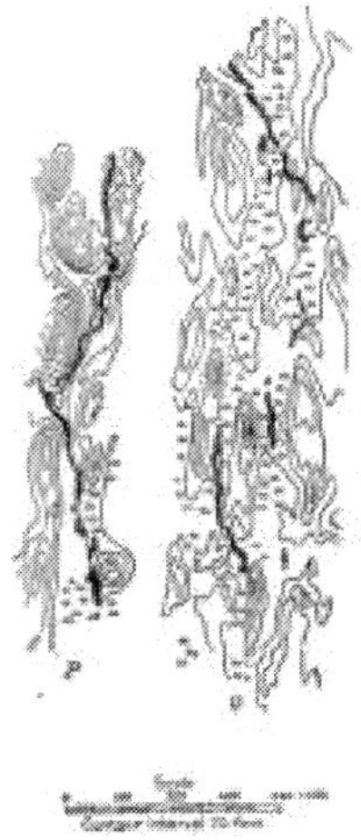

FIG. 342.—Small sketch maps showing the relationships in size, proportions, and orientation of drumlins and eskers in southern Wisconsin. The eskers are in solid black (after Alden).

Eskers.—Intra-morainal features, or those developed beneath the glacier but relatively near its margin, include the "serpentine kame", *esker*, or, as it is called in Scandinavia, the *os* (plural *osar*) (Fig. 340). These diminutive ridges have a width seldom exceeding a few rods, and a height a few tens of feet at most, but with slightly sinuous undulations they may be followed for tens or even hundreds of miles in the general direction of the local ice movement (Fig. 341). They are composed of poorly stratified, thick-bedded sands, gravels, and "worked over" materials, and are believed to have been formed by subglacial rivers which flowed in tunnels beneath the ice. Inasmuch as the deposits were piled against the ice walls, the beds were disturbed at the sides when these walls disappeared, and the stratification, which was somewhat arched in the beginning, has been altered by sliding at both margins. As already stated, eskers have not a general distribution within the glaciated area, but are often found in great numbers at specially favored localities. Formed as they are beneath the ice, it is believed that many have their materials redistributed so soon as uncovered at the glacier margin, because of the vigorous drainage there. They are thus to be found only at those favored localities where for some reason border drainage is less active, or where the ice ended in a body of water.

Drumlins.—A peculiar type of small hill likewise found behind the marginal moraine in certain favored districts has the form of an inverted boat or canoe, the long axis of which is parallel to the direction of ice movement, as is that of the esker (Fig. 342). Unlike the esker, this type of hill is composed of till, and from being found in Ireland it is called a *drumlin*, the Irish word meaning a little hill (Fig. 343). Drumlins are usually found in groups more or less radial and not far behind the outermost moraine, to which their radiating axes are perpendicular. The manner of their formation is involved in some uncertainty, but it is clear that they have been formed beneath the margin of the glacier, and have been given their shape by the last glacier which occupied the district.

The mutual relationships of nearly all the molded features resulting from continental glaciation may be read from Fig. 344.

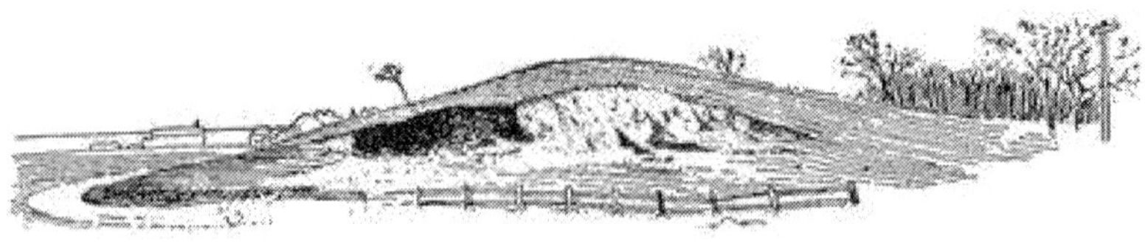

FIG. 343.—View of a drumlin, showing an opening in the till. Near Boston, Massachusetts (after Shaler and Davis).

The shelf ice of the ice age.—Shelf ice, such as we have become familiar with in Antarctica as a marginal snow-ice terrace floating upon the sea, no doubt existed during the ice age above the Gulf of Maine (see Fig. 324, p. 298), and perhaps also over the deep sea to the westward of Scotland. Though the inland ice probably covered the North Sea, and upon the American side of the Atlantic the Long Island Sound, both these basins are so shallow that the ice must have rested upon the bottom, for neither is of sufficient depth to entirely submerge one of the higher European cathedrals.

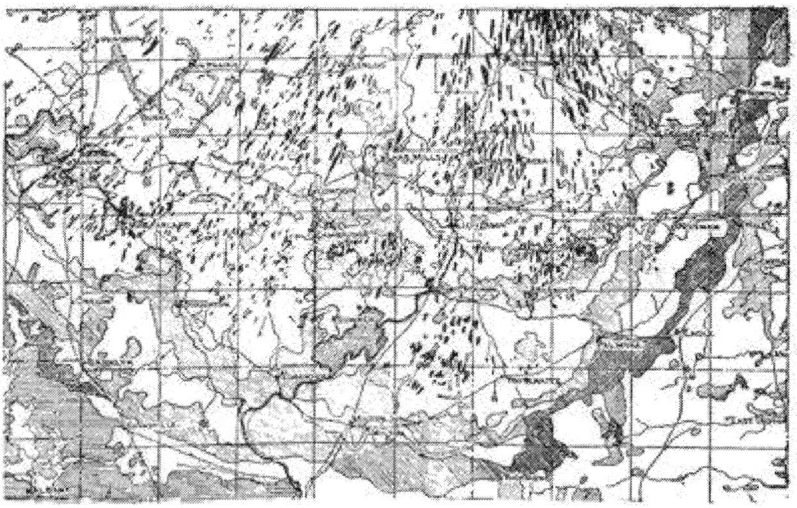

FIG. 344.—Outline map of the front of the Green Bay lobe of the latest continental glacier of the United States. Drumlins in solid black, moraines with diagonal hachure, outwash plains and the till plain or ground moraine in white (after Alden).

Character profiles.—All surface features referable to continental glaciers, whether carved in rock or molded from loose materials, present gently flowing outlines which are convex upward (Fig. 345). The only definite features carved from rock are the *roches moutonnées*, with their flattened shoulders, while the hillocks upon moraines and kames, and the drumlins as well, approximate to the same profile. The esker in its cross sections is much the same, though its serpentine extension may offer some variety of curvature when viewed from higher levels.

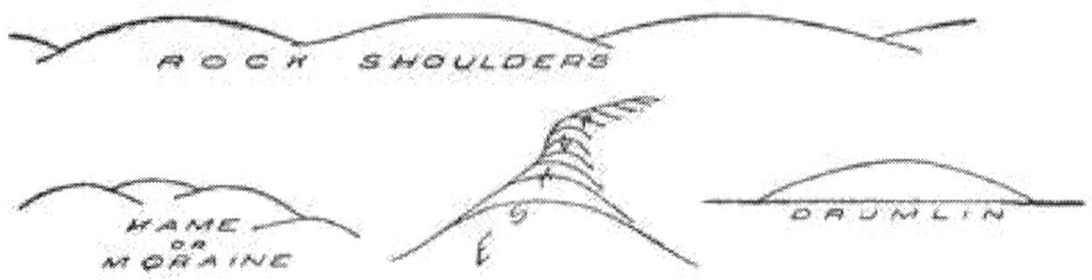

FIG. 345.—Character profiles referable to continental glacier.

READING REFERENCES FOR CHAPTER XXII

General:—

JAMES GEIKIE. The Great Ice Age. 3d ed. London, 1894, pp. 850, maps 18.

CHAMBERLIN and SALISBURY. Geology, vol. 3, 1906, pp. 327-516.

FRANK LEVERETT. The Illinois Glacial Lobe, Mon. 38, U. S. Geol. Surv., 1899, pp. 817, pls. 34; Glacial formations and Drainage Features of the Erie and Ohio Basins, Mon. 41, *ibid.*, 1902, pp. 802, pls. 25; Comparison of North American and European Glacial Deposits, Zeit. f. Gletscherk., vol. 4, 1910, pp. 241-315, pls. 1-5.

Former glaciations previous to Ice Age:—

A. STRAHAN. The Glacial Phenomena of Paleozoic Age in the Varanger Fjord, Quart. Jour. Geol. Soc., London, vol. 53, 1897, pp. 137-146, pls. 8-10.

BAILEY WILLIS and ELIOT BLACKWELDER. Research in China, Pub. 54, Carnegie Inst. Washington, vol. 1, 1907, pp. 267-269, pls. 37-38.

A. P. COLEMAN. A Lower Huronian Ice Age, Am. Jour. Sci. (4), vol. 23, 1907, pp. 187-192.

W. M. DAVIS. Observations in South Africa, Bull. Geol. Soc. Am., vol. 17, 1906, pp. 377-450, pls. 47-54.

DAVID WHITE. Permo-Carboniferous Climatic Changes in South America, Jour. Geol., vol. 15, 1907, pp. 615-633.

Driftless and drift areas:—

T. C. CHAMBERLIN and R. D. SALISBURY. Preliminary Paper on the Driftless Areas of the Upper Mississippi Valley, 6th Ann. Rept. U. S. Geol. Surv., 1885, pp. 199-322, pls. 23-29.

R. D. SALISBURY. The Drift, its Characteristics and Relationships, Jour. Geol., vol. 2, 1894, pp. 708-724, 837-851.

R. H. WHITBECK. Contrasts between the Glaciated and the Driftless Portions of Wisconsin, Bull. Geogr. Soc., Philadelphia, vol. 9, 1911, pp. 114-123.

Glacier gravings:—

T. C. CHAMBERLIN. The Rock Scorings of the Great Ice Invasions, 7th Ann. Rept. U. S. Geol. Surv., 1888, pp. 147-248, pl. 8.

The dispersion of the drift:—

R. D. SALISBURY. Notes on the Dispersion of Drift Copper, Trans. Wis. Acad. Sci., etc., vol. 6, 1886, pp. 42-50, pl.

N. S. SHALER. The Conditions of Erosion beneath Deep Glaciers, based upon a Study of the Bowlder Train from Iron Hill, Cumberland, Rhode Island, Bull. Mus. Comp. Zoöl. Harv. Coll., vol. 16, No. 11, 1893, pp. 185-225, pls. 1-4 and map.

WILLIAM H. HOBBS. The Diamond Field of the Great Lakes, Jour. Geol., vol. 7, 1899, pp. 375-388, pls. 2 (also Rept. Smithson. Inst., 1901, pp. 359-366, pls. 1-3).

Glacial features:—

T. C. CHAMBERLIN. Preliminary Paper on the Terminal Moraine of the Second Glacial Epoch, 3d Ann. Rept. U. S. Geol. Surv., 1883, pp. 291-402, pls. 26-35.

G. H. STONE. Glacial Gravels of Maine and their Associated Deposits, Mon. 34, U. S. Geol. Surv., 1899, pp. 489, pls. 52.

W. C. ALDEN. The Delaven Lobe of the Lake Michigan Glacier of the Wisconsin Stage of Glaciation and Associated Phenomena. Prof. Pap. No. 34, U. S. Geol. Surv., 1904, pp. 106, pls. 15; The Drumlins of Southeastern Wisconsin, Bull. 273, U. S. Geol. Surv., 1905, pp. 46, pls. 9.

W. M. DAVIS. Structure and Origin of Glacial Sand Plains, Bull. Geol. Soc. Am., vol. 1, 1890, pp. 196-202, pl. 3; The Subglacial Origin of Certain Eskers, Proc. Bost. Soc. Nat. Hist., vol. 35, 1892, pp. 477-499.

F. P. GULLIVER. The Newtonville Sand Plain, Jour. Geol., vol. 1, 1893, pp. 803-812.

CHAPTER XXIII

GLACIAL LAKES WHICH MARKED THE DECLINE OF THE LAST ICE AGE

FIG. 346.—The Illinois River where it passes through the outer moraine at Peoria, Illinois, showing the flood plain of the ancient stream as an elevated terrace into which the modern stream has cut its gorge (after Goldthwait).

Interference of glaciers with drainage.—Every advance and every retreat of a continental glacier has been marked by a complex series of episodes in the history of every river whose territory it has invaded. Whenever the valley was entered from the direction of its divide, the effect of the advancing ice front has generally been to swell the waters of the river into floods to which the present streams bear little resemblance (Fig. 346). Because of the excessive melting, this has been even more true of the ice retreat, but here *when the ice front retired up the valley* toward the divide. A sufficiently striking example is furnished by the Wabash, Kaskaskia, Illinois, and other streams to the southward of the divide which surrounds the basin of the Great Lakes (Fig. 347).

FIG. 347.—Broadly terraced valleys outside the divide of the St. Lawrence basin, which remain to mark the floods that issued from the latest continental glacier during its retreat (after Leverett).

Wherever the relief was small there occurred in the immediate vicinity of the ice front a temporary diversion of the streams by the parallel moraines, so that the currents tended to parallel the ice front. This temporary diversion known as "border drainage" was brought to a close when the partially impounded waters had, by cutting their way through the moraines, established more permanent valleys (Fig. 348).

Temporary lakes due to ice blocking.—Whenever, on the contrary, the advancing ice front entered a valley from the direction of its mouth, or a *retreating ice front retired down the valley*, quite different results followed, since the waters were now impounded by the ice front serving as a dam. Though the histories of such blocking of rivers are often quite complex, the principles which underlie them are in reality simple enough. Of the lakes formed during advancing hemicycles of glaciation, and of all save the latest receding hemicycle, no satisfactory records are preserved, for the reason that the lake beaches and the lake deposits were later disturbed and buried by the overriding ice sheets. We have, however, every reason to suppose that the histories of each of these hemicycles were in every way as complex and interesting as that of the one which we are permitted to study.

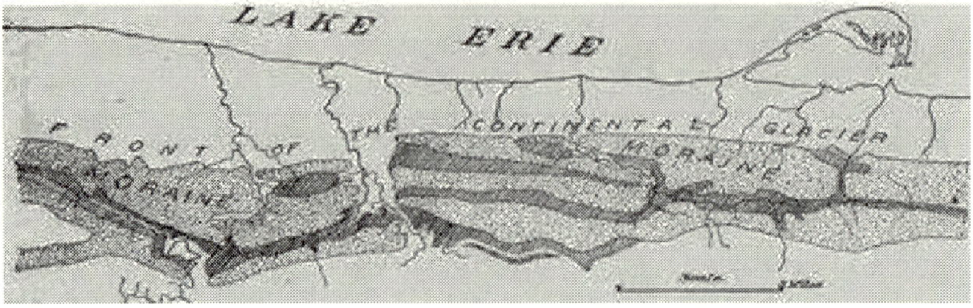

FIG. 348.—Border drainage about the retreating ice front south of Lake Erie. The stippled areas are the morainal ridges and the hachured bands the valleys of border drainage (after Leverett).

As an introduction to the study of the ice-blocked lakes of North America, and to set forth as clearly as may be the fundamental principles upon which such lakes are dependent, we shall consider in some detail the late glacial history of certain of the Scottish glens, since their area is so small and the relief so strong that relationships are more easily seen; it is, so to speak, a pocket edition of the history of the more extended glacial lakes.

FIG. 349.—The "parallel roads" of Glen Roy in the southern highlands of Scotland (after Jamieson).

The "parallel roads" of the Scottish glens.—In a number of neighboring glens within the southern highlands of Scotland there are found faint terraces upon the glen walls which under the name of the "parallel roads" (Fig. 349) have offered a vexed problem to scientists. Of the many scientists who long attempted to explain them, though in vain, was Charles Darwin, the father of modern evolution. He offered it as his view that the "roads" were beaches formed at a time when the sea entered the glens and stood at these levels. When, however, Jamieson's studies had discovered their true history, Darwin, with a frankness characteristic of some of the greatest scientists, admitted how far astray he had been in his reasoning. Let us, then, first examine the facts, and later their interpretation. The map of Fig. 350 will suffice to set forth with sufficient clearness the course of the several "roads." These "roads" are found in a number of glens tributary to Loch Lochy, and of the three neighboring valleys, Glen Roy has three, Glen Glaster two, and Glen Spean one "road." The facts of greatest significance in arriving at their interpretation relate to their elevations with reference to the passes at the valley heads, their abrupt terminations down-valleyward, and the morainic accumulations which are found where they terminate. The single "road" of Glen Spean is found at an elevation of 898 feet, a height which corresponds to that of the pass or col at the head of its valley and to the lowest of the "roads" in both Glens Glaster and Roy. Similarly the upper of the two "roads" in Glen Glaster is at the height of the pass at its head (1075 feet) and corresponds in elevation to the middle one of the three "roads" in Glen Roy. Lastly, the highest of the "roads" in Glen Roy is found at an elevation of 1151 feet, the height of the col at the head of the Glen. In the neighboring Glen Gloy is a still higher "road" corresponding likewise in elevation to that of the pass through which it connects with Glen Roy.

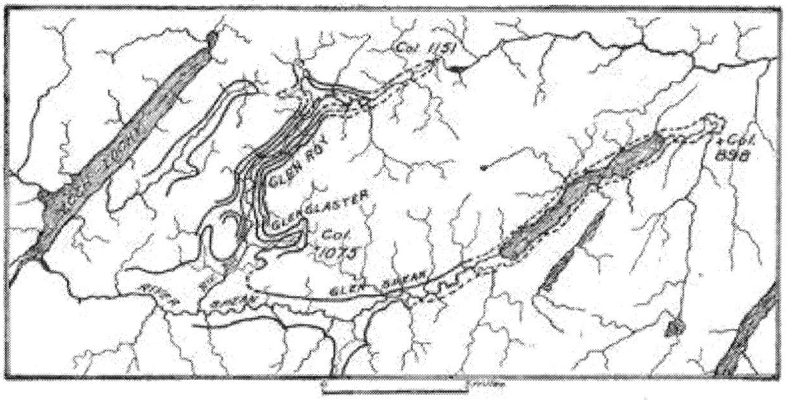

FIG. 350.—Map of Glen Roy and neighboring valleys of the Scottish highlands with the so-called "roads" entered in heavy lines. Glens Roy, Glaster, and Spean have three "roads", two "roads", and one "road", respectively (after Jamieson).

To come now to the explanation of the "roads", it may be said at the outset that they are, as Darwin supposed, beach terraces cut by waves, not as he believed of the ocean, but of lakes which once filled portions of the glens when glaciers proceeding from Ben Nevis to the southwestward were blocking their lower portions. The several episodes of this lake history will be clear from a study of the three successive idealistic diagrams in Fig. 351.

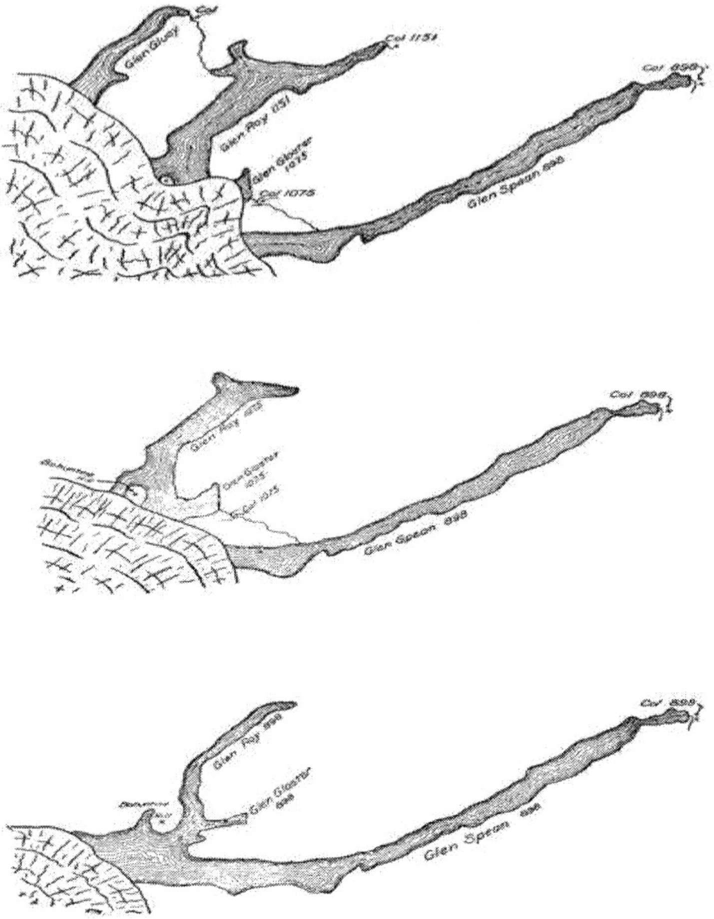

FIG. 351.—Three successive diagrams to set forth in order the late glacial lake history of the Scottish glens.

To derive the principles underlying this history, it is at once seen that *all changes are initiated by the retirement of the ice front to such a point that it unblocks for the waters of a lake an outlet that is lower than the one in service at the time.* This is the principle which explains nearly all episodes of glacial lake history. Thus, when the ice front had retired so as to open direct connections between Glen Roy and Glen Glaster, the col at the head of Glen Roy was abandoned as an outlet, and the waters fell to the level fixed for Glen Glaster. A still further retirement at last opened direct connection between Glen Glaster and Glen Spean, so that the lake common to Glens Glaster and Roy fell to the level of the col which was the outlet of the Spean valley at the time. This stage continued until the ice front had retired so far that the waters drained naturally down the river Spean to Loch Lochy and thence to the ocean.

FIG. 352.—Harvesting time on the fertile floor of the glacial Lake Agassiz (after Howell).

Only in their far grander scale and in the lesser relief of the land over which they formed, do the complex histories of the great ice-blocked lakes of North America differ from these little valley lakes whose beaches may be visited and the relationships worked out, thanks to Jamieson, in a single day's strolling.

FIG. 353.—Map of Lake Agassiz (after Upham).

The glacial Lake Agassiz.—The grandest of the temporary lakes referable to blocking by the continental glaciers of the ice age must be looked for in the largest valleys that lay within the territory invaded and *which normally drain toward the retiring ice front.* In North America these rivers are the Red River of the North in Minnesota, the Dakotas, and Manitoba; and the St. Lawrence River system. To the ice dam which lay across the Red River valley we owe the fertility of that vast plain of lake deposits where is to-day the most intensive wheat farming of the northwest (Fig. 352). Lakes Winnipeg, Winnipegoosis, and Manitoba, and the Lake of the Woods, are all that now remain of this greatest of the glacial lakes, which in honor of the distinguished founder of the glacial theory has been called Lake Agassiz (Fig. 353). With their natural outlet blocked by the ice in northern Manitoba and Keewatin, the waters of the Red were swollen by melting from the retiring glacier and spread over a vast area before finding a southern outlet along the course of the present Lake Traverse and the valley of the Minnesota River. Along this route there flowed a mighty flood which carved out a broad valley many times too large for the Minnesota, its present occupant, and this giant prehistoric river has been called the Warren River (Fig. 354).

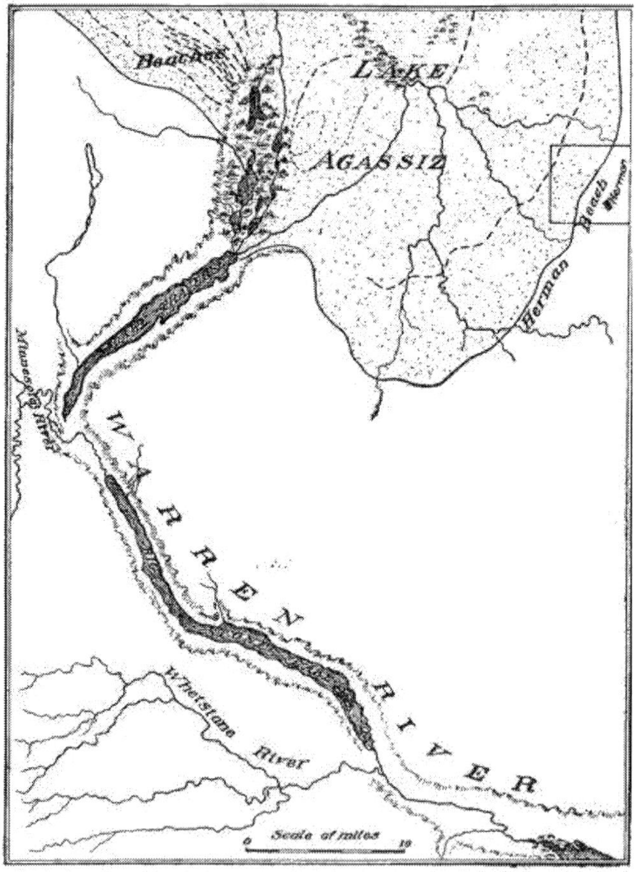

FIG. 354.—Map of the southern end of the Lake Agassiz basin, showing the position of some of the beaches and the outlet through the former Warren River (after Upham).

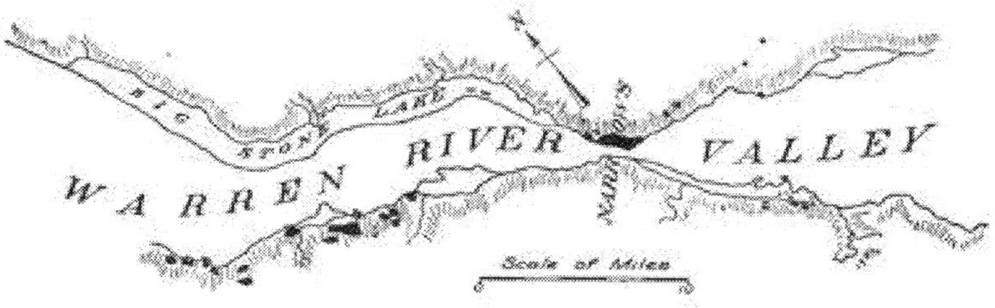

FIG. 355.—Narrows of the Warren River below Big Stone Lake, where it passed between jaws of hard granite and gneiss (after Upham).

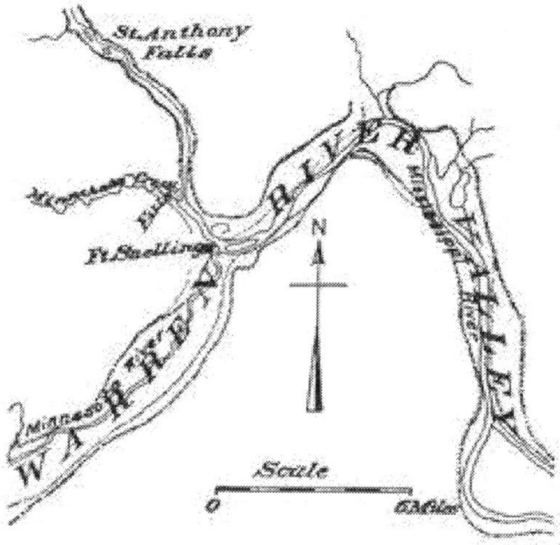

FIG. 356.—Map of the valley of the Warren River in the vicinity of Minneapolis, with the young valley of the Mississippi entering it at Fort Snelling (after Sardeson).

It is interesting to follow this ancient waterway and to discover that, like our normal, present-day streams, it was held up in narrows wherever outcroppings of harder rock had constricted its channel (Fig. 355). The upper end of the Warren River valley is now occupied by the long and relatively narrow Lakes Traverse and Big Stone, each the result of blocking by delta deposits where a tributary stream has emerged into the valley, but this gigantic channel continues down to and beyond Minneapolis, occupied as far as Fort Snelling by the Minnesota River—a mere pygmy compared to its predecessor. To the earnest student of glacial geology there can be few sights more impressive than are obtained by standing at Fort Snelling, just above the confluence of the Minnesota and the Mississippi rivers, and surveying first the steep and narrow valley of the Mississippi above the junction,—a stream fitted to its valley for the simple reason that it has carved it,—and then gazing up and down that broad valley in which the great Warren River once flowed majestically to the sea, now the bed of the Minnesota above the Fort and of the Mississippi below it (Fig. 356).

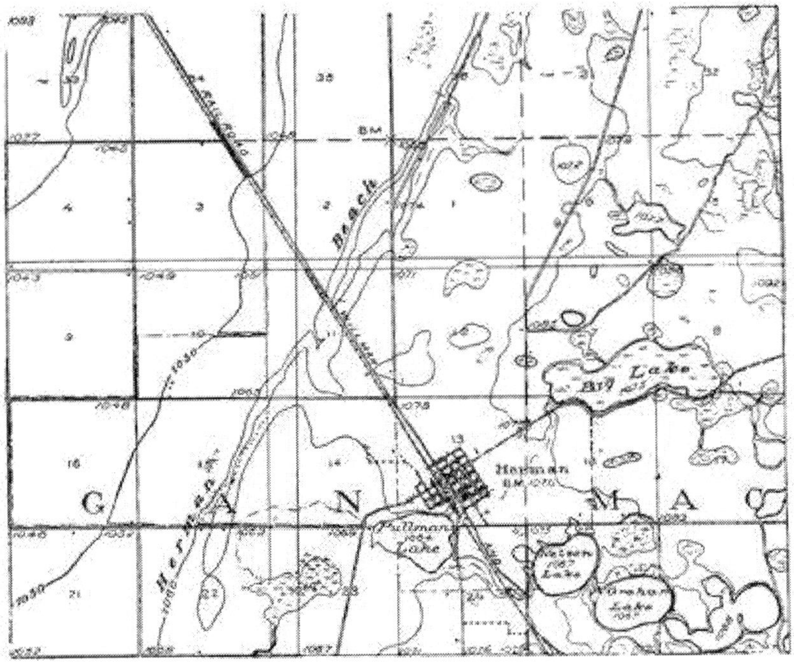

FIG. 357.—Portion of the Herman quadrangle of Minnesota, showing the position of the Herman beach on the shore of the former Lake Agassiz. The lake basin is to the left, and the pitted morainal deposits appear to the right (U. S. G. S.).

Just as the "parallel roads" of Glen Roy, roads in name only, are the beaches of earlier glacial lake stages, so in Lake Agassiz we have parallel beaches of the barrier type which are often roads in fact as well as in name, and which mark the stages of successive lakes within this vast basin. The Herman beach, corresponding to the highest level of the lake, is thus a sharp topographic boundary between lake deposits and morainal accumulations, and is further itself a well-marked topographic feature composed of wave-washed and hence well-drained materials (Fig. 357). Farmers of the district have been quick to realize that these level and slightly elevated ridges lack the clay which would render them muddy in the wet seasons, and are thus ideally adapted for roads. They have in many sections been thus used over long stretches and are known as the "ridge roads."

Episodes of the glacial lake history within the St. Lawrence valley.—Within this great drainage basin it has apparently been possible to read the records of each stage in the latest lake history—complex as this has been. We have only to recall the lake stages cited from the Scottish glens and remember that each new stage was begun in a retirement of the glacier front which unblocked an outlet of lower level than the last. This sequence might, however, have been varied by a temporary readvance of the ice, as indeed once occurred in the Huron-Erie lobe of the great North American glacier.

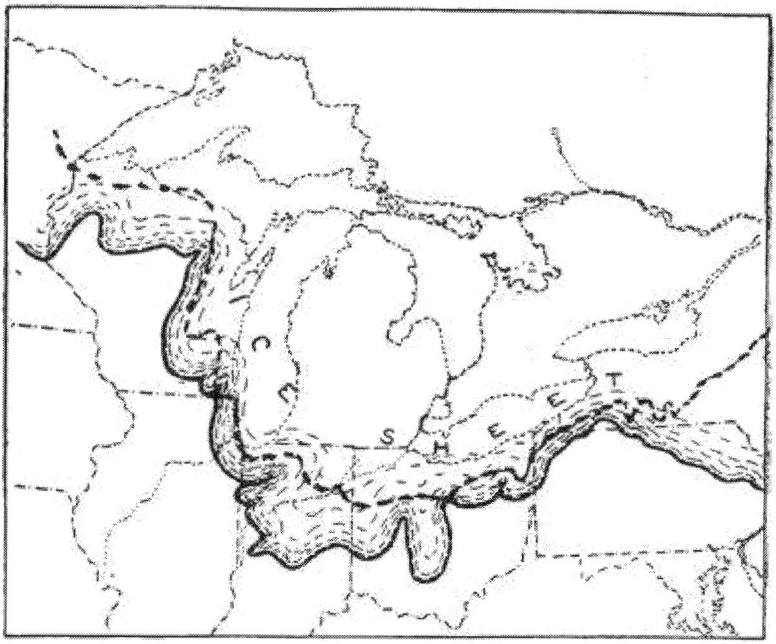

FIG. 358.—The continental glacier of North America in an early stage of its recession, when it covered the entire St. Lawrence drainage basin. The dashed line is the approximate position of the divide (based on a map by Goldthwait).

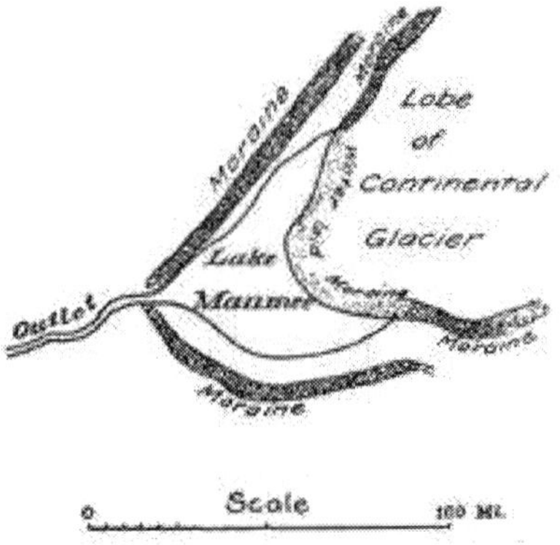

FIG. 359.—Outline map of the early Lake Maumee, with the bordering moraine and the water-laid moraine remaining on the site of the former ice cliff.

The crescentic lakes of the earlier stages.—So long as the glacier covered the entire drainage basin of the St. Lawrence River system, all water was freely drained away by streams which flowed *away from* the ice front (Fig. 358). So soon, however, as at any point the front had retired behind the divide, impounding of the waters must locally have occurred. Lakes of this type are to-day to be seen in Greenland and in the southern Andes; and though upon a diminutive scale, some idea of their aspect may be obtained from the appearance of the Märjelen Lake of Switzerland, here blocked by a mountain glacier (Fig. 446, p. 411). Within all areas of small relief, such as the prairie country surrounding the present Laurentian lakes, the earlier and smaller stages of such ice-blocked lakes are generally crescentic in outline. This is because a moraine in most cases forms the land margin of the lake, and because the ice cliff upon the opposite border, although somewhat straightened, as a consequence of wave-cutting and iceberg formation, still retains the convex outlines characteristic of ice lobes (Fig. 359).

FIG. 360.—Map to show the first stages of the ice-dammed lakes within the St. Lawrence basin (after Leverett and Taylor).

Within each of the Great Lake basins a crescentic lake early appeared at that end of the depression which was first uncovered by the glacier: Lake Duluth in the Superior basin, Lake Chicago in the Michigan basin, and Lake Maumee in the Huron-Erie basin (Fig. 360).

We may now, with profit, trace the successive episodes of the glacial lake history, considering for the earlier stages those changes which occurred within the Huron-Erie basin, since, these are in essential respects like those of the Michigan and Superior basins, although worked out in greater detail. Lake Chicago must, however, be brought into consideration, since in all save the earliest and the later stages, the waters from the Huron-Erie depression were discharged through the Grand River into this lake and thence by the so-called "Chicago outlet" into the Mississippi (plate 20 A).

The early Lake Maumee.—The area, outline, and outlet of this lake are indicated upon Fig. 360. Its ancient beaches have been traced, as well as the water-laid moraine beneath its former ice cliff; and no observant traveler who should take his way down the ancient outlet from Fort Wayne, Indiana, past the town of Huntington, could fail to be impressed by its size, suggesting as it does the great volume of water which must once have flowed along it. Now a channel a mile or more in width, its bed for the twenty-five miles between Fort Wayne and Huntington may be seen from the tracks of the Wabash Railway as a series of swamps merely, while at Huntington the Wabash river enters by a young V-shaped valley at the side, much as the Mississippi emerges into the old channel of the Warren River at Fort Snelling, Minnesota (see p. 327).

The Huron River of southern Michigan, which now discharges into Lake Erie, then found its lower course blocked by the glacier and was thus compelled to find a southerly directed channel now easily followed to the northern horn of the crescent of Lake Maumee.

The later Lake Maumee.—When the ice lobe had retired its front sufficiently, an outlet lower than that at Fort Wayne was uncovered past the city of Imlay, Michigan, into the Grand River, and thence through Lake Chicago and its outlet into the Mississippi. This old outlet south of Chicago follows the course of the present Drainage Canal and the line of the Chicago & Alton Railway. The traveler journeying southward by train from Chicago has thus the opportunity of observing first the beaches of the former lake, and then the several channels which were joined in the main outlet at the station of Sag (plate 20 A).

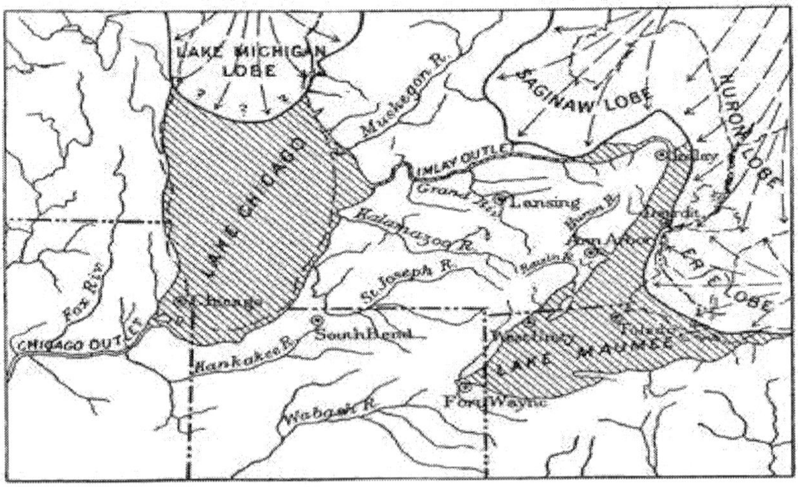

FIG. 361.—Outline map of the later Lake Maumee and of its "Imlay outlet" to Lake Chicago (after Leverett).

In this stage of our history Lake Maumee pushed a shrunk arm up past the site of Ypsilanti in Michigan (Fig. 361), the well-marked beach being found on Summit Street opposite the State Normal College. The Huron River, which in the first lake stage had followed the valley now occupied by the Raisin River southward into Indiana, now discharged directly into a bay upon this arm of Lake Maumee, and so formed a delta at Ann Arbor.

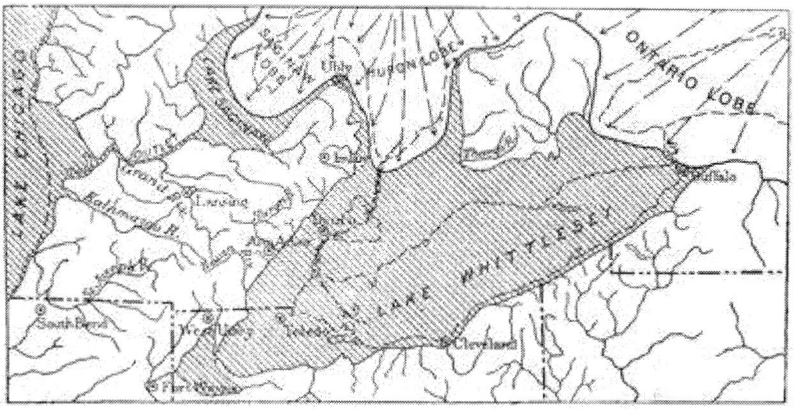

FIG. 362.—Outline map of Lakes Whittlesey and Saginaw (after Leverett).

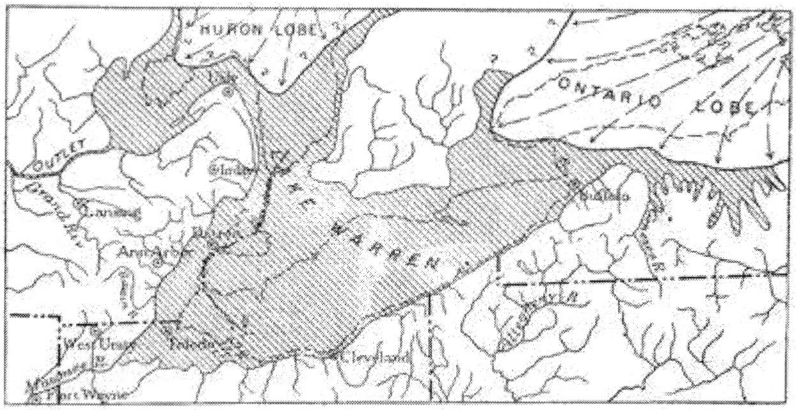

FIG. 363.—Map of the glacial Lake Warren, the last of the lakes in the Huron-Erie basin, which discharged through the "Grand River outlet" into the Mississippi (after Leverett).

Lakes Arkona and Whittlesey.—The ice front in the Huron-Erie basin now retired so far that the impounded waters, instead of following the more direct "Imlay outlet" to the Grand, passed at a lower level completely around "the thumb" of Michigan into the Saginaw basin. Meanwhile a crescent-shaped lake had developed in that basin, so that now the waters of the Maumee basin were joined to those in the Saginaw basin as a common lake, just as the lowering of the waters in Glen Roy caused a union with those of Glen Glaster in the example cited for illustration. Our records of this third North American lake stage, referred to as Lake Arkona, are however most imperfect, for the reason that it was followed by a readvance of the ice front which closed the passage around "the thumb" and raised the level of the waters until an outlet was found past the town of Ubly at a lower level than the "Imlay outlet." When the waters of a lake are thus rising, strong beach formations result, and those of this stage, which is known as the Lake Whittlesey stage, are much the strongest that are found within the Huron-Erie basin. Traced for some three hundred miles entirely around the southern and western margins of Lake Erie, this beach is for much of the distance the famous "ridge road" (Fig. 362).

Lake Warren.—As the ice advance which had produced Lake Whittlesey came to an end, the normal recession was resumed and a lake once more formed as a body common to the Saginaw and Erie basins. This lake, known as Lake Warren, extended a shrunk arm far eastward along the ice front into western New York, though it was still blocked from entering the great Mohawk valley (Fig. 363).

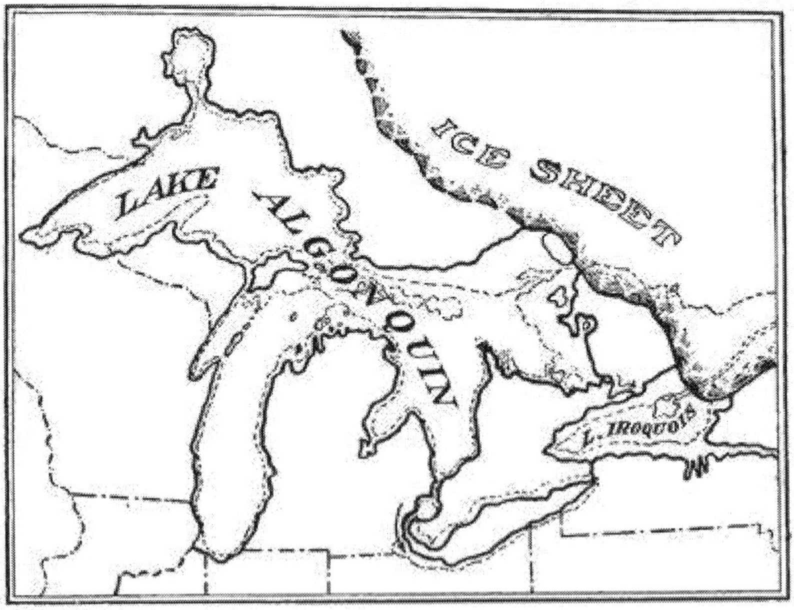

FIG. 364.—Map of the Glacial Lake Algonquin (after Leverett).

Lakes Iroquois and Algonquin.—It must be evident that toward the close of the Lake Warren stage a profound change was imminent—a transfer of the glacial waters from their course to the Mississippi and the Gulf to the trench which crosses New York State and enters the Atlantic. So soon as the ice front had retired sufficiently to lay bare the bed of the Mohawk, an outlet was found by this route and its continuation down the Hudson valley to the sea. The Lake Ontario basin now became occupied by a considerably larger water body known as Lake Iroquois, and the three upper lakes, then joined as Lake Algonquin, discharged their combined waters into Lake Iroquois at first through a great channel now strongly marked across Ontario in the course of the Trent River and Lake Simcoe, the so-called "Trent outlet." At this time a smaller Lake Erie probably occupied the basin of that lake, and later the Trent outlet was abandoned for the Port Huron outlet (Fig. 364).

FIG. 365.—Outline map of the Nipissing Great Lakes with their outlet past North Bay into the Champlain Sea.

The Nipissing Great Lakes.—We have now followed the ice front step by step in its retreat across the valley of the St. Lawrence system. The successive unblocking of outlets offers but one further possibility—the opening of the French River-Nipissing Lake-Ottawa River, or "North Bay outlet." Though not so to-day, the bed of this ancient channel was then much lower than that of the "Mohawk outlet", and so soon as the glacier had in its retreat uncovered this northern channel, the waters of the upper lakes discharged through it past the site of Ottawa and into an arm of the sea which then occupied the lower St. Lawrence valley and has been called the Champlain Gulf or Sea (Fig. 365). The level of the waters was lowered and the area of the lakes correspondingly reduced.

The reader who has had no opportunity to observe these ancient channels which carried the swollen waters of the former glacier lakes, will find it interesting to consider that every one of them has been fixed upon by engineers for improvement as artificial waterways. Thus we have the Illinois Drainage Canal and projected ship canal along the "Chicago outlet", the projected Mississippi-Lake Erie Canal along the "Fort Wayne outlet", the Grand River canal project to connect Lake Michigan and Saginaw Bay along the course of the "Grand River outlet", the Trent Canal along the "Trent outlet", the Erie Canal along the "Mohawk outlet", and, lastly, the proposed Georgian Bay ship canal to the ocean along the "North Bay" or "Nipissing outlet."

Summary of lake stages.—We have omitted in this summary of late lake history in the Laurentian basin all the less important lake stages, including some of a transitional nature which were represented by beaches and outlets easily traced to-day. This is because it is an outline only which it seems best to present, and the episodes of this abridged history may be tabulated as follows:

EPISODES OF GLACIAL LAKE HISTORY

Mississippi Drainage

Lake Maumee (early), Fort Wayne outlet.

Lake Maumee (late), Imlay City outlet.

Lake Arkona, "thumb" outlet.

Lake Whittlesey (with readvance of glacier), Ubly outlet.

Lake Warren, "thumb" outlet.

Atlantic Drainage

Lakes Iroquois and Algonquin (early), Trent and Mohawk outlets.

Lakes Iroquois and Algonquin (late), Port Huron and Mohawk outlets.

Nipissing Great Lakes, North Bay outlet.

Permanent changes of drainage affected by the glacier.—While the lake history which we have sketched is made up of episodes which endured only while the ice front lay between certain stations upon its retreat, there were none the less brought about the profoundest of permanent modifications in the drainage of the region. It is possible to restore upon maps in part only the preglacial drainage of the north central states, but we know at least that it was as different as may be from that which we find to-day. The Missouri and the Ohio take their courses to-day along the margin of the glaciated area as an inheritance from the border drainage of the ice age. Within the glaciated regions rivers have in many cases been compelled by morainal obstructions to enter upon new courses, or even to travel in the opposite direction along their former channels. In districts of considerable relief these diversions have sometimes caused the streams to plunge over the walls of deep valleys, and it may truthfully be said that we owe much of our most beautiful scenery in part to the carving and molding of glaciers, but especially to the cascades and waterfalls directly due to their interference with drainage.

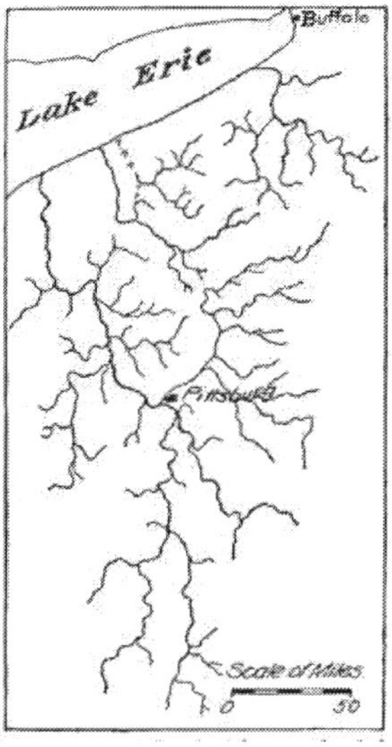

FIG. 366.—Probable preglacial drainage of the upper Ohio region (after Chamberlin and Leverett).

Many diversions or reversals of former drainage lines, through the influence of the continental glacier, are at once suggested by the abnormal stream courses, which appear upon our maps, and the correctness of these suggestions may often be confirmed by very simple observations made upon the ground. The map of Fig. 366 shows how different was the preglacial drainage of the upper Ohio region from that of to-day.

An interesting additional example is furnished by the Still River which in Connecticut is tributary to the Farmington, and is no less remarkable for its abnormal northerly course and sluggish current perpetuated in its name, than for the way in which it is joined to the Farmington system (Fig. 367 *A*). A careful study of the district has shown that the Still River was once a part of the Naugatuck and flowed southward toward Long Island Sound like other rivers of the district (Fig. 367 *B*). It possessed, however, an advantage in a narrow belt of softer rock along its course, and because of this advantage it captured a portion of one of the tributaries to the Farmington (Fig. 367 *C*). The continental glacier later covered the region, and on its retreat laid down morainal obstructions directly across this river and also at the head of the severed arm of the Farmington tributary (Fig. 367 *D*). The now impounded waters found their lowest outlet near Sandy Brook, and in waterfalls and cascades the now reversed river falls one hundred feet to the bed of that stream. With the aid of the excellent topographic maps which are now supplied by a generous government at a merely nominal price, such bits of recent history may be read at many places within the glaciated region.

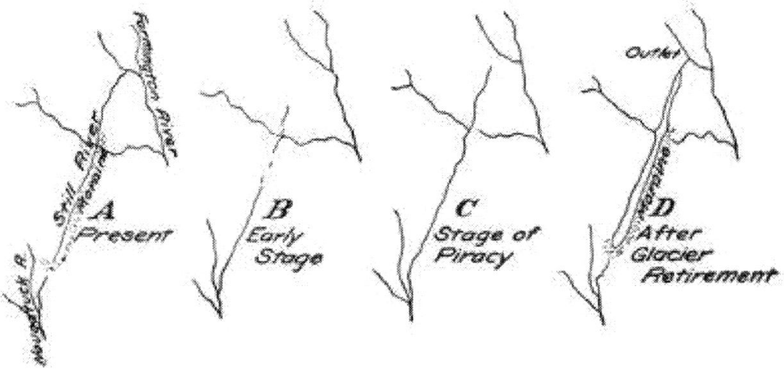

FIG. 367.—Diagrams to illustrate the episodes in the recent history of the Still River tributary to the Farmington in Connecticut. *A*, present drainage; *B*, early stage; *C*, after capture of a tributary to the Farmington; *D*, after blocking by morainal obstructions of the ice age.

Glacial Lake Ojibway in the Hudson Bay drainage basin.—When by passing over the "height of land" in northern Ontario the greatly reduced continental glacier had vacated the basin of St. Lawrence drainage, it was in a position to impound those waters which normally drained to Hudson Bay. The lake which then came into existence has been called Lake Ojibway and was the latest of the entire series. Though of but recent discovery in a country till lately a trackless wilderness, its extension seems to have been that of the clay beds suited for farming. The beaches and outlets remain to be mapped when the country has been made more easily accessible.

READING REFERENCES FOR CHAPTER XXIII

Parallel roads of Glen Roy:—

CHARLES DARWIN. Observations on the Parallel Roads of Glen Roy and of Other Parts of Lochaber in Scotland, with an attempt to prove that they are of Marine Origin, Phil. Trans., vol. 8, 1839, pp. 39-82.

LOUIS AGASSIZ. Geological Sketches, Boston, 1876, vol. 2, pp. 32-76.

T. T. JAMIESON. On the Parallel Roads of Glen Roy and their Place in the History of the Glacial Period, Quart. Jour. Geol. Soc. Lond., vol. 19, 1863, pp. 235-259.

Glacial Lake Agassiz:—

WARREN UPHAM. The Glacial Lake Agassiz. Mon. 25, U. S. Geol. Surv., pp. 658, pls. 38.

F. W. SARDESON. Beginning and Recession of St. Anthony's Falls, Bull. Geol. Soc. Am., vol. 19, 1908, pp. 29-36.

Glacial lakes in the St. Lawrence valley:—

CHAMBERLIN AND SALISBURY. Geology, vol. 3, pp. 394-405.

FRANK LEVERETT. Outline of the History of the Great Lakes (Presidential Address), 12th Rept. Mich. Acad. Sci., 1910, pp. 19-42. The Pleistocene Features and Deposits of the Chicago Area. Chicago, 1897, pp. 86, pls. 8 (Chicago Outlet).

H. L. FAIRCHILD. Glacial Lakes in Western New York, Bull. Geol. Soc. Am., vol. 6, 1895, pp. 353-374, pls. 18-23; Glacial Waters in Central New York. Bull. 127, N. Y. State Mus., 1909, pp. 66, pls. 42, and maps in cover.

Early lakes in the Erie basin:—

FRANK LEVERETT. On the Correlation of Moraines with Raised Beaches of Lake Erie, Am. Jour. Sci. (3), vol. 43, 1892, pp. 281-301.

F. B. TAYLOR. The Great Ice Dams of Lakes Maumee, Whittlesey, and Warren, Am. Geol., vol. 24, 1899, pp. 6-38, pls. 2-3; Relation of Lake Whittlesey to the Arkona Beaches, 7th Rept. Mich. Acad. Sci., 1905, pp. 30-36.

FRANK LEVERETT. The Ann Arbor Folio, Folio No. 155, U. S. Geol. Surv., 1908, pp. 10-12.

CHAPTER XXIV

THE UPTILT OF THE LAND AT THE CLOSE OF THE ICE AGE

The response of the earth's shell to its ice mantle.—There is now good reason to believe that the earth's outer shell makes a response by oscillations of level due to the loading by ice, on the one hand, and to the removal of this burden upon the other. We know, at least, that both in northern Europe and in North America areas which have undergone depression during and elevation after the ice age, correspond closely to the regions which were ice covered. Wherever in these regions there was high relief before the advent of the ice, river valleys were drowned at the land margins and were also gouged out into troughs through erosion by the outlet tongues upon the margin of the ice sheet. Such furrowed and half-submerged valleys have a characteristic U-shaped section, so that their walls rise precipitously from the sea. From their typical occurrence in Scandinavian countries the name *fjord* has been applied to them.

It is now no less clear that the removal of the ice blanket brought from the earth a relatively quick response in uplift, which began before the ice front had retired across the present international boundary of the United States, and that this uplift continued until the final disappearance of the ice. A far slower elevation of a somewhat different nature has continued, even to the present day.

It is obvious that at the time of their formation all shore lines referable to the work of waves must have been horizontal, and hence any variations from a perfect level which they reveal to-day must indicate that a tilting movement of the ground has occurred since the waters departed from their basins. We have thus provided for us in the positions of these ancient water planes, particularly because of their wide extent, a complete record the refinement of which is not easily overstated. Interpreting this record, we find that it was the uptilt of the land to the northward which brought the glacial lake history to an end and inaugurated the present system of St. Lawrence drainage. The outlet of the Nipissing Great Lakes is to-day more than a hundred feet above the level of the outlet at Port Huron, where the upper lakes are now discharging their waters, and this difference in level can only be ascribed to an upward tilting of the land since the latest of the glacial lake stages.

The abandoned strands as they appear to-day.—The traveler by steamer upon the upper lakes, as he comes within view of each rocky headland, may note how the profile against the horizon is notched by a series of steps or terraces (Fig. 368), and if he has followed the discussion in previous chapters, he will suspect that these terraces mark the now abandoned shore lines which have come to their present position through a series of uplifts of the ground accompanied by earthquake shocks. As his steamer skirts the shore he may chance to note a cave within the rock cliff which represents the now elevated sea-arch of an ancient shore.

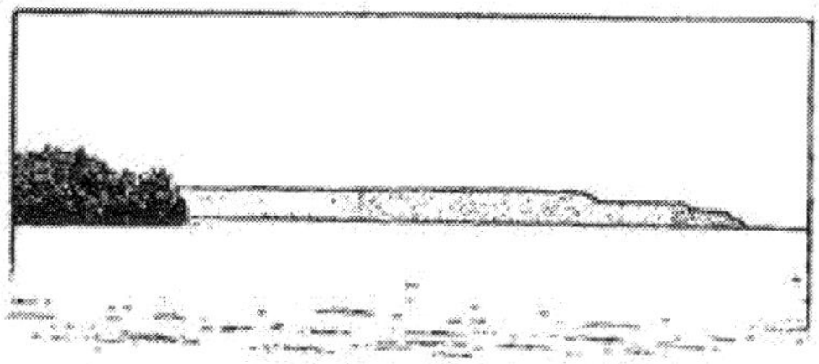

FIG. 368.—The notched rock headland of Boyer Bluff between Green Bay and Lake Michigan (after Goldthwait).

Disembarking from the steamer and traveling inland at any point where the shores are high, the traveler is certain to come upon still more convincing proofs of the ancient strands; perhaps in a storm beach of the unmistakable "shingle", half buried though it may be under dunes of newly drifted sand, or possibly at higher levels the highway has been cut through a shingle barrier as fresh and unmistakable as though formed upon the present shore. Sometimes it is the rock cliff and terrace, at other times barrier ridges of shingle, or, again, it is the sloping cliff and terrace cut in the drift deposits; but of whatever sort, if studied with proper regard to the topography of the district, the evidence is clear and unmistakable.

The records of uplift about Mackinac Island.—Nowhere are the records of the recent uplift of the lake region more easily read than about Mackinac Island in the straits connecting Lake Michigan with Lake Huron. Approaching the island by steamer from St. Ignace, its profile upon the horizon is worthy of remark (Fig. 369). From a central crest broken by minor irregularities and bounded on all sides by a cliff, the island profile slopes gently away to a still lower cliff, below which is another terrace.

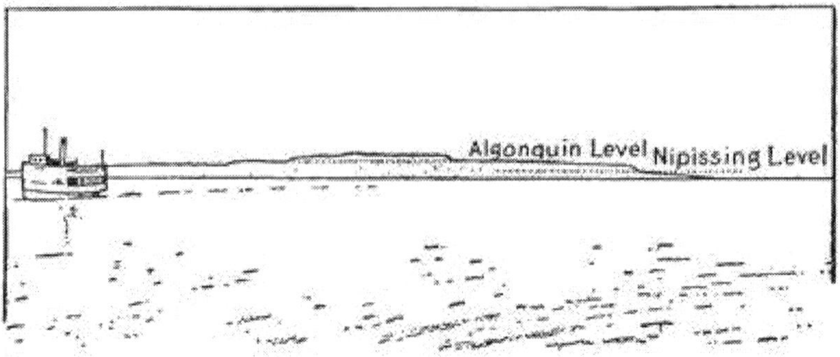

FIG. 369.—View of Mackinac Island from the direction of St. Ignace. The irregular central portion is the only part of the island that was not submerged in Lake Algonquin. The terrace at its base is the old shore line of Lake Algonquin, and the lower terrace the strand of Lake Nipissing (after a photograph by Taylor).

FIG. 370.—The "Sugar Loaf", a stack near the shore of Lake Algonquin, as it is seen from the observatory upon Mackinac Island (after a photograph by Taylor).

When we have reached the island and have climbed to the summit, we there find the surface which is characteristic of erosion by running water, whereas at lower levels are found the forms carved or molded by the action of waves. This central "island", superimposed upon the larger island, is all that rose above Lake Algonquin, the earliest of the glacial lakes in this northern district; and as we look out from the observatory upon the summit, it is easy to call up a picture of the country when the lake stood at the base of this highest cliff. To the northward one sees the "Sugar Loaf" rise out of a sea of foliage, as it formerly did from the waters of Lake Algonquin (Fig. 370). It is a huge stack near the former island shore. If we turn now to the southward and direct our gaze toward the Fort, we encounter a veritable succession of beach ridges formed of shingle and ranged like a series of waves within the cleared space of the "Short Target Range" (Fig. 371). These ridges mark each a stage within a series of successive uplifts which have brought the island to its present height.

FIG. 371.—View from the observatory upon Mackinac Island across the "Short Target Range" toward the Fort. Beach ridges appear in succession within the cleared space (after a photograph by Rossiter).

FIG. 372.—Notched stack of the Nipissing Great Lakes at St. Ignace (after a photograph by Taylor).

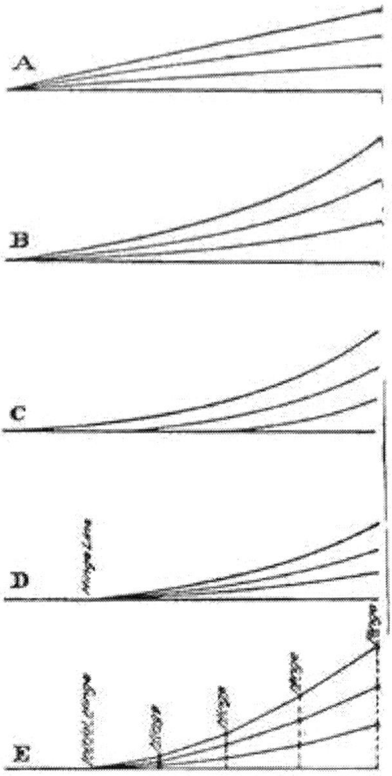

FIG. 373.—Series of diagrams to illustrate the evolution of ideas concerning the uplift of the lake region since the ice age. *A*, simple northerly up-canting (Gilbert): *B*, northerly acceleration of the up-canting (Spencer and Upham); *C*, northerly "feathering out" of beaches (Spencer and Upham); *D*, hinge, line of up-canting found within the lake region (Leverett); *E*, multiple and northwardly migrating hinge lines of up-canting (Hobbs).

If now we descend from our position and visit the "battlefield", we find there a great ridge of level crest, behind which the British force was stationed in its defense of the island in 1812. Near by in the woods is Pulpit Rock, a strikingly perfect stack of the Nipissing Lake. Across the straits at St. Ignace is an even finer example of the notched stack (Fig. 372). Other less prominent beaches, but all later than the Nipissing Lakes, intervene between this level and the present shore to mark the stages in the continued uplift of the land.

The present inclinations of the uplifted strands.—It is not enough that we should have recognized the marks of former shores now at considerable elevations above the existing lakes; if we are to know the nature of the uplift, we must prepare accurate maps based upon measurements by precise leveling at many localities. Such methods are, however, of comparatively recent application in this field; and, as in the investigation of so many other problems, the earlier observations were largely of the nature of reconnaissances with the elevation of beaches estimated by comparatively crude methods only. The evolution of ideas concerning the uptilt has, therefore, been a gradual one.

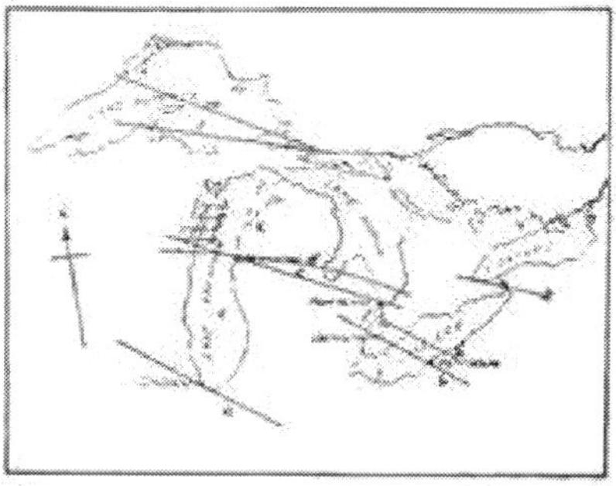

FIG. 374.—Map of the Great Lakes region to show isobases and hinge lines of uptilt. *a*, isobase of the Chicago outlet; *b*, main hinge line of the Lake Whittlesey beach (Leverett); *b*[1], hinge line of the Lake Warren beach (Taylor); *c*, isobase of the Port Huron outlet; *d*, main hinge line of highest Algonquin beach (Goldthwait); *e*, *f*, *g*, *h*, additional hinge lines of Algonquin beaches in Door County peninsula (Hobbs); *l*, isobase of the Lake Superior outlet for the Algonquin beaches (Leverett): *m*, isobase of the same outlet for the Nipissing beaches (Leverett).

It was early observed that the beaches corresponding to a given lake stage were higher to the northward and northeastward, and the natural conclusion from this was that the earth's crust had here been canted like a trap door (Fig. 373, *A*). As we are to see, this but half-correct assumption has led to a striking prophecy relating to future changes within the lake region which we now know to be without warrant in the facts. Later it was learned that the uptilt of the lake beaches is much accelerated to the northward (Fig. 373, *B*), and that new beaches make their appearance from beneath others as we proceed in this direction—there is a "feathering out" of beaches to the northward (Fig. 373, *C*).

The hinge lines of uptilt.—Still later in the study of the region, it was learned that the axis or fulcrum about which the region has been uptilted, instead of lying to the southward of the lake district, as had been assumed by Gilbert, lay within the region and about halfway up the basin of Lake Michigan (Fig. 373, *D*, and Fig. 374). Similarly, in the uptilt which followed the ice retreat in northern Europe a definite hinge line of movement has been discovered.

Lastly, it has been shown, as a result of the use of precise leveling methods, that not one but several hinge lines of movement lie within the region, and that the separate sections into which they divide the area are each in turn characterized by increased up-cant as we proceed to the northward (Fig. 373, *E*. and Fig. 374).

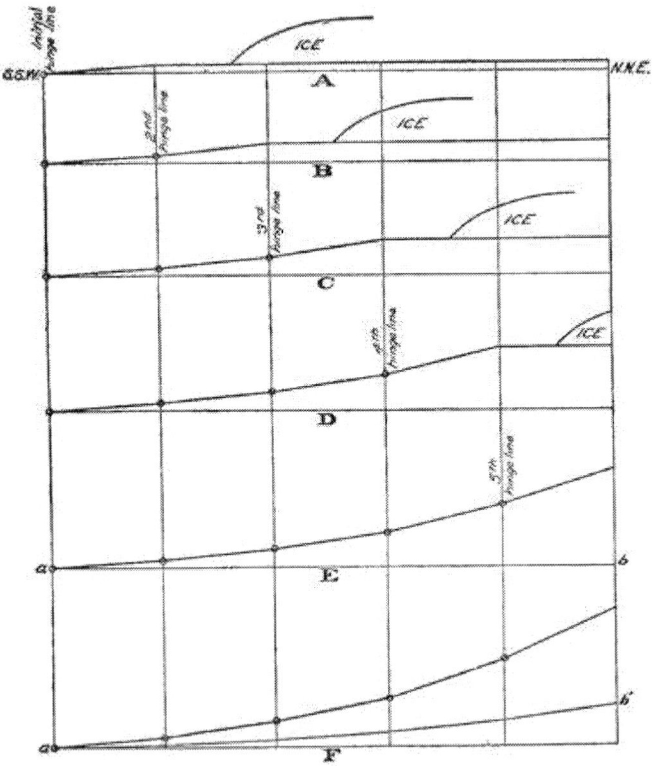

FIG. 375.—Series of idealistic diagrams to indicate the nature of the quick recovery of the crust by uplift in blocks unloaded of the ice in succession. A further and slower uptilt, added after the completion of the first movement, is brought out in the last diagram (*b′*).

The beaches of Lake Maumee, the earliest of the series of lakes within the Huron-Erie lobe and within the extreme southern portion of the Great

Lakes area, show only the slightest possible northerly uptilt, and the well-marked hinge line disclosed in the Whittlesey beach is evidence that the elastic recoil, as it were, from the weight of the mantling glacier did not begin until after the draining of Lake Whittlesey. The determination by Taylor that there is a similar initial hinge line in the Warren beach—that this strand begins its uptilt some fifteen miles farther northeast than does the Whittlesey beach—is one of the greatest importance in obtaining a correct idea of the recent uplift; for it shows that the draining of Lake Whittlesey was followed by a period of quick uplift and seismic activity, that the stage of Lake Warren was one of comparative stability of the land, and, lastly, that the draining of Lake Warren was followed by a second period of rapid uplift and earthquake disturbance. The strongly marked hinge lines, additional to the initial one indicated for the Algonquin beaches in the profiles by Goldthwait from the west shore of Lake Michigan, when considered in the light of this northeasterly migration of the still earlier hinge line in the southern district, are best explained through the assumption of a succession of quick recoveries of the crust by uplift, separated by periods of relative stability, and brought on by the removal in turn of the ice burden from successive blocks of the shell which are separated by the several hinge lines (Fig. 375).

The elaborate study of erosion in the outlet of Lake Agassiz had indicated identical interruptions in the up-canting process for that basin.

Future consequences of the continued uptilt within the lake region.—One of the most distinguished of American geologists, Dr. G. K. Gilbert, in order to determine whether the uptilt revealed by canted beach lines is still in progress, carried out an elaborate study upon the gauge records preserved at the various gauging stations about the Great Lakes. Upon the basis of these studies, he concluded that the uplift continues, that the axes of equal uplift (isobases) take their course about fifteen degrees north of west, so that the lines of greatest uptilt should be perpendicular to this direction, or fifteen degrees east of north. He further believed that the basin was undergoing an up-cant in the simple manner of a trap door, the hinge of which lay to the southward of Chicago, and the study of the gauge records led him to believe that "the rate of change is such that the two ends of a line one hundred miles long and lying in a south-southwest direction are relatively displaced four tenths of a foot in one hundred years."

Gilbert's prophecy of a future outlet of the Great Lakes to the Mississippi.—The *natural* rock sill, over which the waters of Lake Chicago once flowed to the Mississippi, is to-day but eight feet above the common mean level of Lakes Michigan and Huron, and if the tilting of the lake region were to continue upon Gilbert's assumption of a canting plane with the hinge of the movement to the south of Chicago, a time must come when the "Chicago outlet" will again come into use and the lakes once more drain to the Mississippi and the Gulf. Upon the basis of his measurements, Gilbert ventured the prophecy that the first high-water discharge into the Mississippi should occur in from five hundred to six hundred years, and for continuous discharge in fifteen hundred years. In twenty-five hundred years Niagara Falls should at low water stages be dry from this cause, and in thirty-five hundred years it should have become extinct.

This prophecy, emanating from a high scientific authority and relating to changes of such profound economic and commercial importance, has been often quoted and has taken a firm hold upon the popular imagination. Obviously, it depends upon the now exploded theory that the lake basin has been canted *as a plane* and that the axis of uptilt lies somewhere to the southward of the lake region, or, in any event, to the southward of the present Port Huron outlet. We know to-day that instead of being uniformly distributed over the entire lake region, the uptilting goes on at a much higher rate within the northern areas, and that since the early stage of Lake Whittlesey the hinge line of uplift has been steadily migrating northward with the retreat of the ice and is now well to the northward of the present outlet. There is, therefore, no known uptilt of the district which separates the present from the former Chicago outlet, and there is no apparent natural cause which should result in the reoccupation of the old outlet to the Mississippi. The prophecy must be regarded as one that has been outgrown with the progress of science.

Geological evidences of continued uplift.—It has recently been claimed, on the basis of a reëxamination of Gilbert's study of the lake gauge records, that his methods are open to serious criticism and that in reality the figures afford no evidence of continued uplift of the region. However this may be, there are not lacking geological evidences which do not admit of doubt, and these are in a striking way confirmatory of the latest conclusions upon the manner of the recent uplift.

If our conclusions have been correct, the several lake basins should now be behaving in different ways as regards the changes upon their shores. If it is true that the lines of greatest uptilt run north-northeasterly, there should be, speaking broadly, a "spilling over" of waters upon the south-southwesterly shores and a laying bare of the north-northeasterly shore terraces of the basins. This should, however, be true only of basins whose outlets are to the northeastward of the existing main hinge line of uptilt. Lake Huron, having its outlet at the southern margin of its basin, should not have its waters encroaching upon the southern shore, for the simple reason that any continued uptilt of the basin can only have the effect of pouring more water through the outlet. Lake Michigan and Saginaw Bay, which are arms of the Huron basin, ought, however, to become flooded upon their southern shores, *were it not that the hinge line of uptilt to-day lies to the northward of the outlet at*

Port Huron, and, further, that the two connecting channels still have their beds lower than the sill of the outlet channel. Now the evidence goes to show that no encroachment of waters is occurring upon the Chicago shore of Lake Michigan, and although the shores of Saginaw Bay are so excessively flat as to reveal slight changes of level by large migrations of the strand, yet the ancient meander posts fixed by the early surveys are still found near the water's edge.

Drowning of southwestern shores of Lakes Superior and Erie.—Within the basins occupied by Lakes Superior and Erie, a wholly different condition is found. In each case the outlet is found to the northeastward (Fig. 374, p. 345), and the northwesterly trend of the isobases from these outlets is responsible for a continued elevation from uptilt of the outlets with reference to the western and southern shores. In consequence, the waters are encroaching upon these shores, and rivers which there enter the lake are drowned at their mouths, with the formation of estuaries. Upon Lake Superior these changes are very marked near Duluth and particularly in the St. Louis River, within which, since the early treaty with the Indians, certain rapids have disappeared and submerged trunks of trees are now found in the channel of the river. As far east as Ontonagon essentially the same conditions are found.

Upon the shores within the Porcupine Mountain district, the waters are clearly rising. Here old cedar trees may be seen, in some cases dead but still upright and standing in from six to eight inches of water a number of feet out from the present shore, while others near the shore, but upon the land and still living, are washed by the waves, and losing their lower bark in consequence. An old road along the shore has had to be abandoned because of the encroaching water.

Upon the opposite or northeastern shore of the lake, on the other hand, the land is everywhere rising out of the water, and the waves are now building storm beaches well out upon the wave-cut terrace. Here the streams, instead of forming estuaries by drowning, drop down in rapids to the level of the lake.

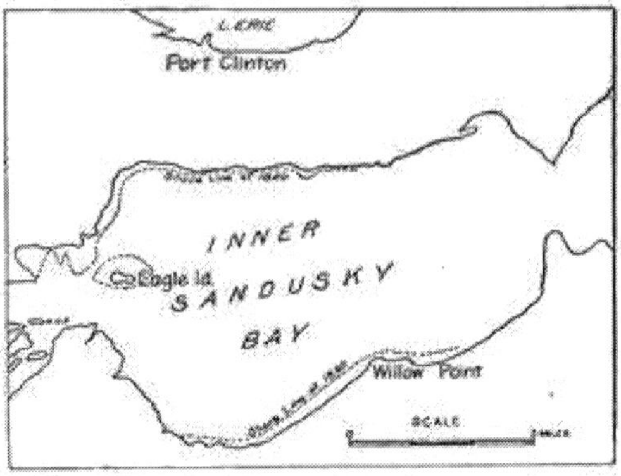

FIG. 376.—Portion of the Inner Sandusky Bay, to afford a comparison of the shore line of 1820 with that of to-day (after Moseley).

At the southwestern margin of Lake Erie there is everywhere evidence of a rapid encroachment by the water. In the caves of South Bass Island stalactites, which must obviously have formed above the lake level, are now permanently submerged. It is, however, about Sandusky Bay upon the southwest shore that the most striking observations have been made. Moseley has collected historical records of the killing of forest trees through a submergence which was the result of an advance of the water upon the shores. It seems to be proven from his studies that the water is now rising in Sandusky Bay at a rate of about 2.14 feet per century. In Fig. 376 there is a comparison of the shores of the inner bay separated by an interval of about ninety years.

READING REFERENCES FOR CHAPTER XXIV

Uptilt in basin of Lake Agassiz:—

WARREN UPHAM. The Glacial Lake Agassiz, Mon. 25, U. S. Geol. Surv., pp. 474-522.

Uptilt in Laurentian Basin:—

G. K. GILBERT. Recent Earth Movement in the Great Lakes Region, 18th Ann. Rept. U. S. Geol. Surv., 1898, Pt. ii, pp. 595-647.

J. W. SPENCER. Deformation of the Algonquin Beach, etc., Am. Jour. Sci. (3), vol. 41, 1891, pp. 14-16.

F. B. TAYLOR. The Highest Old Shore Line of Mackinac Island, Am. Jour. Sci. (3), vol. 43, 1892, pp. 210-218.

A. C. LAWSON. Sketch of the Coastal Topography of the North Side of Lake Superior, with reference to the abandoned strands, etc., 20th Ann. Rept. Geol. and Nat. Hist. Surv. Minn., 1893, pp. 181-289, pls. 7-12.

J. B. WOODWORTH. Ancient Water Levels of the Champlain and Hudson Valleys, Bull. 84, N.Y. State Mus., 1905, pp. 265, pls. 28.

E. L. MOSELEY. Formation of Sandusky Bay and Cedar Point, Proc. Ohio State Acad. Sci., vol. 4, 1905, Pt. v, pp. 179-238.

F. E. WRIGHT. Rept. Geol. Surv. Mich. for 1903, 1905, p. 37.

J. W. GOLDTHWAIT. The Abandoned Shore Lines of Eastern Wisconsin, Bull. 17, Wis. Geol. and Nat. Hist. Surv., 1907, pp. 134, pls. 37; A Reconstruction of Water Planes of the Extinct Glacial Lakes in the Lake Michigan Basin, Jour. Geol., vol. 16, 1908, pp. 459-476; Isobases of the Algonquin and Iroquois Beaches and their Significance, Bull. Geol. Soc. Am., vol. 21, 1910, pp. 227-248, pl. 5; An Instrumental Survey of the Shore Lines of the Extinct Lakes Algonquin and Nipissing in Southwestern Ontario, Mem. 10, Dept. of Mines, Canada, 1910, pp. 57, pls. 4.

WILLIAM H. HOBBS. The Late Glacial and Post-glacial Uplift of the Michigan Basin, Pub. 5, Mich. Geol. and Biol. Surv., 1911, pp. 68, pls. 2.

LAWRENCE MARTIN. [Post-glacial Modifications in and Around the Great Lakes], Mon. 52, U. S. Geol. Surv., 1911, pp. 455-459.

Uptilt in northern Europe:—

G. DE GEER. Quaternary Changes of Level in Scandinavia, Bull. Geol. Soc. Am., vol. 3, 1892, pp. 65-68, pl. 2.

H. MUNTHE. Studies in the Late Quaternary History of Southern Sweden, paper No. 25, Livret Guide, Cong. Géol. Intern., 1910, pp. 96, many plates and maps.

CHAPTER XXV

NIAGARA FALLS A CLOCK OF RECENT GEOLOGICAL TIME

Features in and about the Niagara gorge.—A striking example of those permanent alterations of drainage which have resulted from the presence of the late continental glacier in North America is to be found in the Niagara gorge between Lakes Erie and Ontario. With the aid of borings many of the now buried channels of the region have been followed out, and in a later paragraph we shall refer to some of the stronger lines of the earlier drainage system. Before undertaking the study of Niagara history, it is essential that one become somewhat familiar with the present topography in and about the Niagara gorge.

Below the present cataract the river flows through a deep gorge for about seven miles before issuing at the Lewiston Escarpment (Fig. 381, p. 355). This gorge has been cut in beds of rock sediments which dip at a gentle angle southward toward Lake Erie. The capping of the rock series is a compact and relatively resistant limestone which is known as the Niagara limestone, beneath which there are alternating beds of shale with thinner limestone and sandstone. The plain formed by the upper surface of the limestone capping terminates in the Lewiston Escarpment, which is transverse to the direction of the gorge and seven miles distant below the Falls. The depth of the gorge varies markedly, the above-water portion being represented at the upper end by the height of the cataract, one hundred and sixty-five feet, while at its lower end near Lewiston it is twice that amount. Halfway down the gorge a sharp turn is made at an angle of more than ninety degrees, and the upstream arm is extended to form a basin which contains the famous whirlpool. This visible extension of the upper gorge is continued in a buried channel, the St. Davids Gorge, which extends to the escarpment, broadening as it does so in the form of a trumpet. The materials which fill this earlier channel are notably coarse glacial deposits (Fig. 389).

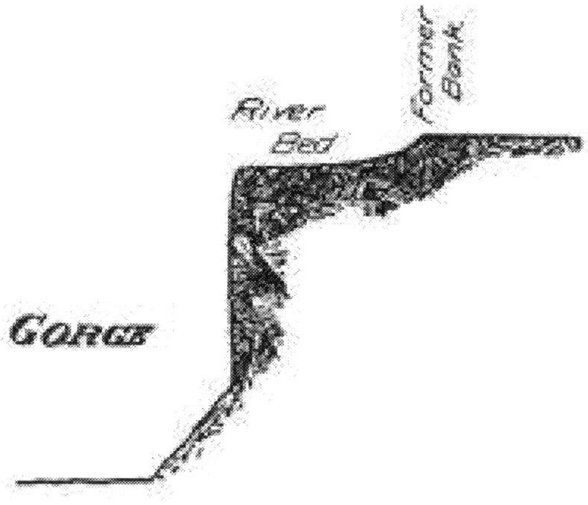

FIG. 377.—Ideal cross section of the Niagara gorge to show the marginal terrace.

Directly above the whirlpool the Niagara gorge is first contracted, but almost immediately swells out into the form of a sausage, which under the name of the Eddy Basin extends to the constricted channel occupied by the Whirlpool Rapids. This Gorge of the Whirlpool Rapids extends to and a little above the railroad bridges, where it again suddenly widens and deepens and with surprisingly uniform cross section now continues as far as the cataract. This uppermost section is known as the Upper Great Gorge. About a mile below the whirlpool is that remarkable projection into the gorge from the Canadian wall which is known as Wintergreen Flats, below which and nearer the river are Fosters Flats. Almost throughout its entire length the Niagara gorge is bordered on either side by a narrow and gently incurving terrace eroded below the general level of the plain and meeting the gorge in a sharp angle (Fig. 377).

The features immediately about the cataract show that the Falls are to-day in a condition which, so far as we know, has occurred but once before in their entire history—the waters of the river are divided unequally by an island, and for this reason, as we shall see, the cataract enters over the *side wall* of the gorge instead of at its *end* (Fig. 381), although the turning of the channel from this cause is combined with a bend of the river.

FIG. 378.—View of the bed of the Niagara River above the cataract, where water has been drained off in installing a power plant. Some separated blocks of limestone are still in place (after J. W. Spencer).

The drilling of the gorge.—There appear to be two important processes which are responsible for the recession of the Falls, the rate of which is determined largely by the resistance of the limestone capping and the tenacity of the looser shale beneath it. One of the eroding processes operates from below and undermines the cap until the unsupported cornice falls in blocks to the bottom of the gorge; the other makes its attack directly from above, selecting for the purpose the lines of jointing of the rock which it widens by solution and corrasion until the included blocks are in so far separated that they are torn out and go over the brink of the Falls (Fig. 378). This process of overhead attack in the powerful currents just above a cataract is even better illustrated by the Falls of St. Anthony near Minneapolis, which have had a similar history of recession to that of the Niagara Falls (Fig. 379).

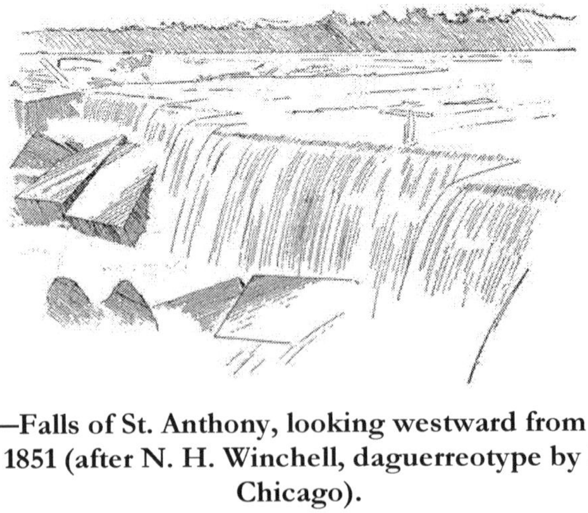

FIG. 379.—Falls of St. Anthony, looking westward from Hennepin Island in 1851 (after N. H. Winchell, daguerreotype by Hessler of Chicago).

The blocks of the capping limestone at Niagara Falls are to some extent fixed in size by the joint planes present in them, and as they fall to the bottom of the gorge, they promote or retard the further recession of the Falls according as they can or cannot be moved about by the churning currents beneath the cataract. Of the retarding effect there is an illustration in the accumulation of the blocks below the American and the intermediate Luna Falls (plate 23 A), which the weaker currents upon the American side find too heavy to handle.

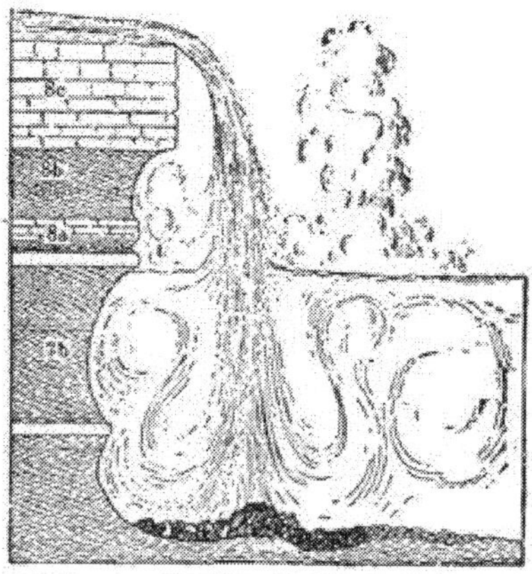

FIG. 380.—Ideal section to show the nature of the drilling process beneath the cataract.

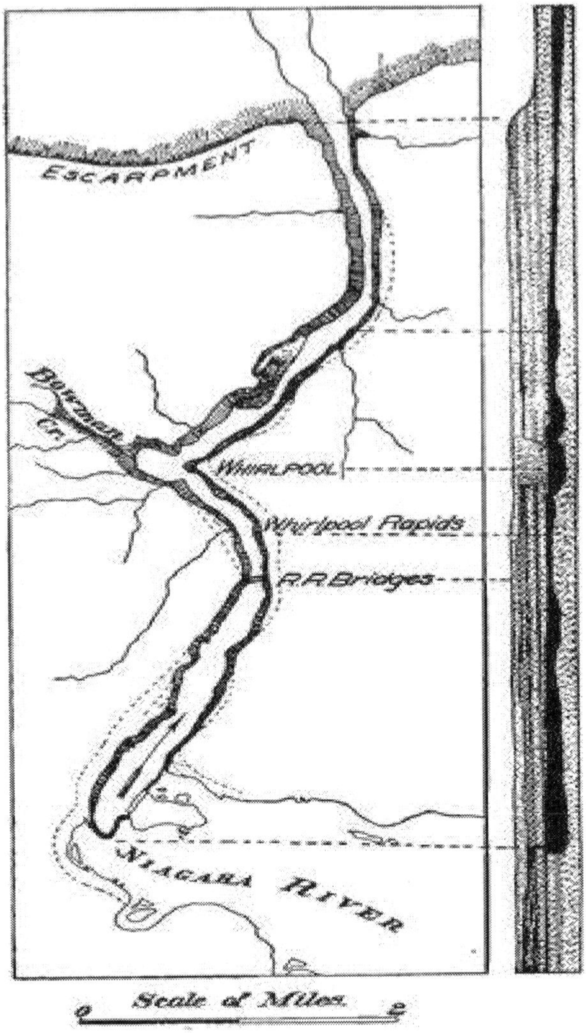

FIG. 381.—Plan and section of the Niagara gorge, showing how in each section the depth is proportional to the width, except in the lowest section where subsequent river action of the normal type has modified the bed of the channel (plan after Taylor and section after Gilbert).

The Canadian Fall, with its much greater power, is an example of the promotion of recession through the churning about of the blocks at the base of the cataract. We have here to do with a churn drill which bores its way into the bottom of the gorge with increasing radius of rotary motion with each increase in volume of the falling water. Under this rotary churning the soft shales are torn out near the bottom and in succession the harder layers above until the capping is reached (Fig. 380). The conditions appear now to be such that the effective work is largely concentrated, as it usually has been, near the middle of the channel; and so the gorge recedes with a margin of the earlier river bed remaining as a terrace on either side and extending to the former river bank (Fig. 377).

As must have been noted, one peculiarity of the operation of the churn drill beneath the cataract is that the depth of the gorge will bear a direct proportion to its width, and if the volume of water has varied during the process of recession, these changes in volume will be registered in the width and also in the depth of that section of the gorge which was drilled at the time—the cross section of the gorge at any place is proportional to the volume of the water falling in the cataract which produced it, modified, however, by the competency to handle the joint blocks of definite size (Fig. 381).

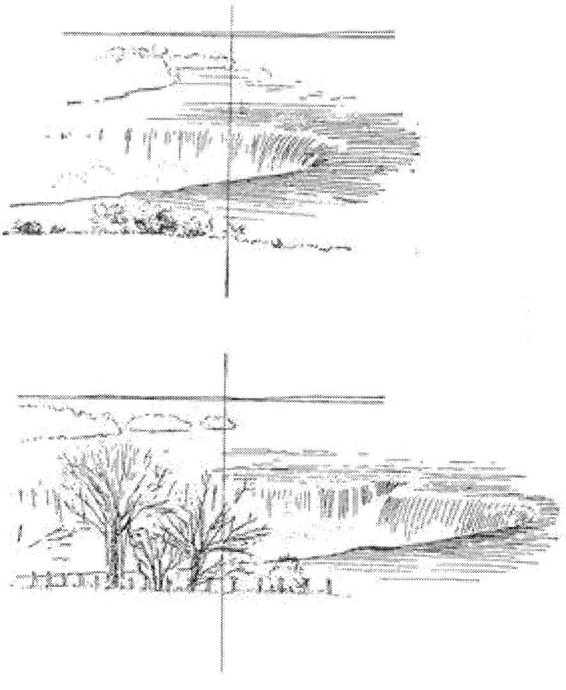

FIG. 382.—Comparison of a sketch of the Canadian Fall made with the aid of a camera lucida in 1827 with a photograph taken from the same view point in 1895 (after Gilbert).

The present rate of recession.—There are various sketches, more or less accurate, made in the early part of the nineteenth century, and from the later period there are daguerreotypes, photographs, and maps, which refer especially to the Canadian Fall; and which, taken together, render possible a comparison of the earlier with the later brinks. By comparing the earliest with the recent, views it is seen at a glance that the Falls are receding, and at a quite appreciable rate (Fig. 382). A careful comparison of the maps made in 1842, 1875, 1886, 1890, and 1905 of the brink of the Canadian Fall (Fig. 383) indicates that for the period covered the rate of recession has been about five feet per year, and similar studies made of the American Fall show that it has been receding at the rate of only three inches per year, or one twentieth the rate of the recession of the Canadian Fall.

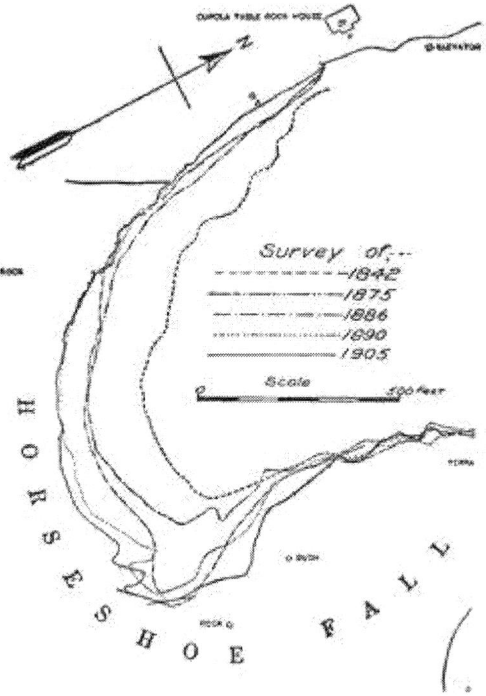

FIG. 383.—Map to show the recession of the brink of the Canadian Fall, based upon maps of different dates (after Gilbert).

Future extinction of the American Fall.—It is because of this many times more rapid recession of the Canadian Fall that the Niagara cataract, instead of lying athwart the gorge, enters it from its side. The Canadian Fall is thus in reality swinging about the American, and the time can already be roughly estimated when this more effective drilling tool will have brought about a capture, so to speak, of the American Fall through the cutting off of its water supply. It will then be drained and left literally "high and dry", an enduring witness to the geological effect of an island in making an unequal division of the waters for the work of two cataracts.

As already pointed out, the inefficiency of the American Fall as an eroding agent is amply attested by the wall of blocks already appearing above the water below it. The tourist who a thousand years hence pays a visit to the Niagara cataract, provided the water flow is allowed to remain as it has been, will find above this rampart of blocks a bare cliff in part undermined, and surmounted by a nearly flat table surface which is cut off from the existing cataract by a higher section of the gorge (Fig. 384). It is quite likely that this table will furnish the most satisfactory viewpoint of the future cataract of that date.

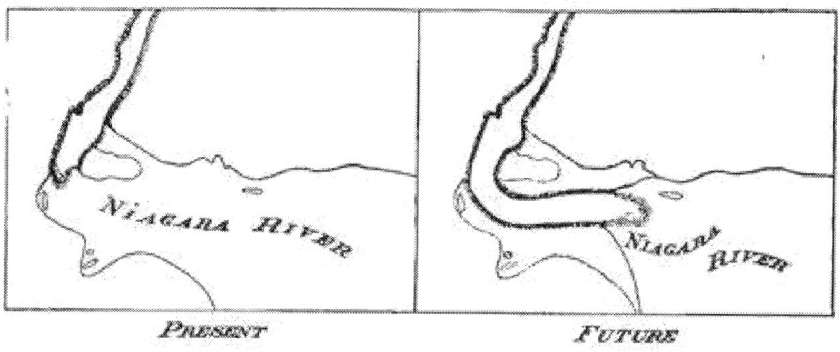

FIG. 384.—Comparison of the present with the future falls.

The captured Canadian Fall at Wintergreen Flats.—What we have predicted for the future of the present American Fall will be the better understood from the study of a monument to earlier capture made long before the upper section of the gorge had been cut or the whirlpool had come into existence. The tables were then turned, for it was a fall upon the Canadian side of the gorge that was captured by one upon the American. The locality is known as Wintergreen Flats, or sometimes as Fosters Flats; though the first name properly applies to a higher surface near the brink of the gorge, and Fosters Flats to a lower plain near the level of the river (see Fig. 381, p. 355). The peculiar topographic features at this locality are well brought out in Gilbert's bird's-eye view of the locality (Fig. 385); in fact, in some respects better than they appear to the tourist upon the ground, for the reason that the abandoned channel and the Flats on the site of the since undermined island are both heavily forested and so not easy to include in a single view. For one who has studied the existing cataract this early monument is full of meaning. Standing, as one may, upon the very brink of the former cataract, it is easy to call up in imagination the grandeur of the earlier surroundings and to hear the thunder of the falling water. A particularly vivid touch is added when, in digging over the sand about the great blocks of fallen limestone underneath the brink, one comes upon the shells of an animal still living in the Niagara River, though only in the continual spray beneath the cataract.

FIG. 385.—Bird's-eye view of the captured Canadian Fall at Wintergreen Flats, showing the section of the river bed above the cliff and the blocks of fallen Niagara limestone strewn over the abandoned channel below (after Gilbert).

The Whirlpool Basin excavated from the St. Davids Gorge.—It has already been pointed out that a rock channel now filled with glacial deposits extends from the Whirlpool Basin to the edge of the escarpment at St. Davids (Fig. 389, p. 363). In plan this buried gorge has a trumpet form, being more than two miles wide at its mouth and narrowing to the width of the upper gorge before it has reached the Whirlpool. Near the Whirlpool it has been in part excavated by Bowman Creek, thus revealing walls that are well glaciated. Different opinions have been expressed concerning the origin of this channel, one being that it is the course either of a preglacial river or one incised between consecutive glacial invasions; and another that it is a cataract gorge drilled out between glacial invasions after the manner of the later Niagara gorge. In either case its contours have been much modified by the later glacier or glaciers, whose work of planing, polishing, and widening is revealed in the exposed surfaces; and it is not improbable that a cataract has receded along the course of an earlier river valley.

As we shall see, there are facts which point rather clearly to an earlier cataract which ended its life immediately above the present Whirlpool. When the later Niagara cataract had receded to near the upper end of the Cove section, or near the present Whirlpool, the falling water must have been separated from this older channel and its filling of till deposits by only a thin wall of rock, and this must have been constantly weakened as its thickness was further reduced.

When this weakened dam at last gave way, it must have produced a debacle grand in the extreme. It is hardly to be conceived that the "washout" of the ancient channel to form the Whirlpool Basin could have occupied more than a small fraction of a day, though it is highly probable that the broken rock partition below the Whirlpool was not immediately removed entire. The mandible-like termination of the Eddy Basin immediately above the Whirlpool has led Taylor to believe that the cataract quickly reëstablished itself at this point upon the last site of the extinct St. Davids cataract. If reduced in power for a short interval, as a result of the obstructions still remaining in the lately broken dam below the Whirlpool, the remarkable narrowing of the gorge at this point would be sufficiently accounted for.

Being compelled to turn through more than a right angle after it enters the Whirlpool Basin, the swift current of the Niagara River is forced to double upon itself against the opposite bank and dive below the incoming current before emerging into the Cove section below the Whirlpool (Fig. 386).

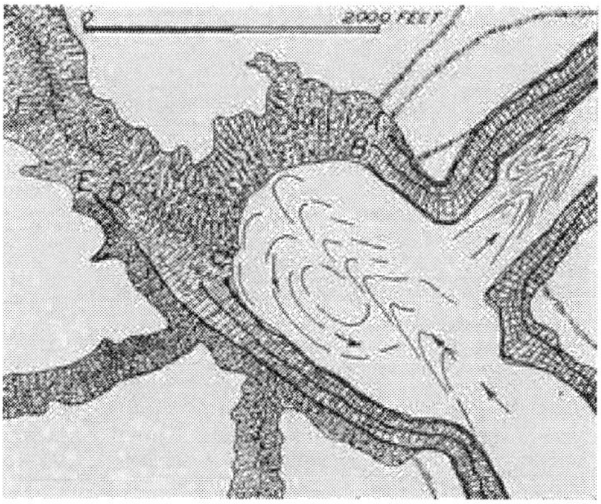

FIG. 386.—Map of the Whirlpool Basin, showing the rock side walls like those of the Niagara Gorge, and the drift bank which forms the northwest wall (after Gilbert).

In tearing out the loose deposits which had filled this part of the buried St. Davids Gorge, many bowlders of great size were left which slid down the slope and in time produced an armor about the looser deposits beneath, so as to protect them and prevent continued excavation. Thus it is found that the submerged northwestern wall of the basin is sheathed with bowlders large enough to retain their positions and so stop a natural process of placer outwashing upon a gigantic scale (Fig. 386).

The shaping of the Lewiston Escarpment.—To understand the formation of the Lewiston Escarpment cut in the hard Niagara limestone, it is necessary to consider the geology of a much larger area—that of the Great Lakes region as a whole. To the north of the Lakes in Canada is found a most ancient continent which was in existence when all the area to the southward lay below the waters of the ocean. In a period still very many times as long ago as the events we have under discussion, there were laid down off the shore of this oldland a series of unconsolidated deposits which, hardened in the course of time, and elevated, are now represented by the shales, sandstone, and limestone which we find, one above the other, in the Niagara gorge in the order in which they were laid down upon the ocean floor. The formations represented in the gorge are but a part of the entire series, for other higher members are represented by rocks about Lake Erie and even farther to the southward. These strata, having been formed upon an outward sloping sea floor, had a small initial dip to the southward, and this has been probably increased by subsequent uptilt, including the latest which we have so recently had under discussion. At the present time the beds dip southward by an angle of less than four degrees, or about thirty-five feet in each mile.

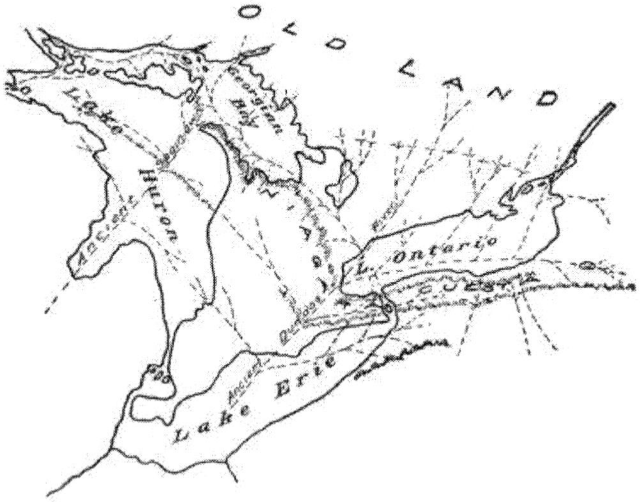

FIG. 387.—Map to show the cuestas which have played so important a part in fixing the boundaries of the Lake basins, and also the principal preglacial rivers by which they have been trenched (based upon a map by Grabau).

When the elevation of the land in the vicinity of this shore had caused a recession of the waters, there was formed a coastal plain on the borders of the oldland like that which is now found upon our Atlantic border between the Appalachians and the sea (Fig. 272, p. 246). The rivers from the oldland cut their way in narrow trenches across thc newland, and because of the harder limestone formations, their tributaries gradually became diverted from their earlier courses until they entered the trunk stream nearly at right angles and produced the type of drainage network which is called "trellis drainage." It is characteristic of this drainage that few tributaries of the second order will flow up the natural slope of the beds, but on the contrary these natural slopes are followed in the softer rock nearly at right angles again to the tributaries of the first order of magnitude (Fig. 387). Thus are produced a series of more or less parallel escarpments formed in the harder rock and having at their base a lowland which rises gradually in the direction of the oldland until a new escarpment is reached in the next lower of the hard formations. Such flat-topped uplands in series with intermediate lowlands and separated by sharp escarpments are known as *cuestas* (see p. 246), and the Lewiston Escarpment limits that formed in Niagara limestone (Figs. 387 and 388).

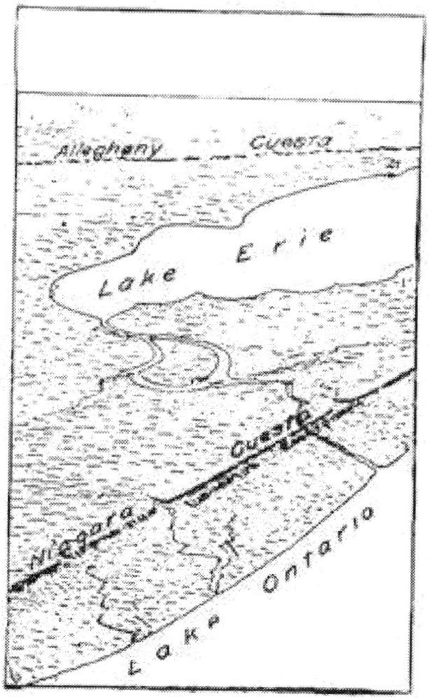

FIG. 388.—Bird's-eye view of the cuestas south of Lakes Ontario and Erie (after Gilbert).

Episodes of Niagara's history and their correlation with those of the Glacial Lakes.—Of the early episodes of Niagara's history, our knowledge is not as perfect as we could desire, but the later events are fully and trustworthily recorded. The birth of the Falls is to be dated at the time when the ice front had here first retired into what is now Canadian territory, thus for the first time allowing the waters from the Erie basin to discharge over the Lewiston Escarpment into the basin of the newly formed Lake Iroquois (Fig. 364, p. 334). Since the level of Lake Iroquois was far above that of the present Lake Ontario, the new-born cataract was not the equivalent in height of the escarpment to-day. The Iroquois waters then bathed all the lower portion of the escarpment, so that the foot of the Fall was upon the borders of the Lake.

In order to interpret the history of the Niagara gorge, we must remember that the effective drilling of this gorge was in each stage dependent mainly upon the volume of water discharged from Lake Erie, a large discharge being recorded by a channel drilled both wide and deep, while that produced by the discharge of a smaller volume was correspondingly narrow and shallow. To-day the gorges of large cross section have, moreover, a relatively placid surface, whereas through the constricted sections the water of the river is unable to pass without first raising its level at the upper end and under the head thus produced rushing through under an increased velocity. The best illustration of such a constricted section is the Gorge of the Whirlpool Rapids.

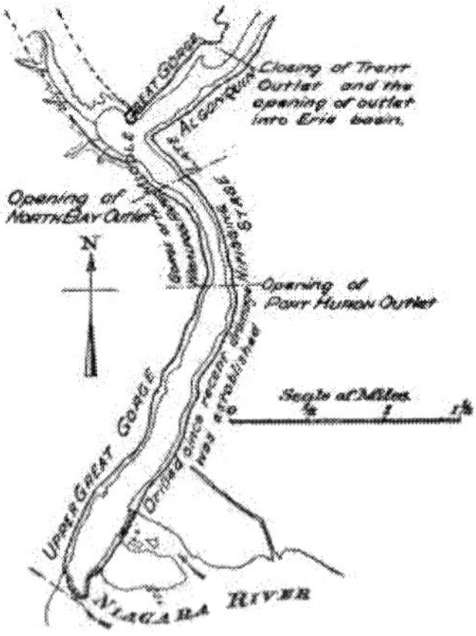

FIG. 389.—Sketch map of the greater portion of the Niagara Gorge to show the changes in cross section in their relations to Niagara history (based upon a map by Taylor).

Our reading of the history should begin at the site of the present cataract, since the records of later events are so much the more complete and legible, and it should ever be our plan to proceed from the clearly written pages to those half effaced and illegible.

As we have learned, the most abrupt change in the cross section of the gorge is found a little above the railroad bridges, where the Upper Great Gorge is joined to the Gorge of the Whirlpool Rapids (Fig. 389). In view of the remarkably uniform cross section of the Upper Great Gorge, there is no reason to doubt that it has been drilled throughout under essentially the same volume of water, and that its lower limit marks the position of the former cataract when the waters from the upper lakes were transferred from the "North Bay Outlet" into the present or "Port Huron Outlet" and Lake Erie. As the upper limit of the Gorge of the Whirlpool Rapids thus corresponds to the closing of the "North Bay Outlet" and the extinction of the Nipissing Great Lakes, so its lower limit doubtless corresponds to the opening of that outlet and the termination of the preceding Algonquin stage; for in the stage of the Nipissing lakes the water of the upper lakes, as we have learned, reached the ocean through the northern outlet.

Mr. Frank Taylor, who has given much study to the problem of Niagaran history, believes that the Middle Great Gorge, comprising the Eddy Basin and the Cove section, represents the gorge drilling which occurred during the later stage of Lake Algonquin after the "Trent Outlet" had been closed and the waters of the upper lakes had been turned into the Erie Basin.

Summarizing, then, the episodes of the lake and the gorge history are to be correlated as follows:—

Glacial Lake	Niagara Gorge
Early Lakes Iroquois and Algonquin.	Drilling of the gorge from the Lewiston Escarpment to the Cove section above the Wintergreen Flats.
Later Lakes Iroquois and Algonquin with upper lakes discharging into Erie basin.	Drilling of Middle Great Gorge.
Nipissing Great Lakes with the upper lake waters diverted from Lake Erie.	Drilling of the narrow Gorge of the Whirlpool Rapids.
Recent St. Lawrence drainage since the waters of the upper lakes were discharged into Lake Erie through occupation of the Port Huron Outlet.	Drilling of Upper Great Gorge to the present cataract.

Time measures of the Niagara clock.—In primitive civilizations time has sometimes been measured by the lapse necessary to accomplish a certain task, such, for example, as walking the distance between two points; and the natural clock of Niagara has been of this type. But men possess differences in strength and speed, and the same man is at some times more vigorous than at others, and so does not work at a uniform rate. The cataract of Niagara, charged with the pent-up energy of the waters of all the Great Lakes, can rush its work as it is clearly unable to do at times when the greater part of this energy has been diverted. Units of distance measured along the gorge are therefore too unreliable for our use, with the unique exception of the stretch from the railroad bridges to the site of the present cataract, within which stretch the gorge cross sections are so nearly uniform as to indicate an approximation to continued application of uniform energy. This energy we may actually measure in the existing cataract, and so fix upon a unit of time that can be translated into years.

In order to secure the normal rate of recession of this Upper Great Gorge, we should add to the volume of water in the Canadian Fall that now passing over the American; and for the reason that the blocks which fall from the cataract cornice and are the tools of the drilling instrument approximate to a definite size fixed by their joint planes, the effect of this added energy it is not easy to estimate. We may be sure, however, that the drilling action would be somewhat increased by the junction of the two Falls, and thus are assured that the average rate of recession within the Upper Great Gorge has been somewhat in excess of the five feet per year determined by Gilbert for the present Canadian Fall. The Upper Great Gorge is about two miles in length, and its beginning may thus be dated near the dawning of the Christian Era. The Whirlpool Gorge was cut when the ice vacated the North Bay Outlet in Canada, and still lay as a broad mantle over all northeastern Canada. For the earlier gorge and lake stages, the time estimates are hardly more than guesses, and we need not now concern ourselves with them.

The horologe of late glacial time in Scandinavia.—A glacial timepiece of somewhat different construction and of greater refinement has been made use of in Scandinavia to derive the "geochronology of the last 12,000 years." Instead of retreating over the land and impounding the drainage as it did so, the latest continental glacier of Scandinavia ended below sea level, and as it retired, its great subglacial river laid down a giant esker known as the Stockholm Os, which was bordered by a delta and fringed on either side by water-laid moraines of the block type. These recessional moraines are upon the average less than 1000 feet apart, and are believed to have each been formed in a single season. The delta deposits which surround the esker are of thin-banded clay, and as an additional uppermost band is found outside every moraine, these bands are also believed to represent each the delta deposit of a single year. In studies extending over many years, Baron de Geer, with the aid of a large body of student helpers, has succeeded in completing a count of moraines and clay layers, and so in determining the time to be 12,000 years since the ice front of the latest continental glacier lay across southern Sweden. The fertility of conception and the thoroughness of execution of this epoch-making investigation recommend its conclusion to the scientific reader.

Reading References for Chapter XXV

G. K. Gilbert. Niagara Falls and their History, Nat. Geogr. Soc. Mon., vol. 1, No. 7, 1895, pp. 203-236.

F. B. Taylor. Origin of the Gorge of the Whirlpool Rapids at Niagara, Bull. Geol. Soc. Am., vol. 9, 1898, pp. 59-84.

A. W. Grabau. Guide to the Geology and Paleontology of Niagara Falls and Vicinity, Bull. N. Y. State Mus., vol. 9, No. 45, 1901, pp. 1-85, pls. 1-11.

J. W. Spencer. The Falls of Niagara, etc. Dept. of Mines, Geol. Surv. Branch, Canada, 1907, pp. 490, pls. 43.

G. K. Gilbert. Rate of Recession of Niagara Falls, etc. Bull. 306, U. S. Geol. Surv., 1907, pp. 31, pls. 11.

G. de Geer. Quaternary Sea Bottoms of Western Sweden. Paper 23, Livret Guide Cong. Géol. Intern., 1910, pp. 57, pls. 3.

CHAPTER XXVI

LAND SCULPTURE BY MOUNTAIN GLACIERS

Contrasted sculpturing of continental and mountain glaciers.—In discussing in a previous chapter the rock pavement lately uncovered by the Greenland glacier, we learned that this surface had been lowered by the processes of plucking and abrasion, the combined effect of which is always to reduce the irregularities of the surface, soften its outlines, and from sharply projecting masses to develop rounded shoulders of rock—*roches moutonnées.*

Though the same processes act in much the same manner beneath mountain glaciers, though here upon all parts of the bed, they are, in the earlier stages at least, subordinated to a third process more important than the two acting together. Sculpture by mountain glaciers, instead of reducing surface irregularities and softening outlines, increases the accent of the relief and produces the most sharply rugged topography that is known. In nearly all places where Alpinists resort for difficult rock climbing, mountain glaciers are to be seen, or the evidence for their former presence may be read in unmistakable characters.

Wind distribution of the snow which falls in mountains.—Until quite recently students of glaciation have concerned themselves but little with the work of the wind in lifting and redistributing the snow after it has fallen. We have already seen that, for the continental glaciers, wind appears to be the chief transporting agent, if we except the marginal lobes where glacier flow assumes large importance. In the case of mountain glaciers, also, we are to find that for the earlier stages particularly wind is of the first importance as a redistributing agent. In the higher levels snow is swept up from the ground by all high winds, and does not find a resting place until it is dropped beneath an eddy in some irregularity of the surface; and if the inherited surface be relatively smooth, this will be found in most cases upon the lee of the mountain crest.

In normal cases at least the inherited irregularities of the higher zones of mountain upland are the gentle depressions which develop at the heads of streams. These become, then, the sites of snowdrifts that are augmented in size from year to year, though at first they melt away in the late summer.

FIG. 390.—Snowdrift hollowing its bed by nivation and building a delta (at the left). Quadrant Mountain, Yellowstone National Park.

The niches which form on snowdrift sites.—Wherever a drift is formed, a process is set in operation, the effect of which is to hollow out and lower the ground beneath it, a process which has been called *nivation*. The drift shown in Fig. 390 was photographed in late summer at an elevation of some 9000 feet in the Yellowstone National Park. The very gently sloping surface surrounding the drift is covered with grass, but within a zone a few feet in width on the borders of the drift no grass is growing, and in its place is found a fine brown soil which is fast becoming the prey of the moving water derived by melting of the drift. This is explained by the water permeating the crevices of the rock and being rent by the nightly freezing. Farther from the drift the ground is dry, and no such action is possible. With each succeeding spring the augmented drift as it melts carries all finely comminuted rock material down slopes beneath the snow to emerge at the lowest margin and be there deposited in the form of a delta. By the operation of this process of nivation the higher parts of the drift site are lowered as deposition goes on upon the lower. The combined effect is thus to produce a *niche* or faintly etched amphitheater upon the slope of the mountain (Fig. 391).

FIG. 391.—Amphitheater formed on a drift site in northern Lapland (after a photograph by G. von Zahn).

The augmented snowdrift moves down the valley—birth of the glacier.—In still lower air temperatures the drifts enlarge with each succeeding year until they endure throughout the summer season. From this stage on, an increment of snow is left from each succeeding season. No longer entirely wasted by melting, the time soon comes when the upper snow layers will by their weight compress the lower into ice, and the mass will begin to creep down the slope along the course of the inherited valley. The enlarged snowdrift which feeds this ice stream is called the *névé* or *firn*.

Against the sloping cliff which had been shaped by nivation at the upper margin of the snowdrift, that snow which is not of sufficient depth to begin a movement towards the valley separates from the moving portion, opening as it does so a cleft or crevasse parallel to the wall. This crack in the snow is called by its German name *Bergschrund* or *Randspalte*, and may perhaps be referred to as the marginal crevasse (Fig. 392).

FIG. 392.—The marginal crevasse or Bergschrund on the highest margin of a glacier (after Gilbert).

The excavation of the glacial amphitheater or cirque.—It has been found that the marginal crevasse plays a most important rôle in the sculpture of mountains by glaciers, for the great amphitheater which is everywhere the collecting basin for the nourishment of mountain glaciers is not an inherited feature, but the handiwork of the ice itself. This was the discovery of Mr. W. D. Johnson, an American topographer and geologist, who, in order to solve the problem of the amphitheater allowed himself to be lowered into such a crevasse upon the Mount Lyell glacier of the Sierra Nevadas in California.

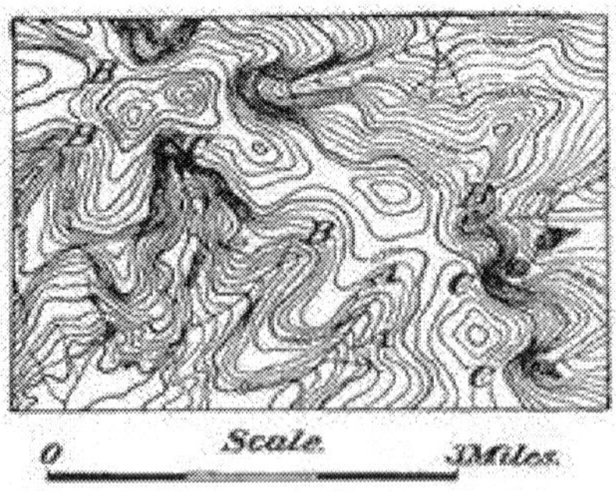

FIG. 393.—Niches and cirques in the same vicinity in the Bighorn Mountains of Wyoming. *A*, *A*, unmodified valleys; *B*, *B*, niches on drift sites; *C*, *C*, cirques on small glacier sites (after map by F. E. Mathes, U. S. G. S.).

Let down a distance of a hundred and fifty feet, he reached the bottom of the crack, and in a drizzling rain of thaw water stood upon a floor composed of rock masses in part dislodged from a wall which extended some twenty feet upwards upon the cliff side of the crevasse. It was evident that the warm air of the day produced the thaw water which was constantly dripping and which filled every crack and cranny of the rock surface. With the sinking of the sun below the peaks the sudden chill, so characteristic of the end of the day in high mountains, causes this water to freeze and thus rend the rock along its planes of jointing. Broad and thin plates of ice, loosened by melting at the walls, could be extracted from the crevices of the rock as mute witnesses to the powerful stresses developed by this most vigorous of weathering processes.

FIG. 394.—Subordinate small cirques in the amphitheater on the west face of the Wannehorn above the Great Aletsch Glacier of Switzerland.

In short, the rock wall above the glacier, which in its initial stage was the upper wall of the niche hollowed beneath the snowdrift, is first steepened and later continually both recessed and deepened by an intensive frost rending which is in operation at the base of the marginal crevasse. The same process does not go on as rapidly above the surface of the névé for the reason that *the necessary wetting of the rock surface does not there so generally result from the daily summer thaw.* At the bottom of the marginal crevasse alone is this condition fully realized. Intensive frost action *where the rock is wet with thaw water daily* is thus a fundamental cause, both of the hollowing of the early drift site to form the niche, and of the later enlargement of this niche into an amphitheater or cirque when the drift has been transformed into the névé of a glacier. Inasmuch as the crevasse forms where the snow and ice pull away from the rock toward the middle of the depression, the cirque wall in its early stage has the outline of a semicircle. In the Bighorn Mountains of Wyoming, all stages, from the unmodified valley heads to the full-formed cirque, may be seen near one another (Fig. 393). It will be noted that wherever a glacier has formed, as indicated by the cirque, there is a series of lakes which have developed in the valley below (see p. 412).

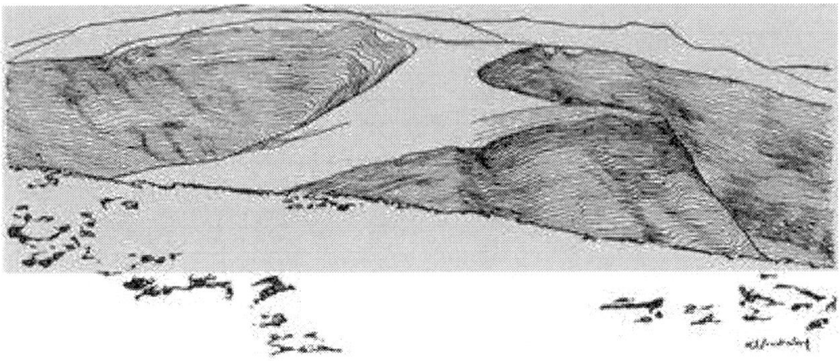

FIG. 395.—"Biscuit cutting" effect of glacial sculpture in the Uinta Mountains of Wyoming (after Atwood).

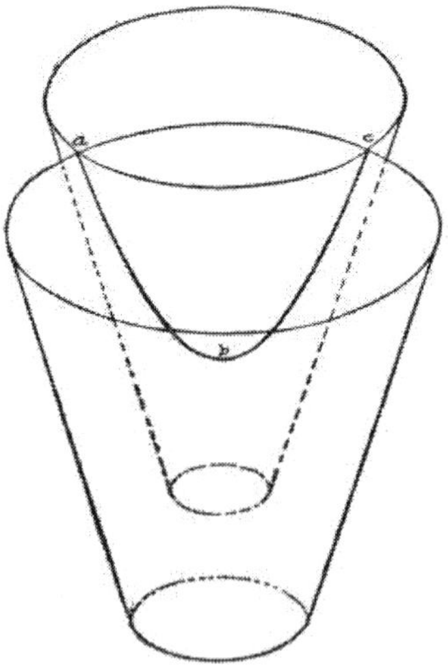

FIG. 396.—Two intersecting inverted cones representing glacial cirques of different sizes, to show that their intersection is the arc of a hyperbola, the curve to which the col approximates.

Life history of the cirque.—In its earliest stage the cirque is more or less uniformly supplied with snow from all sides, and so it enlarges by recession in a manner to retain its early semicircular outline. In a later stage a larger proportion of the snow reaches the cirque at its sides so that its further enlargement causes it to broaden and to flatten somewhat that part of its outline which represents the head of the valley (Fig. 389, p. 364). As the territory of the upland is still further invested by the cirques, their nourishment becomes still more irregular, and the circular outline gives place to a scalloped border, as the amphitheater becomes differentiated into subordinate smaller cirques, each of which corresponds to a scallop of the outline (Fig. 398 and Fig. 394).

Grooved and fretted uplands.—The partial investment by cirques of a mountain upland yields a type of topography quite unlike that produced by any other geological process. The irregularly connected remnants of the inherited upland resemble nothing so much as a layer of dough from which biscuits have been cut (Fig. 395). The surface as a whole, furrowed as it is below the cirques, may be described as a *grooved upland* (plate 19 A). A further continuation of the process removes all traces of the earlier upland, for the cirques intersect from opposite sides and thus yield palisades of sharp rock pinnacles which rise on precipitous walls from a terraced floor. This ultimate product of cirque sculpture by glaciers is called a *fretted upland* (plate 18 A and 19 B).

PLATE 18.

A. **Fretted upland of the Alps seen from the summit of Mount Blanc.**

B. **Model of the Malaspina Glacier and the fretted upland above it (after model by L. Martin).**

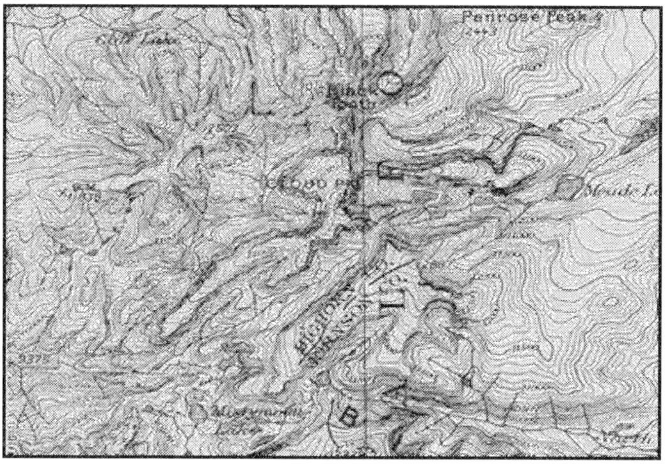

A. **Contour map of a grooved upland, Bighorn Mountains, Wyoming (U. S. Geol. Survey).**

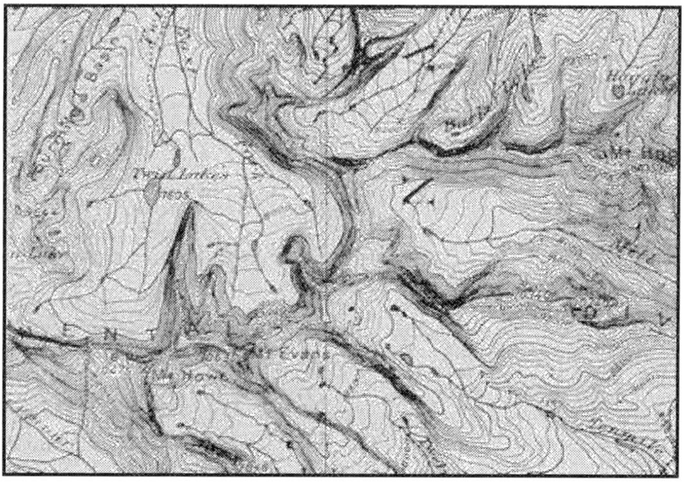

B. **Contour map of a fretted upland, Philipsburg Quadrangle, Montana (U. S. Geol. Survey).**

The features carved above the glacier.—The ranges of pinnacles carved out by mountain glaciers have become known by various names of foreign derivation, such as *arête*, *grat*, *aiguille* mountains, "files of *gendarmes*", etc. They may, perhaps, be best referred to as *comb ridges*, and according to their position they are differentiated into main and lateral comb ridges, as will be clear from the second map of plate 19.

FIG. 397.—A col shaped like a hyperbola between Mount Sir Donald and Yogo Peak in the Selkirks (after a plate by the Keystone Plate Co.).

With the gradual invasion of the upland upon which the cirques have made their attack, the area from which winds may gather up the snow is steadily diminished, and hence cirque recession is correspondingly retarded. Cirques which have approached each other from opposite sides of the ridge until they have become tangent at one point may, however, still receive nourishment at the sides and so continue to cut down the intervening rock wall to form a pass or *col*. The theoretical curve which results from this intersection is that known as the hyperbola, of which an illustration is afforded by Fig. 396. An approximation to this form is clearly furnished by most of the mountain passes in glaciated mountain districts, and a particularly good illustration is furnished from the vicinity of Glacier on the line of the Canadian Pacific Railway (Fig. 397).

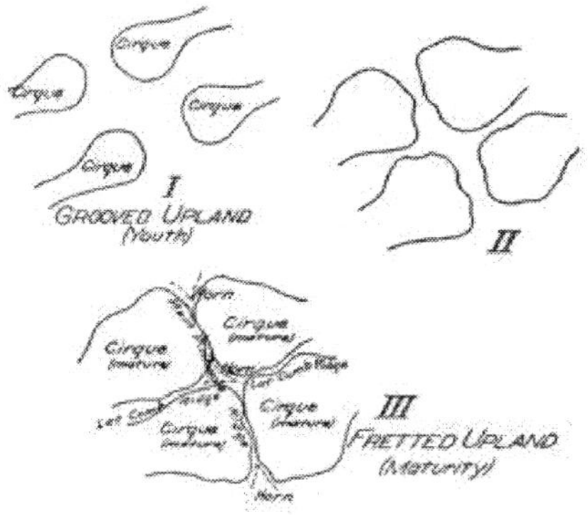

FIG. 398.—Diagrams to illustrate the progressive investment of an upland by cirques with the formation of comb ridges, cols, and horns. I, early stage, youth; II, intermediate stage; III, late stage, maturity.

Upon either side of the col the land mass is left in high relief, rising from a more or less triangular base (Fig. 398, III) into a sharp horn or tooth. An illustration of such a *horn* is furnished by the Matterhorn in the Swiss Alps, or by Mount Sir Donald in the Selkirks, though less noteworthy examples may be found in every maturely glaciated mountain district.

The features shaped beneath the glacier.—Those features which are carved above the glacier—the comb ridge, the col, and the horn—are all shaped as a result of intensive weathering upon the cirque wall. The shaping at lower levels is accomplished by processes in operation below the glacier surface, where weathering is excluded and where plucking and abrasion work together to tear away and grind off the rock surface. By their joint action the valley is both deepened and widened, directly to the height of the glacier surface, and indirectly through undermining as far up as rock extends. Thus the valley is transformed into one of broad and flat bed and precipitous side walls—the U-shaped section illustrated by valleys of the Swiss Alps and in fact in all districts which have been strongly glaciated by mountain glaciers (Fig. 399).

FIG. 399.—The U-shaped Kern valley in the Sierra Nevadas of California (after W. B. Scott).

As high up in the valley as it has been occupied by the glacier, the bed is rounded, smoothed, and polished, and marked by the characteristic glacial scorings or striæ which point down the valley. Above the level of the glacier's upper surface, on the other hand, erosion is accomplished through undermining or sapping, a process which always leaves precipitous slopes of ragged surface made up of the joint planes on which the fallen blocks have separated from the cliff. Thus there is found a sharp line which separates the smoothly rounded rock surface below from the jagged and precipitous one above (Fig. 400). Inasmuch as this boundary usually separates the scalable from the inaccessible slopes above, snow is apt to lodge at this level and make it strikingly apparent.

FIG. 400.—Glaciated valley wall in the Sierra Nevadas of California, showing the sharp line which separates the abraded from the undermined rock surface (after a photograph by Fairbanks).

FIG. 401.—View of the Vale of Chamonix from the séracs of the Glacier des Bossons. The alb of the opposite side is well brought out.

If uplift of the land occurs while glaciers occupy the valleys of mountains, an increased capacity for deepening the valley is imparted to these ice streams, and we find, as a result, a deep central valley of U cross section excavated within a relatively broad trough visible above the shoulder on either side of the later furrow. Save only for its characteristic curves, such a valley bears close resemblance to a mature stream valley which has been rejuvenated (see p. 173). The remnants of the earlier glacier-carved valley are, as already stated, gently curving high terraces so common in Switzerland, where they are known as *albs* or high mountain meadows. These albs may be seen to special advantage on the sides of the Chamonix valley (Fig. 401), the Lauterbrunnen valley, or in fact almost any of the larger Alpine valleys.

The cascade stairway in glacier-carved valleys.—If now, instead of giving our attention to the cross section, we follow the course of the valley that has been occupied by a glacier, we find that it descends by a series of steps or terraces having many backwardly directed treads (plate 19), whereas a normal and well-established river valley has only forward grades. Because of these backward grades the stream waters are impounded, and so lakes are found strung along the valley in chains as the larger beads are found in a rosary, and these are the characteristic *rock basin lakes* sometimes referred to as "Paternoster Lakes" (see p. 412 and Fig. 402).

PLATE 20.

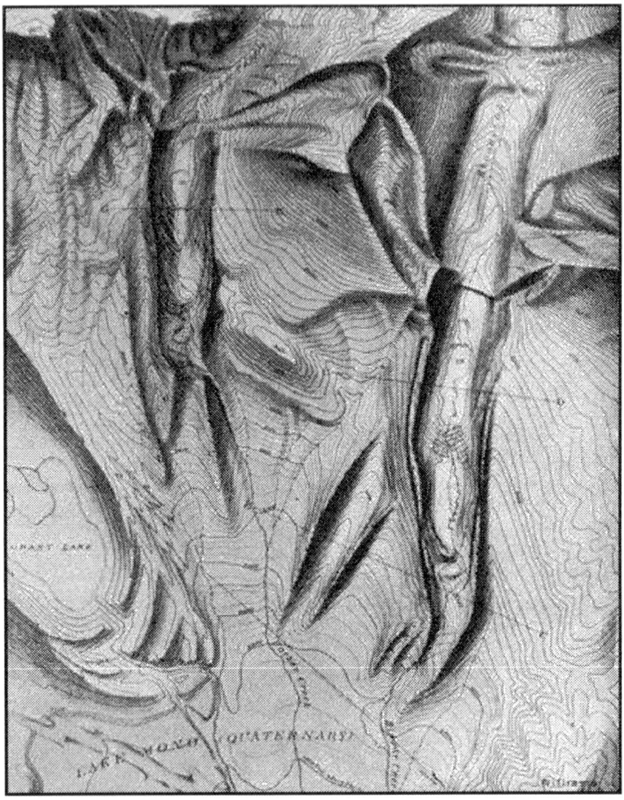

Map of the surface modeled by mountain glaciers in the Sierra Nevadas of California (after I. C. Russell).

When the backward grades upon the valley floor are especially steep, the rock step becomes a *rock bar*, or *Riegel*, of which nearly every Alpine valley has its examples. In a walk from the Grimsel to Meiringen many such bars are passed. Carrying in suspension the sharp rock sand from the glacier deposits along its bed, the stream which succeeds to the glacier as it vacates its valley saws its way through these obstructions with a rapidity that is amazing, thus producing narrow defiles, of which the Gorge of the Aar near

Meiringen and that of the Gorner near Zermatt are such well-known examples (Fig. 403).

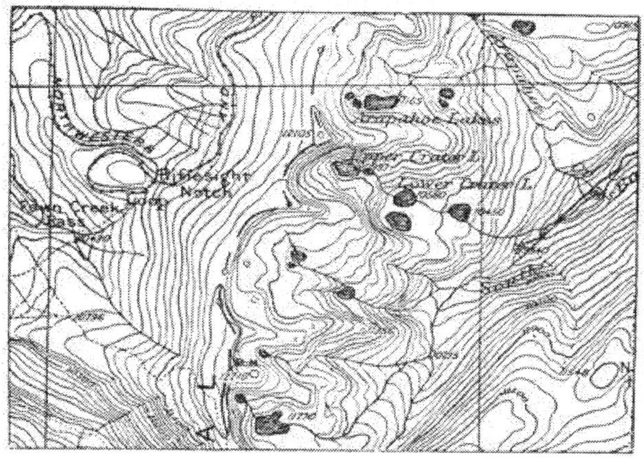

FIG. 402.—Map of an area near the continental divide in Colorado, showing an unglaciated surface to the west of the divide, where the westerly winds have cleared the ground of snow, and the glacier-carved country to the eastward. Note the regular forms of the youthful cirque, the glacier stairway, and the rock basin lakes (U. S. G. S.).

FIG. 403.—Gorge of the Albula River near Berkum in the Engadine, cut through a rock bar by the river which has succeeded to the earlier glacier.

FIG. 404.—Idealistic sketch showing both glaciated and nonglaciated side valleys tributary to a glaciated main valley (after Davis).

It is characteristic of rivers that the tributaries cut their valleys more rapidly than does the main stream within the neighboring section, though they cannot cut lower than their outlets—the side streams enter *accordantly*. This is easily explained because the grades of the tributary streams are the steeper, and, as we well know, the corrosion of a valley is augmented at a most amazing rate for each increase of its grade. No such law controls the processes of plucking and abrasion by which the glacier lowers its floor, for these processes appear to depend for their efficiency upon the depth of the ice, and the supply of cutting tools, quite as much as upon the grade of the bed. To apply a homely illustration, the hollowing of flagstones upon our walks is dependent more upon the number of persons that pass over them, and upon their size and the number of protruding nails in their boot heels, than upon the grades upon which they are placed. At all events we find that the main glacier valleys are cut deeper than the side valleys, so that the latter become *hanging valleys*—they enter the main valley, not upon its bed, but some distance above it (Fig. 404).

The U-shaped hanging valleys, like the main valley, are much too large for the streams which now fill them, and these diminutive side streams plunge over the steep wall of the main valley in ribbon-like falls so thin that the wind turns them aside and disperses the water in the spray of a "bridal veil." Such falls are found by the hundred in every glaciated mountain district, imparting to it one of the greatest of its scenic charms.

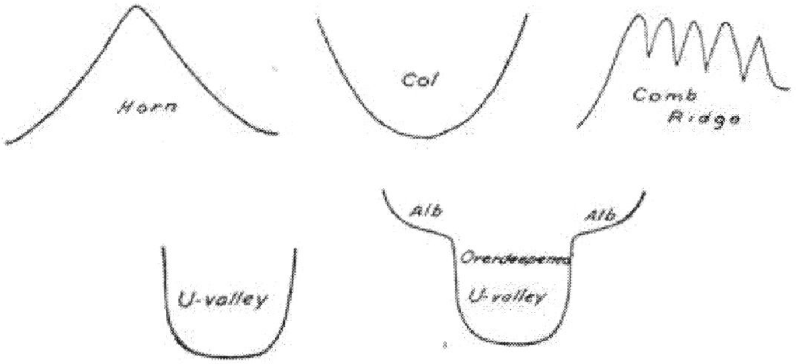

FIG. 405.—Character profiles in landscapes sculptured by mountain glaciers.

FIG. 406.—Flat dome shaped under the margin of a Norwegian ice cap with projecting rock knobs and moraines in foreground.

The character profiles which result from sculpture by mountain glaciers.—The lines which are repeated in landscapes carved by mountain glaciers are easy to recognize (Fig. 405). The highest horizon lines are the outlines of horns which are separated by cols. Minaret-like palisades, or "files of *gendarmes*", often run for long distances as the characteristic comb ridges. Lower down and lacking the lighter background of the sky, we make out with less distinctness the U-valley, either with or without the albs to show that the sculpturing process has been interrupted by uplift.

FIG. 407.—Two views illustrating successive stages in the shaping of tinds or "beehive" mountains.

The sculpture accomplished by ice caps.—In the case of ice caps, the only rock exposed is found in the neighborhood of the margin—the projecting islands known as nunataks. It is essential for the existence of the ice cap that the rock base should have relatively slight irregularities compared to the dimensions of the cap itself. Except in very high latitudes this base must be somewhat elevated, for like mountain glaciers ice caps are nourished by the surface air currents, and their snows are deposited above the snow line.

The Norwegian tind or beehive mountain.—Within temperate or tropical climes the snow line lies so high that only the loftier mountains are able to support glaciers. It follows that those which are formed flow upon relatively high grades with correspondingly high rate of movement and increased cutting power. Within high latitudes the snow is found nearer the sea level, and glaciers are for the most part correspondingly sluggish in their movements as well as less active denuding agents.

To this condition characteristic of high latitude glaciers, there is added in Norway another in the peculiar shape of the basement beneath the recent and the still existing glaciers. The plateau of Norway is intersected by a network of deep and steep walled fjords, and the glaciers have developed as small ice caps perched upon veritable pedestals of rock, over the margins of which their outlet tongues of ice descend on steep slopes into the fjord. The tops of the pedestals thus come to be shaped by the plucking and abrading processes into flat domes (Fig. 406), while the knobs of rock, which as nunataks reach above the surface of the ice, divide the outflowing ice tongues at the margin of the pedestal. These tongues being much more active denuding agents, because of their steep gradients, continually lower their beds, thus transforming the earlier knobs of rock into high and steep mountains of more or less circular base. Such "beehive" mountains upon the margins of the fjords are the characteristic Norwegian *tinds* (Fig. 407).

READING REFERENCES FOR CHAPTER XXVI

I. C. RUSSELL. Quaternary History of Mono Valley, California, 8th Ann. Rept. U. S. Geol. Surv., 1889, pp. 329-371, pls. 27-37.

F. E. MATTHES. Glacial Sculpture of the Bighorn Mountains, Wyoming, 21st Ann. Rept. U. S. Geol. Surv., 1900, Pt. ii, pp. 179-185, pl. 23.

W. D. JOHNSON. Maturity in Alpine Glacial Erosion, Jour. Geol., vol. 12, 1904, pp. 569-578.

G. K. GILBERT. Systematic Asymmetry of Crest Lines in the High Sierras of California, *ibid.*, pp. 579-588.

EMM. DE MARTONNE. Sur la Formation des Cirques, Ann. de Géogr., vol. 10, 1901, pp. 10-16.

W. M. DAVIS. Glacial Erosion in North Wales, Quart. Jour. Geol. Soc. Lond., vol. 65, 1909, pp. 281-350, pl. 14.

ED. BRÜCKNER. Die Glazialen Züge im Antlitz der Alpen, Naturw. Wochenschr., N. F., vol. 8, 1909.

WILLIAM H. HOBBS. Characteristics of Existing Glaciers, pp. 1-96.

CHAPTER XXVII

SUCCESSIVE GLACIER TYPES OF A WANING GLACIATION

Transition from the ice cap to the mountain glacier.—A study of existing glaciers leads inevitably to the conclusion that although subject to short period advances and retreats, yet, broadly speaking, glaciers are now gradually wasting away, surrounded by wide areas upon which are the evidences of their recent occupation. We are thus living in a receding hemicycle of glaciation.

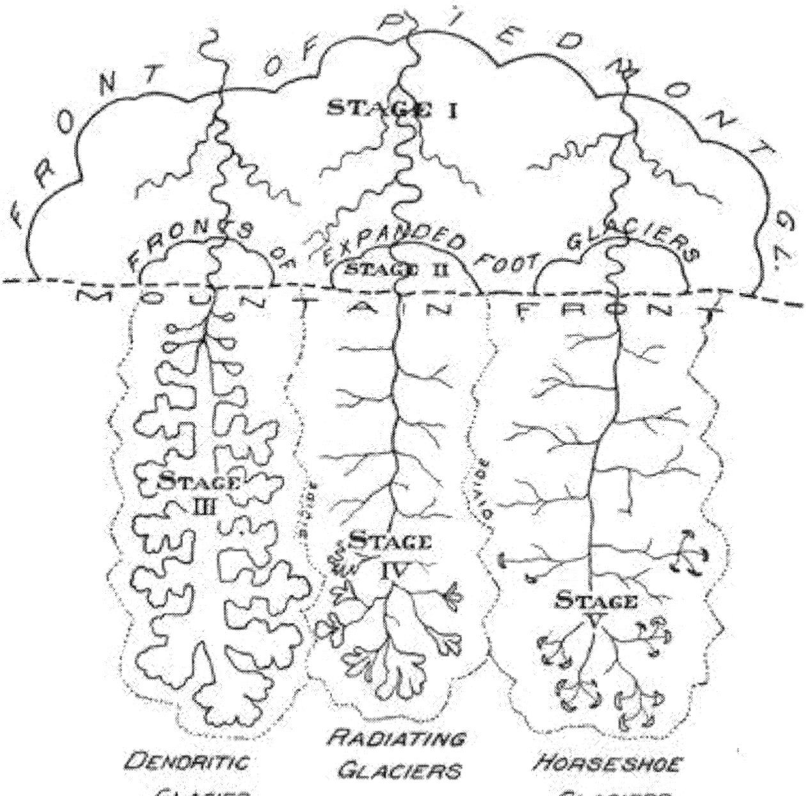

FIG. 408.—Schematic diagram to show the relationships of glacier types formed in succession during a receding hemicycle of glaciation.

Many mountain districts which now support small glaciers only, or none at all, were once nearly or quite submerged beneath snow and ice. If once covered by an ice carapace or cap, our present interest in them begins at that stage of the receding hemicycle *when the rock surface has made its reappearance above the surface of the snow-ice mass.* At this stage intensive frostwork, the characteristic high level weathering, begins, and cirques develop above the scars of those earlier amphitheaters formed in the advancing hemicycle.

The piedmont glacier.—In this early stage of transition from the ice cap to the mountain glacier, the ice flows outward to the mountain front in ill-defined streams divided by the projecting ridges, and upon reaching the mountain front these streams deploy upon it so as to coalesce in a great stagnant ice apron whose upper surface slopes gently forward at an angle of a few degrees at the most (Fig. 408, stage I). This is the *piedmont glacier*, a type found to-day in the high latitudes of Alaska and in the southern Andes (Fig. 409 and pl. 18 B).

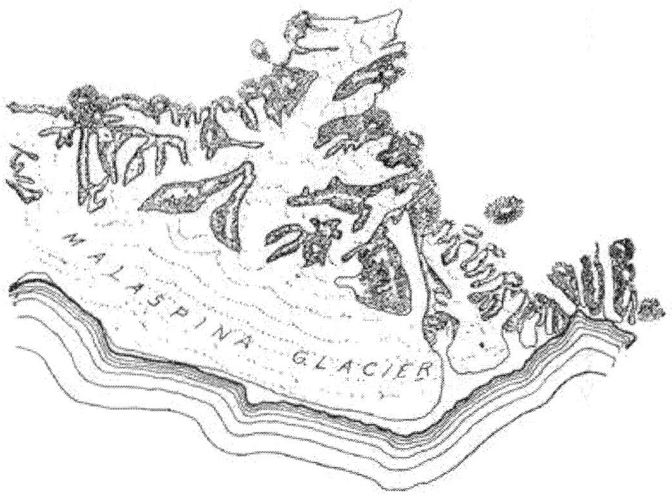

FIG. 409.—Map of the Malaspina glacier of Alaska, the best known of existing piedmont glaciers (after Russell).

During this stage the cirques may be but poorly defined, and ice flows in both directions from rock divides so that the streams transect the range, and later, after the glaciers have disappeared, may expose a pass smoothed and polished upon its floor and with striæ directed in opposite directions from the highest point. The pass of the Grimsel in Switzerland furnishes an excellent illustration of such earlier transection of the range.

FIG. 410.—Map of the Baltoro glacier of the Himalayas, a typical glacier of the dendritic type.

The expanded-foot glacier.—As air temperatures continue to become milder, the glacier streams within the mountains are less deep and hence more clearly defined, and instead of coalescing upon the mountain foreland, they now issue from the mountains to form individual aprons and are described as *expanded-foot glaciers* (Fig. 408, stage II, and Fig. 292, p. 264).

FIG. 411.—The Triest glacier, a hanging glacieret separated from the Great Aletsch glacier to which it was lately a tributary.

The dendritic glacier.—Still later in the hemicycle nourishment of the glaciers is diminished as depletion from melting increases, so that the glacier streams no longer reach to the mountain front. Branches continue to enter the main valley from the several side valleys like the short branches of a tall tree, and because of this arrangement such a glacier may be described as a *dendritic glacier* (Fig. 408, stage III, and Fig. 410).

Inasmuch as the depletion from melting increases at a rapid rate in descending to lower levels, the tributary glacier valleys "hanging" above the main valley in the lower stretches become separated, and may continue to exist as series of hanging glacierets upon either side of the main valley below the glacier front (Fig. 408, stage III, and Fig. 411). It must be clear from this that any attempt to name each separated ice stream without regard to its relationship must lead to endless confusion, for glacier size is in such sensitive adjustment to air temperature that a fall or rise of a few degrees only in the average annual temperature of the district may prove sufficient to fuse many glaciers into one or separate one ice mass into many smaller ones.

When in high latitudes a dendritic glacier descends in fjords to below the level of the sea, it is attacked by the water in the same manner as are the outlets of Greenland glaciers, and is then known as a "tidewater glacier", which may thus be a subtype or variety of the dendritic glacier (Fig. 412).

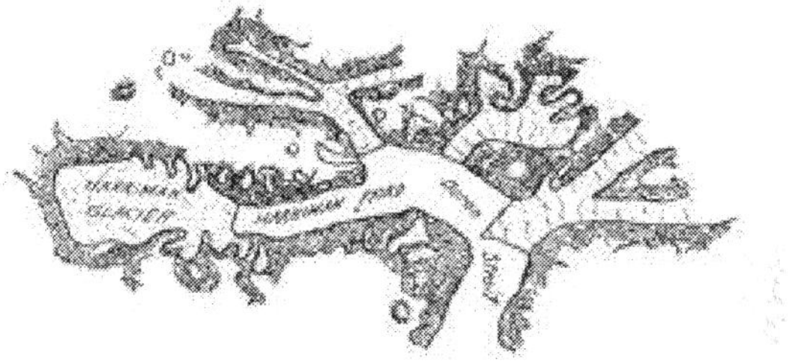

FIG. 412.—The Harriman fjord glacier of Alaska, a tidewater variety of dendritic glacier (after a map by Gannett).

The radiating (Alpine) glacier.—In the progressive wastings of dendritic glaciers, there comes a time when their dendritic outlines give place to radiating ones. Attention has already been called to the division of the cirque into subordinate basins separated by small rock arêtes and yielding a markedly scalloped border (Fig. 394, p. 371). When the ice front retires from the main valley into one of these mature cirques, the now wasted ice stream is broken up into subordinate glacierets, each of which occupies one of the basins within the larger cirque, and these ice streams flow together to produce a glacier whose component elements radiate like the sticks within a lady's fan (Fig. 408, stage IV, and Fig. 413).

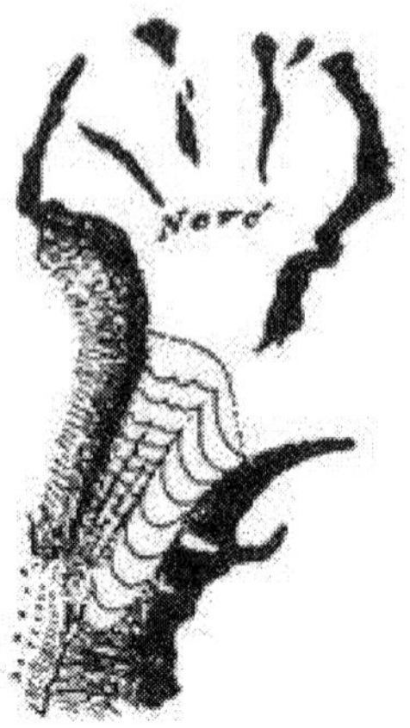

FIG. 413.—Map of the Rotmoos glacier, a radiating glacier of Switzerland (after Sonklar).

The horseshoe glacier.—As the glacier draws near to its final extinction, it is crowded hard against the wall of the amphitheater in which it has so long been nourished. Up to this stage it has offered a swelling front outwardly convex as a direct consequence of the laws controlling its flow. No longer amply nourished, for the first time its front is hollowed, and it awaits its final dissolution curled up against the cirque wall (Fig. 408, stage V, and Fig. 414). Practically all the glaciers of the United States and southern Canada are of this type.

The above classification is one depending directly upon glacier nourishment, and hence also upon size, and upon the stage of the glacial hemicycle. In order to determine the type of any glacier it is necessary to know the outlines of the mountain valley—its divide—and those of the glacier or glaciers within it. It is likely that the types of the advancing hemicycle of glaciation would be much the same, save only for the *new-born* or *nivation glacier*, which would be as different as possible from the horseshoe type, to which in size it corresponds. Upon the continent of Antarctica, where the absence of any general melting of the ice, even in the summer season and near the sea level, introduces special conditions, some additional glacier types are found, which, however, it is not necessary that we consider here.

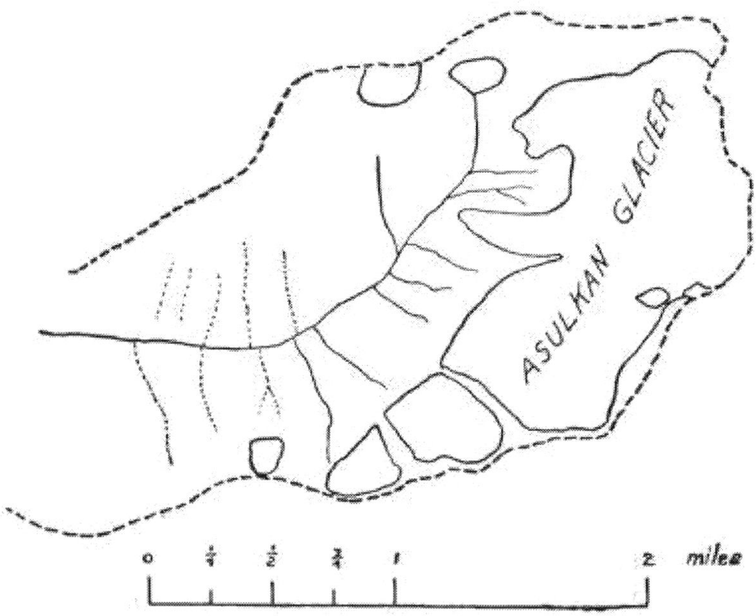

FIG. 414.—Outline map of the Asulkan glacier in the Selkirks, a typical horseshoe glacier.

The inherited-basin glacier.—It may be, however, that glaciers have developed, not upon mountains shaped in a cycle of river erosion, nor yet in succession to an ice cap, as in the normal cases which we have considered. On the contrary, glaciers may develop where basins of one sort or another have been inherited from the preceding period. In such cases inherited depressions may become more important than the auto-sculpture of the glacier. Glaciers which develop under such conditions may be described as *inherited-basin glaciers*.

FIG. 415.—Outline map of the Illecillewaet glacier, an inherited-basin glacier in the Selkirks.

A partly closed basin between ridges may supply a collecting ground for snows carried from neighboring slopes by the wind, and so may yield a broad névé, approaching in size a small ice cap, yet without developing definite ice streams except upon its border. Such a glacier is the Illecillewaet glacier of the Selkirks (Fig. 415).

Again in low latitudes the high and pointed volcanic peaks may push up beyond the snow line into the upper atmosphere, and so become snow-capped. Definite cirques do not develop well under these circumstances, and the loose materials of which such peaks are always composed are attacked in somewhat irregular fashion from the different sides. This is the case of Mount Rainier and similar peaks of the Cascade range of North America.

Summary of types of mountain glacier.—In tabular form the various types of mountain glacier may be arranged as follows:—

Piedmont glacier. Mountain valleys entirely occupied and largely submerged, with overflow upon the foreland to form a common ice apron through coalescence of neighboring streams.

Expanded-foot glacier. Valley entirely occupied and an overflow upon the foreland sufficient to produce individual ice apron.

Dendritic glacier. Valley not completely occupied but with tributary ice streams ranged along the sides of the main stream, and with hanging glacierets separated near the glacier foot.

Radiating glacier. Glacier largely included in a cirque with subordinate glacierets converging below like the sticks in a lady's fan.

Horseshoe glacier. Small glacier remnants hugging the cirque wall and having an incurving front.

Inherited-basin glacier. Of form dependent on a basin inherited and not shaped by the glacier itself.

READING REFERENCE FOR CHAPTER XXVII

WILLIAM H. HOBBS. The Cycle of Mountain Glaciation, Geogr. Jour., vol. 37, 1910, pp. 268-284.

CHAPTER XXVIII

THE GLACIER'S SURFACE FEATURES AND THE DEPOSITS UPON ITS BED

The glacier flow.—The downward flow of the ice within a mountain glacier has been the subject of many investigations and the topic of many heated discussions since the time when Louis Agassiz and his companions set a line of stakes across the Aar glacier and numbered the surface bowlders in preparation for repeated observations. Their first observation was that the line of stakes, which had run straight across the glacier, was distorted into a curve which was convex downstream (Fig. 416, A´), thus showing that the surface layers have more rapid motion in proportion as they are distant from the side margins. Summarizing these and later studies, it may be stated that the glacier increases its rate of motion from its side margin towards its center line, from its bed upwards towards its surface, and below the névé the velocity is greatest where the fall is greatest and also wherever the cross section diminishes. In all these particulars, then, the ice of the glacier behaves like a stream of water. The average rate of flow of Alpine glaciers varies from a few inches to a few feet per day, and is greater during the warm summer season. The Muir glacier of Alaska has been shown to move at the rate of about seven feet per day.

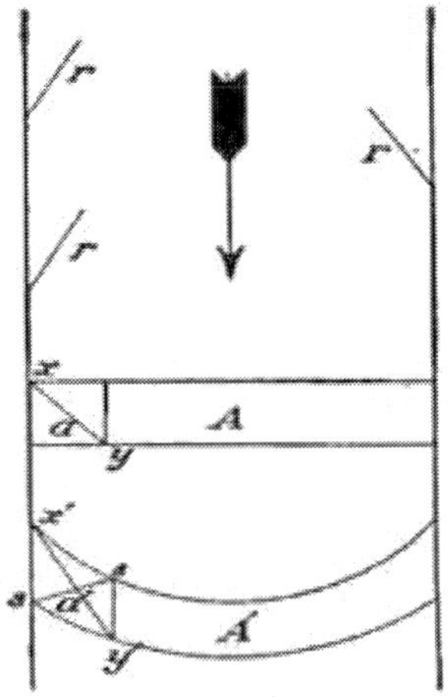

FIG. 416.—Diagram to illustrate the migrations of lines of stakes crossing a glacier, due to its surface movement, *A*, original position of lines; *A* ´, later positions; *a* and *a* ´, original and distorted forms of a square section of the glacier surface near its margin; *r*, *r* ´, diagonal crevasses.

In traveling from the névé downward to the glacier foot, the snow not only changes into ice, but it undergoes a granulating process with continued increase in the size of the nodules until at the foot of the glacier these may be picked out of the partially melted ice as articulating balls the size of the fist or larger. Glacier ice has therefore a structure quite different from that of lake ice, since the latter is developed in parallel needles perpendicular to the freezing surface.

Crevasses and séracs.—Prominent surface indications of glacier movement are found in the open cracks or *crevasses*, which are the marks of its yielding to tensional stresses. Crevasses are apt to run either directly across the glacier, wherever there is a steep descent upon its bed, or diagonally, running in from the margin and directed up-glacier (*r*, *r*, *r*, of Fig. 416), though they occasionally run longitudinally with the glacier when there is a rock terrace at the side of the valley beneath the ice. The diagonal crevasses at the glacier margin are due to the more sluggish movement where the ice is held back by friction upon the walls of the valley, as will be clear from Fig. 416. The square *a* has by this movement been distorted into the lozenge *a*´, so that the line *xy* has been extended into *x*´*y*´, with the obvious tendency to open cracks in the direction *ss*.

Every glacier surface below its névé is marked by steps or terraces, which are well understood to overlie corresponding steps of the cascade stairway to be seen in all vacated glacier valleys (plate 19). The steep risers of these steps are usually marked by parallel crevasses which cross the glacier. Under the rays of the sun, which strike them more from one side than from the other, the slices into which the ice is divided are transformed into sharpened blades and needles which are known as *séracs* (Fig. 401, p. 376, and Fig. 417).

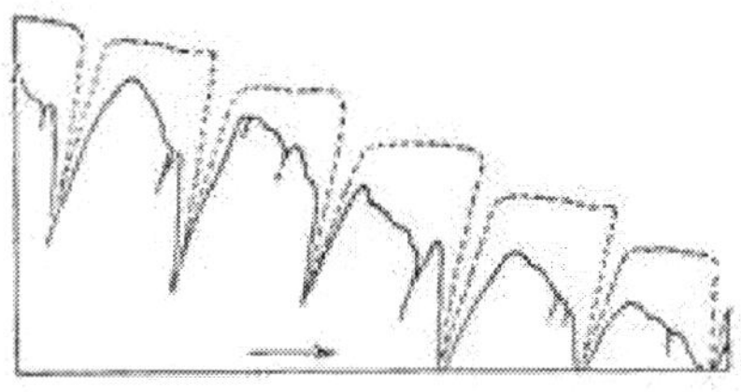

FIG. 417.—Transverse crevasses at the fall below a glacier step transformed by unsymmetrical melting into séracs.

The numerous crevasses tell us that the ice is many times wrenched apart during its journey down the glacier. This has been illustrated by somewhat grewsome incidents connected with accidents to Alpinists, but as they illustrate in some measure both the mode and the rate of motion of Swiss glaciers, they are worthy of our consideration.

Bodies given up by the Glacier *des Bossons*.—In the year 1820, during one of the earlier ascents of Mont Blanc, three guides were buried beneath an avalanche near the *Rochers Rouges* in the névé of the Glacier des Bossons (Fig. 418). In 1858 Dr. Forbes, who had measured the rate of flow of a number of Alpine glaciers, predicted that the bodies of the victims of this accident would be given up by the glacier after being entombed from thirty-five to forty years. In the year 1861, or forty-one years after the disaster, the heads of the three guides, separated from their bodies, with some hands and fragments of clothing, appeared at the foot of the Glacier des Bossons, and in such a state of preservation that they were easily recognized by a guide who had known them in life. Inasmuch as these fragments of the bodies had required forty-one years to travel in the ice the three thousand meters which separate the place of the accident from the foot of the glacier, the rate of movement was twenty centimeters, or eight inches, per day.

FIG. 418.—View of the *Glacier des Bossons* upon the slopes of Mont Blanc showing the position of accidents to Alpinists and the place of reappearance of their bodies.

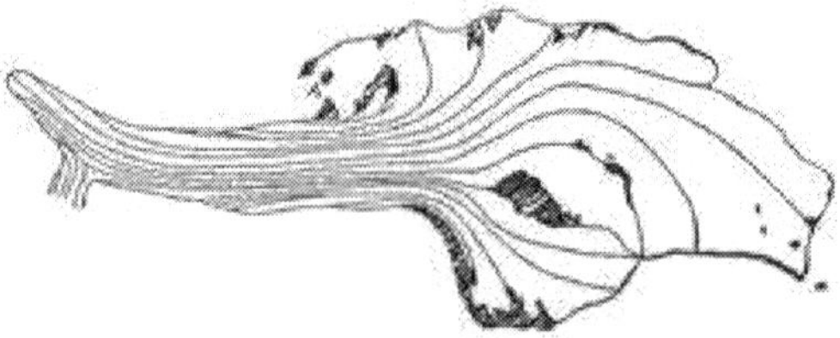

FIG. 419.—Lines of flow upon the surface of the Hintereisferner glacier in the Alps (after Hess).

Various separated parts of the body of Captain Arkwright, who had been lost in 1866 upon the névé of the same glacier, reappeared at its foot after entombment in the ice for a period of thirty-one years. To-day the time of reappearance of portions of the bodies of persons lost upon Mont Blanc is rather accurately predicted, so that friends repair to Chamonix to await the giving up of its victims by the Glacier des Bossons.

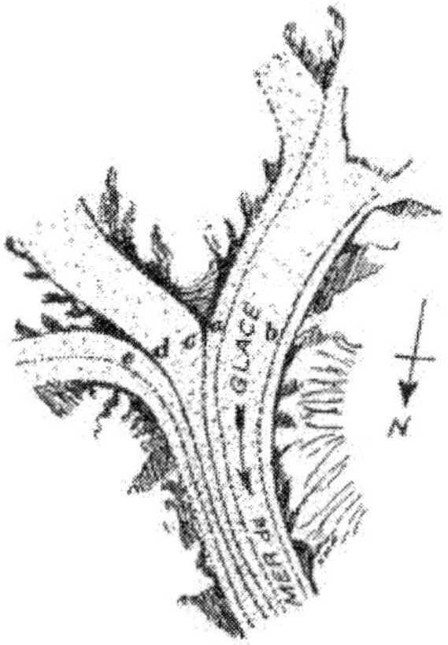

FIG. 420.—Lateral and medial moraines of the *Mer de glace* and its tributary ice streams.

The moraines.—The horns and comb ridges which rise above the glacier surface are continually subject to frost weathering, and from time to time drop their separated fragments upon the glacier. Falling as these do from considerable heights, they reach the ice under a high velocity, and rebounding, sometimes travel well out upon its surface before coming to a temporary rest. Upon a fresh snow surface of the névé their tracks may sometimes be followed with the eye for considerable distances, and their fall is a constant menace to Alpine climbers. Below the névé the larger number of such fragments remain near the cliff, and the lines of flow of the ice within the glacier surface are such that blocks which reach points farther out upon the glacier are later gathered in beneath the cliff at the side (Fig. 419). The ridge of angular rock débris which thus forms at the side of the glacier is called a *lateral moraine* (see Fig. 411, p. 385, and Fig. 420).

At the junction of two glacier streams, the lateral moraines are joined, and there move out upon the ice surface of the resultant glacier as a *medial moraine*. Thus from the number of medial moraines upon a glacier surface it is possible to say that the important tributary glaciers number one more (Fig. 420).

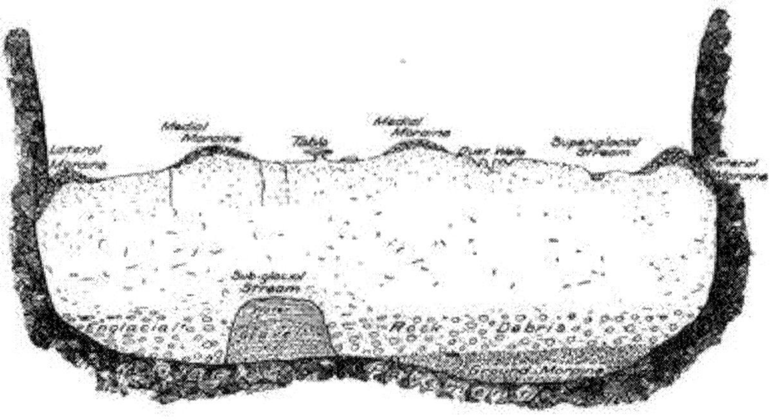

FIG. 421.—Ideal cross-section of a mountain glacier to show the position of moraines and other peculiarities characteristic of the surface of the bed.

The plucking and abrading processes in operation beneath the glacier, quarry the rock upon its bed, and after shaping and smoothing the separated rock fragments, these are incorporated within the lower layers of the ice as *englacial* rock débris. In spaces favorable for its accumulation, a portion of this material, together with much finer débris and rock flour, is left behind as a ground moraine upon the bed of the glacier (see Fig. 421).

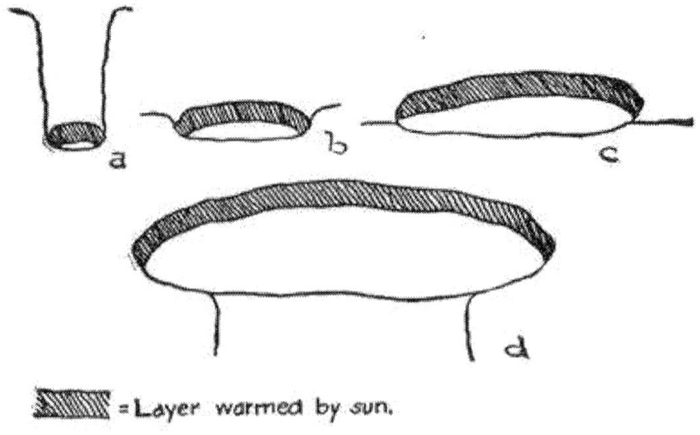

FIG. 422.—Fragments of rock of different sizes, to bring out their different effects upon the melting of the glacier surface.

At the foot of the glacier the relatively angular rock débris, which has been carried upon the surface, and the soled and polished englacial material from near the bottom, are alike deposited in a common marginal ridge known as the *terminal* or *end moraine* (plate 21 B).

PLATE 21.

***A*. View of the Harvard Glacier, Alaska, showing the characteristic terraces (after U. S. Grant).**

***B*. The terminal moraine at the foot of a mountain glacier (after George Kinney).**

Selective melting upon the glacier surface.—The white surface of the glacier generally reflects a large proportion of the sun's rays which reach it, and its more rapid melting is largely accomplished through the agency of rock fragments spread upon its surface. Such fragments, however, promote or retard the melting process in inverse proportion to their size up to a certain limit, and above that size their action is always to protect the glacier from the sun. This nice adjustment to the size of the rock fragments will be clear from examination of Fig. 422, for rock is a poor conductor of heat, and in even the longest summer day a thin outer layer only is appreciably warmed. Large rock blocks, grouped in the medial and lateral moraines, hold back the process of lowering the glacier surface during the summer, so that late in the season these moraines stand fifty feet or more above the glacier as armored ice ridges.

FIG. 423.—Small glacier table upon the surface of the Great Aletsch glacier in 1908.

Isolated and large rock slabs, as the season advances, may come to form the capping of an ice pedestal which they overhang and are known as *glacier tables* (Fig. 423). Such tables the sun attacks more upon one side than upon the other, so that the slab inclines more and more to the south and may eventually slip down until its edges rest against the glacier surface. Rounded bowlders, which less frequently become perched upon ice pedestals, may, from a similar process, slide down upon the southern side and leave a pyramid of ice furrowed upon this side and known as an *ice pyramid.*

Fine dirt when scattered over the glacier surface is, on the other hand, most effective in lowering its level by melting. Use was made of this knowledge to lower the great drifts of snow which had to be removed each season during the construction of the new Bergen railway of southern Norway. Each dirt particle, being warmed throughout by the sun's rays, melts its way rapidly into the glacier surface until the *dust well* which it has formed is so deep that the slanting rays of the sun no longer reach it. When the dirt particles are near together, the thin walls which separate the dust wells are attacked from the sides in the warm air of summer days, thus producing from a patch of dirt upon the glacier surface a *bath tub* (Fig. 424 *d*). At night the water which fills these basins is frozen to form a lining of ice needles projecting inward from the wall, and this, repeated in succeeding nights, may entirely close the basin with water ice and produce the familiar *glacier star* (Fig. 424 *c*).

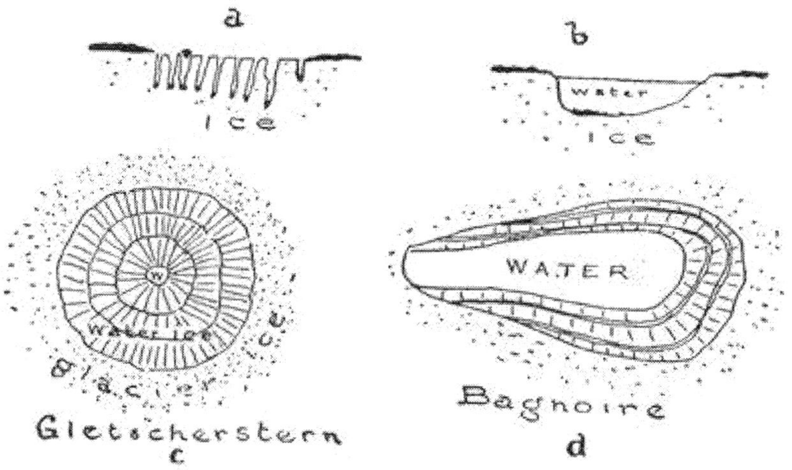

FIG. 424.—Effects of differential melting and subsequent refreezing upon the glacier surface. *a*, dust wells; *b*, glacier *tub* produced by melting about a group of scattered dust particles; *c*, glacier star produced when the inclosed water of the glacier well has frozen in successive nights; *d*, "bath tub."

If the dirt upon the glacier surface, instead of being scattered, is so disposed as to make a patch completely covering the ice to the thickness of an inch or more, the effect is altogether different. Protecting as it now does the ice below, a local ice hillock rises upon its site as the surrounding surface is lowered, and as this grows in height its declivities increase and a portion of the dirt slides down the side. The final product of this shaping is an almost perfectly conical ice hill encased in dirt and known as a *débris*, *sand*, or *dirt cone* (Fig. 425). The novice in glacier study is apt to assume that these black cones contain only dirt, but is rudely awakened to the reality when he attempts to kick them to pieces. Both glacier tubs and débris cones may assume large dimensions; as, for example, in Alaska, where they may be properly described as lakes and hills.

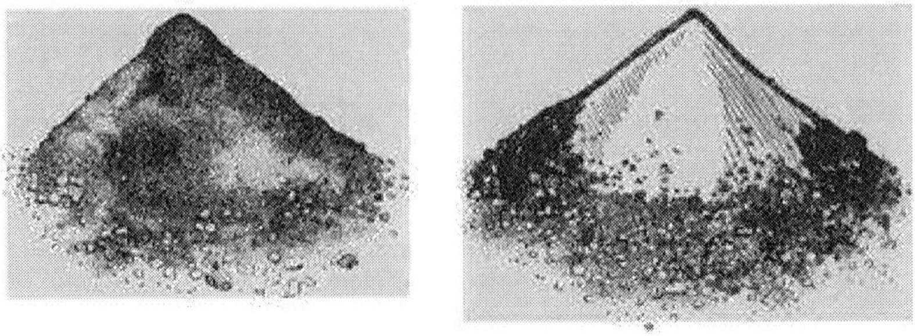

FIG. 425.—Dirt cone and one with its casing in part removed. Victoria glacier (after Sherzer).

A patch of hard and dense snow which is less easily melted than that upon which it rests may lead to the formation of snow cones upon the glacier surface similar in size and shape to the better known débris cones. Such cones of snow have, with doubtful propriety, been designated "penitents", for it is pretty clear that the interesting bowed snow figures, which really resemble penitents and which were first described from the southern Andes under the name of *nieves penitentes*, are of somewhat different character.

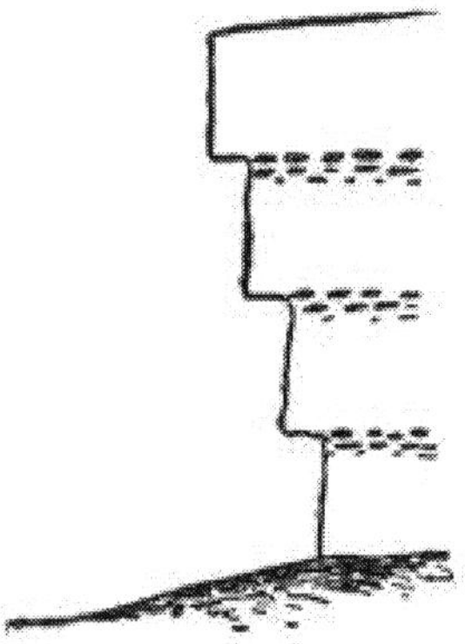

FIG. 426.—Schematic diagram to show the manner of formation of glacier cornices.

One further ice feature shaped by differential melting around rock particles remains to be mentioned. Wherever the seasonal snowfalls of the névé are exposed in crevasses, they are generally found to be separated by layers of dirt, and lines of pebbles similarly separate those ice layers which are revealed at the foot of the glacier. In either case, if the sun's rays can reach these layers in an opened crevasse, the half-buried rock fragments are warmed by the sun upon their exposed surfaces and slowly melt their way down the ice surface, thus removing from it a thin layer of snow or ice and causing that part above the pebble layer to project like a cornice. This process will go on until the overhanging cornice protects the pebbles from any further warming by the sun, but each lower pebble layer that is reached by the sun will produce an additional cornice, so that the original surface may at the bottom have been retired by the process a number of inches. These features are described as *glacier cornices* (Fig. 426).

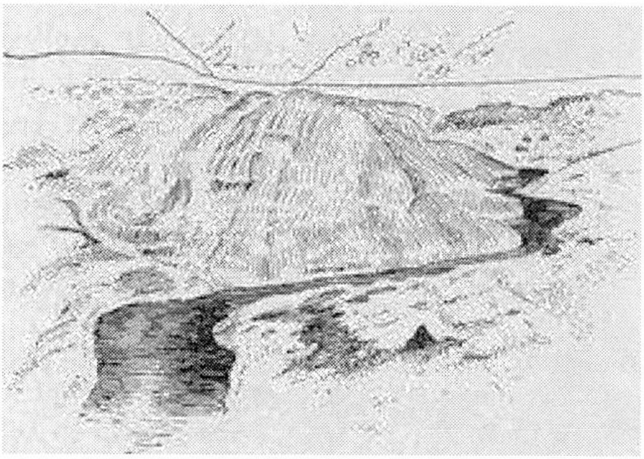

FIG. 427.—Superglacial stream upon the Great Aletsch glacier.

Glacier drainage.—Already in the early morning of every warm summer day, active melting has begun upon the surface of the Swiss glaciers. Rills of icy water soon make their way along depressions upon the surface, and are joined to one another so that they sometimes form brooks of considerable size (Fig. 427). Such streams continue their serpentine courses until these are intersected by a crevasse down which the waters plunge in a whirling vortex which soon develops a vertical shaft of circular section within the ice. Such shafts with their descending columns of whirling water are the well-known *moulins*, or "*mills*", which may be detected from a distance by their gurgling sounds. The first plunge of the water may not reach to the bottom of the glacier, in which case the stream finds a passageway below the surface but above the floor until another crevasse is encountered and a new plunge made, here perhaps to the bottom. Once upon the valley floor the stream is joined by others, and pursues its course within an ice tunnel of its own making (Fig. 421, p. 394) until it issues at the glacier front.

The coarser of the rock débris which was gathered up by the stream upon the glacier surface is deposited within the tunnel in imperfect assortment (gravel and sand), while all finer material and that lifted from the floor (rock flour) is retained in suspension and gives to the escaping stream its opaque white appearance. This *glacier milk* may generally be traced far down the valleys or out upon the foreland, and is often the traveler's first indication that a range which he is approaching supports glaciers.

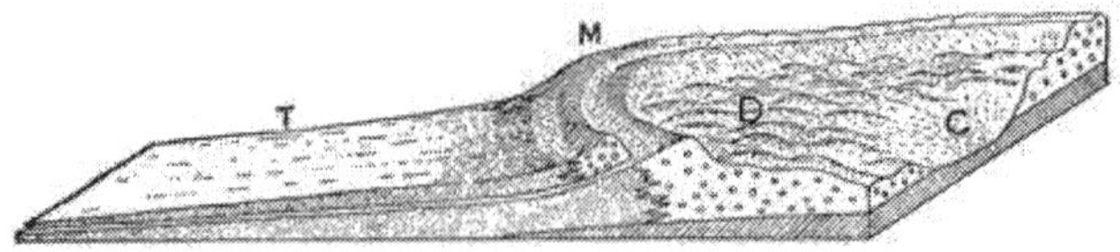

FIG. 428.—Ideal form of the surface left on the site of the apron of a piedmont glacier. *M*, moraine; *T*, outwash; *C*, basin usually occupied by a lake; *D*, drumlins (after Penck).

Deposits within the vacated valley.—For every excavation of the higher portions of the upland through glacial sculpture, there is a corresponding deposit of the excavated materials in lower levels. So far as these materials are deposited directly by the ice, they form the lateral, medial, ground, and terminal moraines already described. A considerable proportion of them are, however, deposited by the water outside the terminal moraine; but as with the shrinking glacier the ice front retires in halting movements over the area earlier ice-covered, the terminal moraines are ranged along the vacated valley as *recessional moraines*, each with a *valley train* of outwash below. About the apron of the piedmont glacier, such deposits are particularly heavy (Fig. 428). During the "ice age" the Swiss glaciers extended down the valleys below the existing ice remnants and spread upon the Swiss foreland as great piedmont glaciers such as may now be seen in Alaska. To-day we find there moraines and glacial outwash, a lake in the middle of the apron site, and sometimes a group of radiating drumlins like those found within the ice lobes of the continental glacier in southern Wisconsin (Fig. 429, and Fig. 344, p. 317).

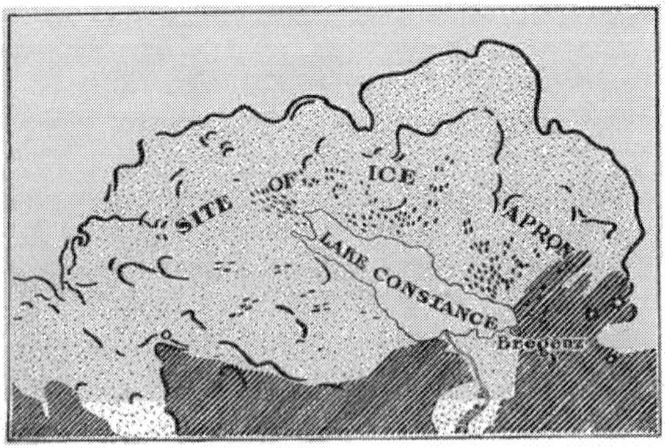

FIG. 429.—Moraines and drumlins about Lake Constance upon the site of the earlier piedmont glacier of the Upper Rhine. The white area outside the outermost moraine is buried in glacial outwash (after Penck and Brückner).

Behind the recessional moraines within the glaciated valley are found the valley moraine lakes (Fig. 448, p. 413), in association with the rock basin lakes due to glacial sculpture (Fig. 447, p. 412). After the glacier has vacated its valley, the precipitous side walls become the prey of frostwork and are the scenes of disastrous avalanches or landslides. Within the cirques, drifts of snow are nourished long after the ice has disappeared, and as a consequence the amphitheater walls succumb to the process of solifluxion (p. 153).

Diversions and reversals of drainage, which are so characteristic of the work of continental glaciers, are hardly less common to glaciated mountain districts. Many of our most beautiful waterfalls have resulted from either the temporary or permanent obstruction of earlier valleys above the falls. The famous Yosemite Falls offers an interesting illustration of the shifting of an earlier waterfall, itself no doubt due to ice blocking in a still earlier glaciation (plate 22 B).

Marks of the earlier occupation of mountains by glaciers.—It is well that we should now bring together within a small compass those evidences by which the existence of earlier mountain glaciers may be proven in any district. These marks are so deeply stamped upon the landscape that no one need err in their interpretation.

MARKS OF MOUNTAIN GLACIERS

High-level sculpture. The grooved upland with its cirques, or the fretted upland with its cirques, cols, horns, and comb ridges.

Low-level sculpture. The U-shaped main valley, the hanging side valleys with their ribbon falls, the glacier staircase with its rock bars and gorges, the rounded, polished, and striated rock floor.

Deposits. The recessional moraines of till and the valley trains of sand and gravel, the soled erratic blocks derived always from higher levels of the valley.

Lakes. The valley moraine lakes and the chains of rock basin lakes.

READING REFERENCES FOR CHAPTER XXVIII

Glacier movement:—

L. AGASSIZ. Nouvelles Études et Expériences sur les Glaciers Actuels, etc., Paris, 1847, pp. 435-539.

H. HESS. Die Gletscher, Braunschweig, 1904, pp. 115-150.

H. F. REID. The Mechanics of Glaciers, Jour. Geol., vol. 4, 1896, pp. 912-928; Glacier Bay and Its Glaciers, 16th Ann. Rept. U. S Geol. Surv., Pt. i, 1898, pp. 445-448.

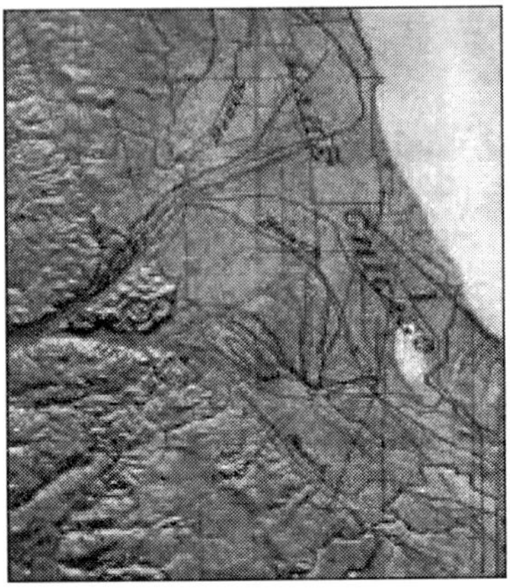

A. **Model of the vicinity of Chicago, showing the position of the ancient beaches and the outlet of the former Lake Chicago.**

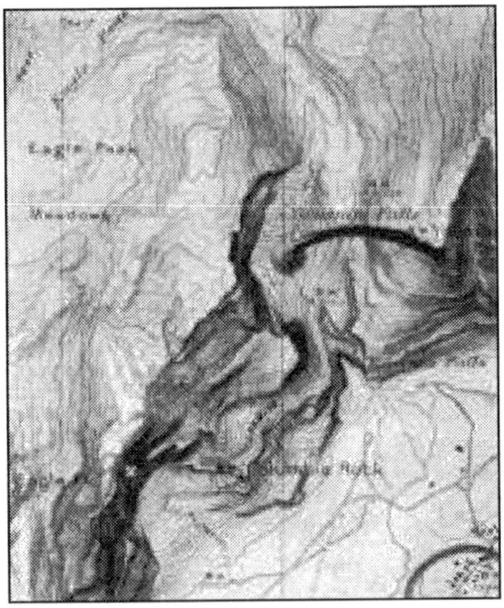

B. **Map of Yosemite Falls and its earlier site near Eagle Peak (after F. E. Matthes).**

CHAPTER XXIX

A STUDY OF LAKE BASINS

Freshwater and saline lakes.—Lakes require for their existence a basin within which water may be impounded, and a supply of water more than sufficient to meet the losses from seepage and evaporation. If there is a surplus beyond what is needed to meet these losses, lakes have outlets and remain fresh; their content of mineral matter is then too slight to be detected by the palate. If, on the other hand, supply is insufficient for overflow, continued evaporation results in a concentration of the mineral content of the water, subject as it is to continual augmentation from the inflowing streams.

As we have seen, there are in areas of small rainfall special weathering processes which tend to bring out the salts from the interior of rock masses, these concentrated salts generally first appearing as a surface efflorescence which is ultimately transferred through the agency of wind and cloudburst to the characteristically saline desert lakes.

Lake basins may be formed in many ways. Depressions of the land surface may result from tectonic movements of the crust; they may be formed by excavating processes; but in by far the greater number of instances they result from the obstruction in some manner of valleys which were before characterized by uniformly forward grades. In relatively few cases loose materials are heaped up in such a manner as to produce fairly symmetrical basins.

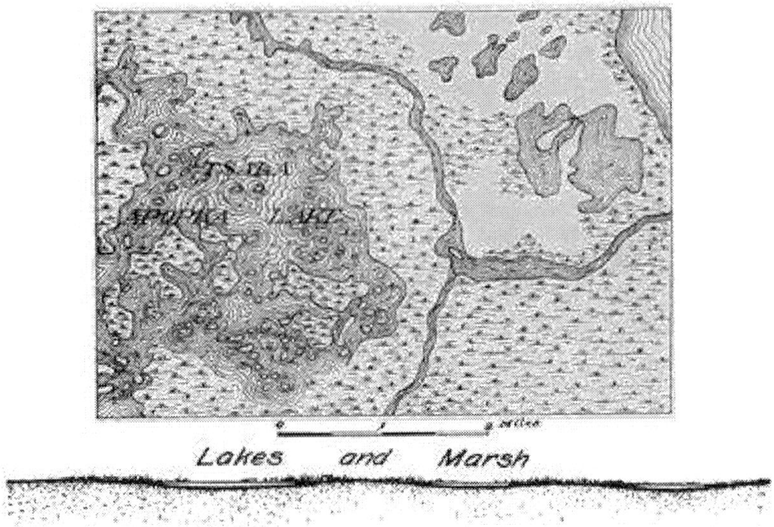

FIG. 430.—Map and diagram to bring out the characteristics of newland lakes.

Newland lakes.—On land recently elevated from the sea, basins of lakes may be merely the inherited slight irregularities of the earlier sea floor, in which case they may be assumed to be largely the result of an irregular distribution of deposits derived from the land. Lakes of this type are especially well exhibited in Florida, and are known as newland lakes (Fig. 430). Such lakes are exceptionally shallow, and are apt to have irregular outlines and extremely low banks. Under these circumstances, they are soon filled with a rank growth of vegetation, so that it is sometimes difficult to properly distinguish lake and marsh.

FIG. 431.—View of the Warner Lakes, Oregon (after Russell).

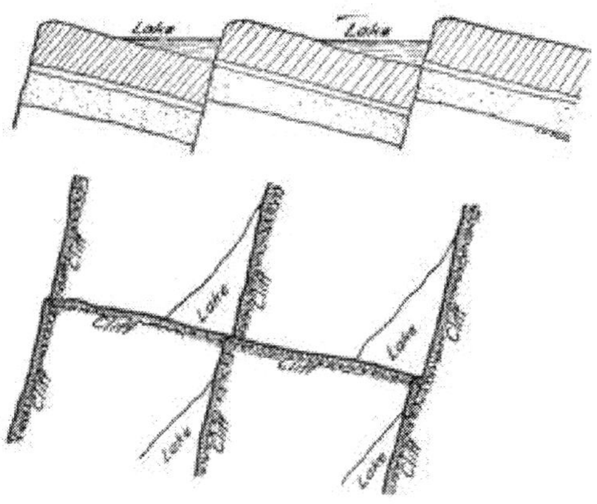

FIG. 432.—Schematic diagrams to illustrate the characteristics of basin-range lakes.

Basin-range lakes.—Newland lakes may be said to have their origin in an uplift of the land and sea floor near their common margin. A lake type dependent upon movements of the earth's crust but within interior areas has been described as the basin-range type and is exemplified by the Warner Lakes of Oregon. In this district great rectangular blocks of the earth's crust, which in their upper portions at least are composed of basaltic lavas, have undergone vertical adjustments in level and have been tilted so that the corresponding corners of neighboring blocks have been given a similar degree of down-tilt (Fig. 431). Lakes formed in this way are of triangular outline, are bounded on the two shorter sides by cliffs, but have extremely flat shores on their longest side. From this shore the water increases gradually in depth and attains a maximum depth at or near the opposite angle. Such lakes naturally betray a tendency to appear in series (Fig. 432), and are unfortunately much too often illustrated on a small scale after a shower by the tilted blocks of imperfectly made cement sidewalks.

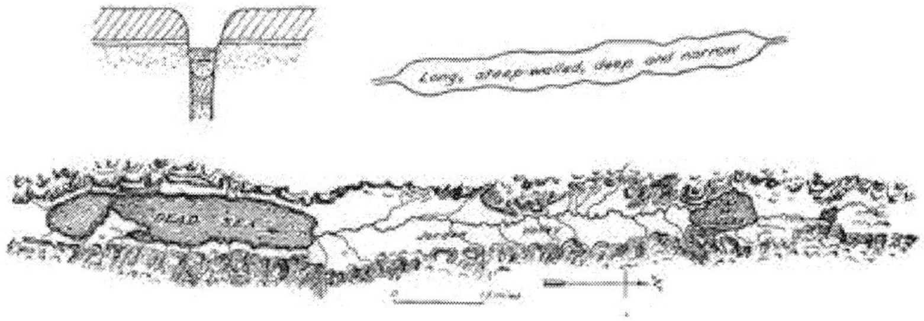

FIG. 433.—Schematic diagrams of rift-valley lakes, and the rift valley of the Jordan with the Dead Sea and the Sea of Galilee as remnants of a larger lake in which their basins were included.

Rift-valley lakes.—Another type of lake basin which has its origin in faulted block movements is known as the rift-valley lake, and is best exemplified by the great lakes of east Central Africa. In this type a strip of crust, many times as long as it is wide, has been relatively sunk between the blocks on either side so as to produce a deep rift, or what in Germany is known as a *Graben* (trench). Such a basin when occupied by water yields a lake which is long, straight, deep, and narrow, and is in addition bounded on the sides by steep rock cliffs. At the ends the shores are generally by contrast decidedly low. If the hard rock at the bottom of the lake could be examined, it would be found to be of the same type as that exposed near the top of the side cliffs. The valley of the Jordan in Palestine is a rift of this character and was at one time occupied by a long and narrow lake of which the Dead Sea and the Sea of Galilee are the existing remnants (Fig. 433).

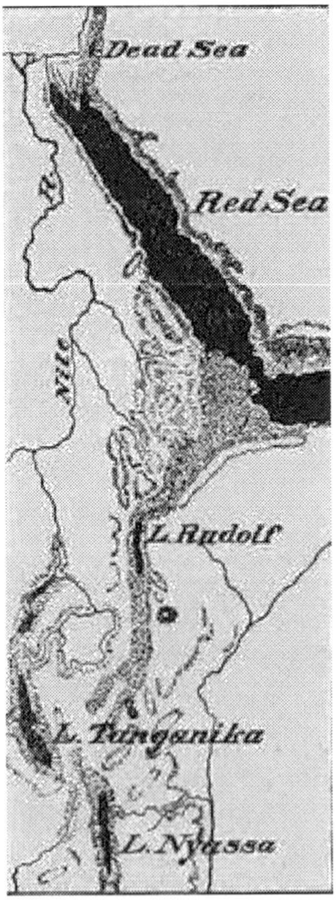

FIG. 434.—Map showing the rift valley lakes of east Central Africa.

One of the most striking examples of a rift valley lake is Lake Tanganyika, while Albert Nyanza, Nyassa, and Rudolf in the same region are similar (Fig. 434).

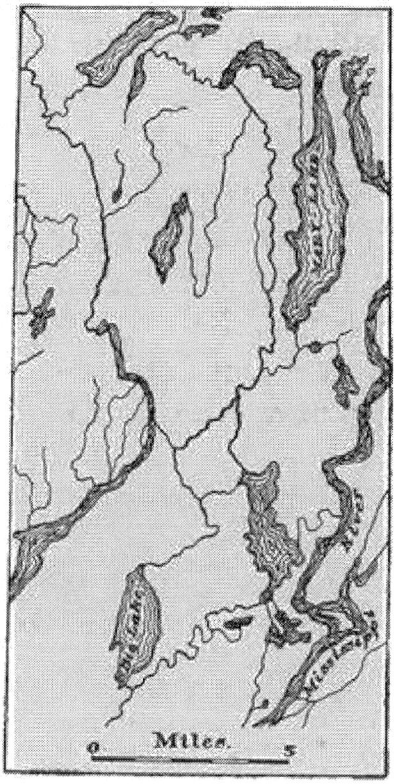

FIG. 435.—Earthquake lakes which were formed in the flood plain of the lower Mississippi during the earthquake of 1811 (after Humphreys).

Earthquake lakes.—The complex adjustments in level of the surface of the ground at the time of sensible earthquakes are many of them made apparent in no other way than by the derangements of the surface water. This is at such times impounded either in pools or in broad lakes, which inasmuch as they date from known earthquakes have been called "earthquake lakes", even though in a strict sense any lake which has originated in earth movements might properly be regarded as an earthquake lake. To avoid unnecessary confusion, the term must, however, be restricted to those lakes which are known to have been formed at the time of definite earthquakes (Fig. 435). Reelfoot Lake in Tennessee, which in late years has acquired undesirable notoriety because of the feuds between the fishermen of the district and the constituted authorities, is a lake more than twenty miles across and came into existence during the great earthquake of the lower Mississippi valley in 1811.

Crater lakes.—The craters of volcanic mountains are natural basins in which surface waters are certain to be collected, provided only the supply is sufficient and seepage into the loose materials is not excessive. Some craters, still visibly more or less active, are occupied by lakes (Fig. 436).

FIG. 436.—View of lake in Poas Crater in Costa Rica, a volcanic crater more than half a mile across and with walls 800 feet deep. At intervals there is an ejection of steam mixed with mud and ash after the manner of a geyser (after H. Pittier).

In the larger number of cases in which craters become occupied by lakes, the evidence of continued activity is lacking, and it would appear in such cases that the lava of the chimney had consolidated into a volcanic plug, closing the bottom of the crater. Notable groups of crater lakes are the *Caldera* of the Roman Campagna (Fig. 437) and the so-called *maare* of the Eifel about the Lower Rhine. Crater lakes are easy to recognize by their circular plan, their steep walls of volcanic materials, and their considerable depth with a maximum near the center.

One of the most remarkable of these water-filled basins is Crater Lake in Oregon, which has a diameter of about six miles and is believed to have resulted from the incaving of a great volcanic cone in the latest stage of its activity. This remarkable feature has now been made a national park and will soon be conveniently reached by tourists and counted one of the greatest nature wonders of the Pacific slope.

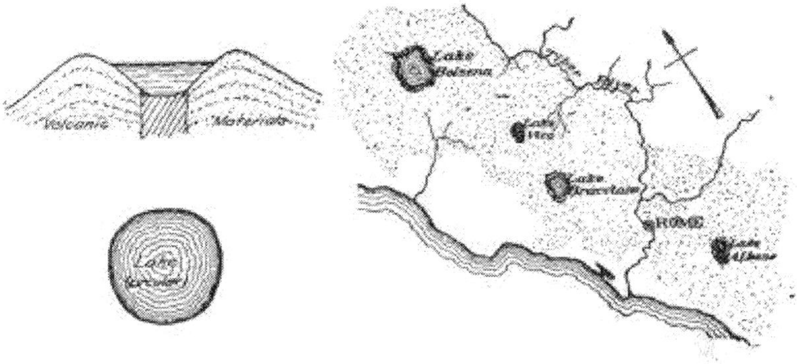

FIG. 437.—Diagrams to illustrate the characteristics of crater lakes. The Roman Campagna is a plain formed of volcanic ash, with the crater lakes of Bracciano, Vico, and Bolseno arranged on a line traversing it.

Coulée lakes.—Far more important as lakes are those volcanic basins which arise from the flow of a stream of lava across the valley of a river so as to impound its waters (Fig. 438).

At the time of the great eruption under Skaptár Jökull in 1783 the river Skaptár and many of its tributaries were blocked by the flow of lava, which it is estimated exceeded in bulk the mass of Mont Blanc.

FIG. 438.—View of Snag Lake, a *coulée* lake with lava dam shown in middle distance (after Fairbanks).

Morainal lakes.—As we have learned, the obstruction of drainage, due to the distribution of rock débris by continental glaciers, has yielded lakes in almost countless numbers. Probably ninety per cent or more of the known lakes have had this origin, and the type is so common within the once glaciated regions that it forms perhaps the best distinguishing mark of former glaciation. The hummocky surface of morainal deposits is so characteristic that the lakes of this type are never very large and are correspondingly irregular in outline. They have often numerous islands, and their banks are formed of the combination of rock flour and ice-worn materials known as till (Fig. 439). The smallest of the morainal lakes are mere kettles on the marginal moraine, and these rapidly become replaced by peat bogs. In contrast with pit lakes, morainal lakes lack the steep surrounding slopes and the encircling plain.

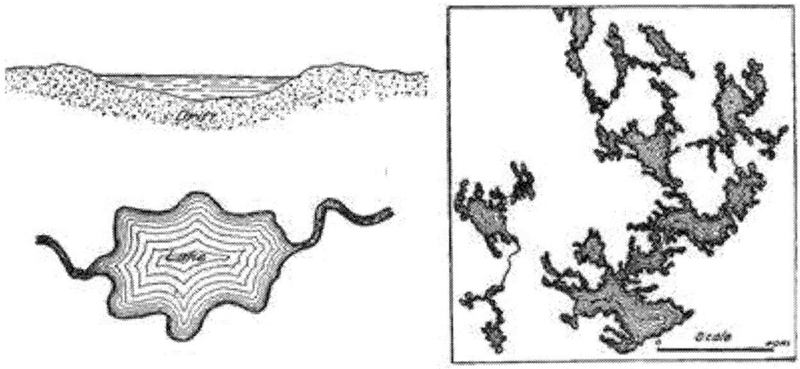

FIG. 439.—Diagrams to illustrate the characteristics of morainal lakes, and a sample map of such lakes from the glaciated region of North America.

Pit lakes.—The so-called pit lakes have their origin in continental glaciation, and are found in groups within broad plains of glacial outwash (mainly sand and gravel), which are for this reason described as "pitted plains" (see p. 314). Those areas which lay between neighboring lobes of the ice sheet were subject to particularly heavy deposits of outwash material, and are, in consequence, particularly likely to be occupied by pit lakes. As has been pointed out in an earlier section, the water derived from surface melting within the marginal portions of a continental glacier descends to the bottom in the crevasses and thereafter flows in an ice tunnel under the same conditions as water flowing in a pipe. Having in most cases a considerable head at the outer margin of the ice, this water may rise and issue well above the lower ice layers and so cover a portion of the ice margin beneath sand

and gravel (Fig. 440). Separated blocks, often of massive proportions, are thus buried beneath nonconducting materials by which they are long protected from further melting. Eventually, however, with the approach of still milder climates they disappear, thus causing the overlying sand and gravel to descend and form a pit of steep walls similar to the sawdust pits over melted ice blocks within our storehouses.

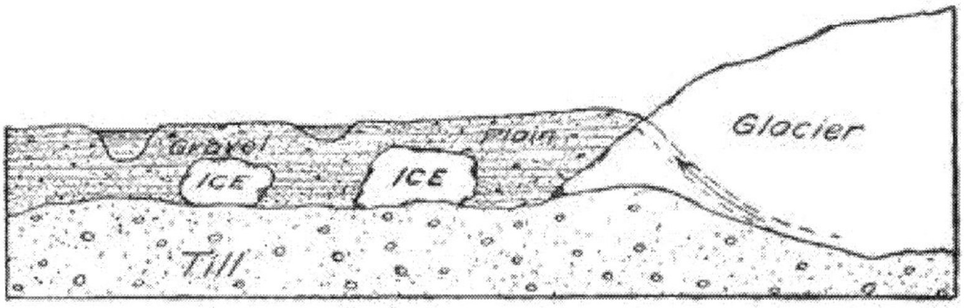

FIG. 440.—Diagram to show the manner of formation of pit lakes.

Pit lakes are thus easily recognized by their occurrence usually in groups within a plain of glacial outwash and by their characteristic banks inclined at the angle of repose of such materials (Fig. 441).

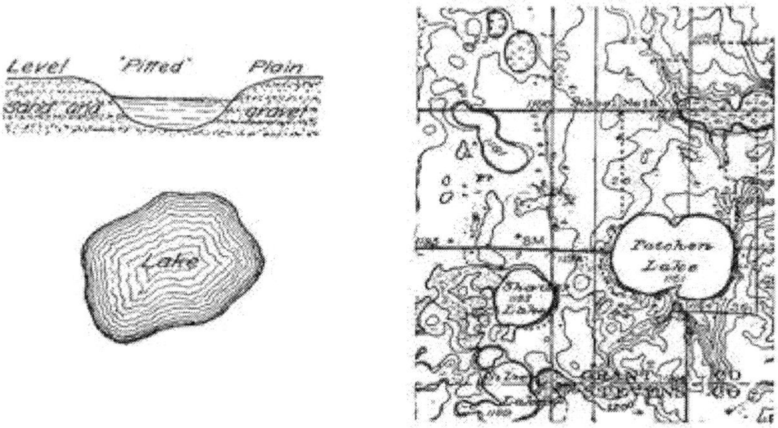

FIG. 441.—Diagrams to illustrate the characteristics of pit lakes and a sample map from the glaciated region of North America.

Glint or colk lakes.—It has been found to be true of existing continental glaciers that where their mass has been held back by a mountain wall, their current at the portals within this rampart becomes greatly accelerated. Though the upper layers of the glacier in the vicinity may move forward with a velocity of but an inch per day, the current within the outlet may be as much as seven hundred or a thousand times as great. In many respects these conditions are similar to those about the raceway of a reservoir where the near-by surface of the water is lowered by the indraught of the outlet and the current in the raceway is so accelerated that, unless protected, the bottom of the race is carried away and a basin excavated which extends a short distance both above and below the position of the dam. In Holland such basins hollowed out beneath breaks in the dykes are known as colks. Basins which were excavated beneath the glacier outlets by a similar process would not be open to our inspection until after the ice had disappeared from the region; but it is most significant that in Scandinavia, where the Pleistocene continental glacier, advancing westward from the Baltic, was held in check by the escarpment at the Norwegian boundary (the *glint*), lake basins have been excavated in hard rock whose walls show the abrading and polishing which are characteristic of glacial sculpture, and whose positions are such that they lie beneath the former outlets partly above and in part below the line of the escarpment. Their position in reference to the rampart and to the former outlets is brought out in Fig. 442. The largest of the glint lakes of this series is Torneträsk in northern Lapland (see p. 277 and Fig. 443).

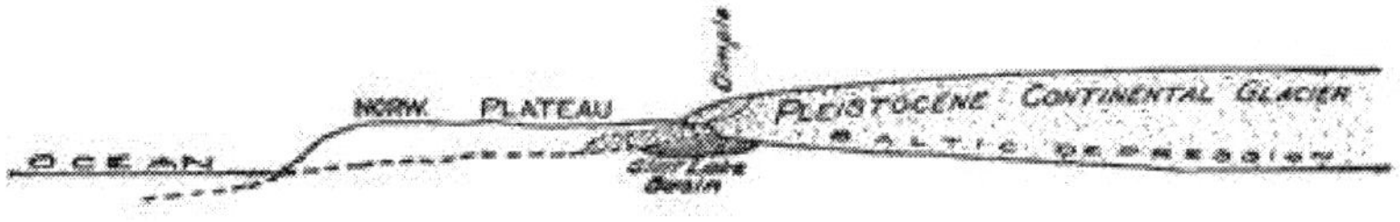

FIG. 442.—Diagram to show the manner of formation of glint or outlet lakes where the continental glacier of Scandinavia issued from the Baltic depression through portals in its mountain rampart.

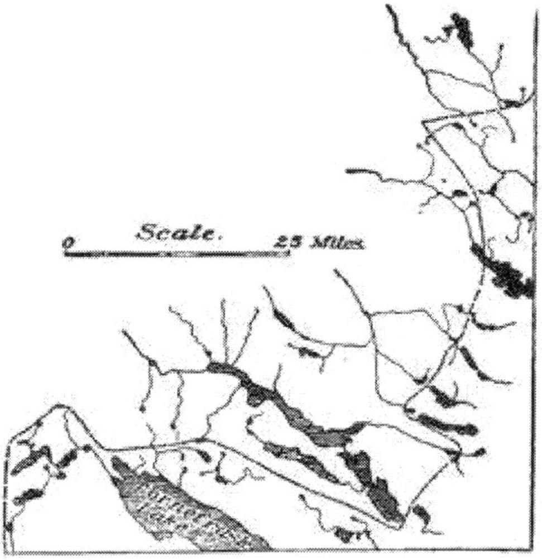

FIG. 443.—Map showing a series of glint lakes which lie across the international boundary of Sweden and Norway.

Ice-dam lakes.—Whenever a continental glacier, either in advancing its front or in retiring, lies across the lines of drainage upon their downstream side, water is impounded along the ice front so as to form ice-dam lakes. Such lakes are found to-day in Greenland and in the southern Andes, and similar bodies of water of far greater size and importance came into existence in Pleistocene times each time that the continental glaciers of northern North America and Europe advanced upon or retired from suitably directed river systems. Thus above the Baltic depression, when the ice front lay to the eastward of the main watershed, each easterly sloping valley was obstructed by the ice and occupied by an ice-dam lake (Fig. 444), the beaches of which may all be traced to-day (Fig. 445).

FIG. 444.—Ice-dam lakes (in black) between the front of the late Pleistocene glacier of northern Europe and the divide near the Norwegian boundary (after G. de Geer).

One side of each ice-dam lake is formed by an ice cliff at the glacier front, and if the region is relatively flat, the remaining shores are likely to be formed by a marginal moraine which the glacier has abandoned in its retreat. In their smaller stages, therefore, ice-dam lakes on prairie country have the form of a crescent, which is the more pronounced because the waves by their attack upon the ice front flatten the curvature of its outline (see Fig. 360, p. 330).

The life of an ice-dam lake is begun and ended in important changes of glacier outline, and after the draining of lakes by this process the land shores may be traced in beaches, and the ice margin by a water-laid moraine of low relief (Fig. 359, p. 330).

A much smaller but in many respects similar ice-dam lake is to-day to be seen at the side of the Great Aletsch glacier, a mountain glacier of Switzerland. The traveler who makes the easy ascent of the Eggishorn may look directly down upon this crescent-shaped lake with its ice cliff on the glacier side (see Fig. 446).

FIG. 445.—Wave-cut terrace at an elevation of 177.5 meters above sea on the southern slope of the northern Dala valley north of Baggedalen in Sweden. To the right in the foreground is a peat bog (after Munthe).

FIG. 446.—View of the Márjelen Lake at the side of the Great Aletsch glacier, seen looking directly down from the summit of the Eggishorn (after a photograph by I. D. Scott).

Glacier lobe lakes.—Upon the sites of the former lobes of the Pleistocene glacier of North America are found the basins of the Laurentian River system, the largest freshwater lakes in the world. There has been much controversy concerning the manner of formation of these lakes, but the view which has seemed to have the largest following is that they were excavated by the eroding action of the continental glacier over the drainage basins of former rivers. It is but one phase of the long controversy between opposing schools, which have advocated on the one hand the efficiency of glacier ice as an eroding agent, and upon the other its supposed protection from the weathering processes. The positions and the outlines of the several lakes of the series sufficiently proclaim their connection with the former glacial lobes, and the name which we have adopted leaves the exact manner of their formation a still open question. The recognition of the importance of the glacial anticyclone, in giving shape to the glacier surface and in effecting a transfer of snow from the central to the marginal portions, has had the effect of emphasizing the relative importance of erosion under the marginal and lobate portions. Thus the importance of ice lobes has been greatly accentuated, though this applies only to the shaping of the basins and not in any important way to the impounding of the present waters. The present Laurentian Lakes owe their existence to the elevation by successive uplifts of the country to the northward and eastward, since the glacier retired from the lake region. When the ice front lay to the northward of the Ottawa River, the discharge of the upper lakes was by a channel through Nipissing River and Lake and thence down the Ottawa River to a gulf in the lower St. Lawrence. The uplift of the land has had the effect of raising a barrier where the former outlet existed, and diverting the waters to a roundabout channel by way of Detroit and Lake Erie (see Fig. 365, p. 335).

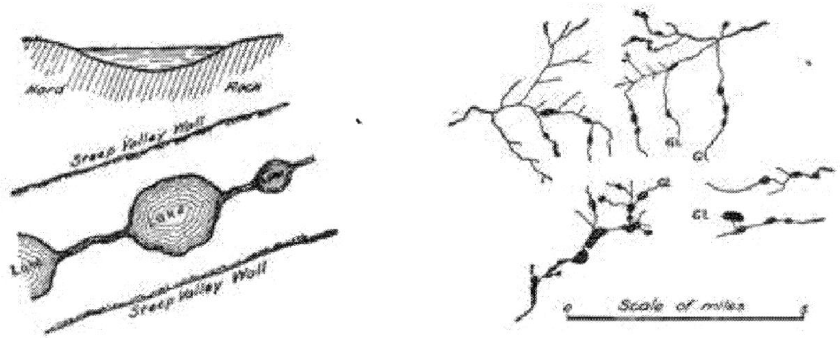

FIG. 447.—Diagrams to illustrate the arrangement and the characters of rock-basin lakes, together with a map of such lakes from the Bighorn Mountains in Wyoming.

Rock-basin lakes.—The reversed grades which develop in a valley deepened by mountain glaciers—the back-tilted treads of the cascade stairway (see p. 376)—furnish a series of basins hollowed in rock which are strung along the course of the valley like pearls upon a thread, or, far better, like the larger beads in a rosary (Fig. 447). This characteristic arrangement accounts for the name "Paternoster Lakes" which has sometimes been applied to them in Europe. Their positions in series within U-shaped mountain valleys, and their rock shores with characteristically smoothed and striated surfaces, make them easy of determination. In the higher portions of the valley, where the treads of the cascade stairway are relatively narrow, such lakes are often approximately circular in outline, but in the lower levels and upon wider treads they may be ribbon-like, though lakes of this type are to a large extent replaced in the lower levels by the valley moraine type or a combination of the two.

FIG. 448.—Convict Lake, a lake behind a moraine dam within a glaciated valley of the Sierra Nevadas, California (after a photograph by Fairbanks).

Valley moraine lakes.—The recessional moraines which mark the halting stations of mountain glaciers, while retiring up their valleys, form dams in the later river and so produce a type of lake which is in contrast with the morainal lakes which result from continental glaciation. They may, therefore, be distinguished by the name *valley moraine lakes*. Their positions on the bed of a U-shaped mountain valley, and the glacial materials which compose the dams, are sufficient for their identification (Fig. 448). Moraine Lake and Lake Louise in the Canadian Rockies are typical examples. Rock basin and valley moraine lakes may occur in alternation or combined in mountain valleys.

FIG. 449.—Lake basins produced by successive slides from the steep walls of a glaciated mountain valley (after Russell).

Landslide lakes.—The sheer-walled valleys which are carved by mountain glaciers are too steep to long retain their perpendicularity when the support of the glacier has been removed. Aided by the ever present joint planes, which admit water to the rock, they succumb to frost action, and further give way in avalanches whenever the rock is of sufficiently porous material to become saturated with water. Landslides sometimes occur successively until the original valley wall has been replaced by a terraced slope. The treads of the steps in this terrace have generally a backward-sloping grade, so that basins are formed to be filled by relatively long and narrow lakes or by successions of small pools (Fig. 449 and plate 23 B).

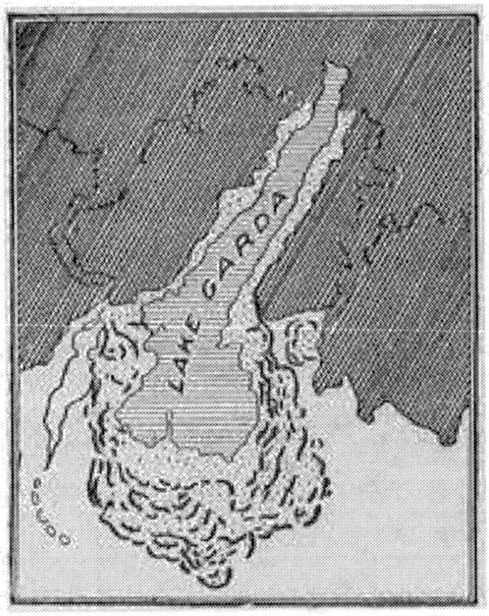

FIG. 450.—Lake Garda, a border lake upon the site of a piedmont apron at the margin of the Alpine highland (after Penck and Brückner).

When the avalanched material is so disposed as to dam the valley, much larger lakes of this type come into existence. During an earthquake which occurred on January 25, 1348, there was a landslide within the valley of the Gail, Carinthia, which destroyed seventeen villages and produced a lake which even to-day is represented by a great marsh.

Border lakes.—Whenever mountain glaciers push out their fronts beyond the borders of the mountain range by which they are nourished, they spread upon the foreland in broad aprons about which morainic accumulations are particularly heavy. This elevation of morainal walls about the margins of the aprons yields natural basins that are occupied by lakes so soon as the glacier retires its front within the valley. Because such lakes are found at the borders of upland districts they have been called *border lakes*. The beautiful Lakes Constance, Lucerne, Maggiore, Lugano, Como, and Garda (Fig. 450), on the borders of the Alpine highland, are all of this type.

PLATE 23.

***A*. View of the American Fall at Niagara, showing the accumulation of rocks beneath (after Grabau).**

***B*. Crystal Lake, a landslide lake in Colorado.**
(Photograph by Howland Bancroft.)

Ox-bow lakes.—The cutting off of a meander within the flood plain of a river yields a lake which is of horseshoe (ox-bow) outline and lies generally with low banks within a plain composed of river silt. Before separating from the parent stream the meander had begun to silt up, especially at the ends. Ox-bow lakes are, however, relatively deep near the convex shore and correspondingly shallow toward the concave margin (Fig. 451).

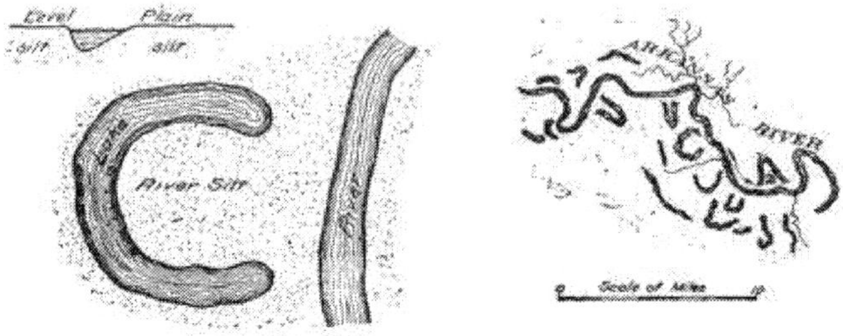

FIG. 451.—Diagrams to bring out the characteristics of ox-bow lakes, together with a map of such lakes from the flood plain of the Arkansas River.

FIG. 452.—Diagrammatic section to illustrate the formation of saucer-like basins between the levees of streams flowing in a flood plain.

Saucer lakes.—As we have learned, a river meandering in its flood plain has banks which are higher than the average level of the plain, for the reason that at flood time the main current of the stream still persists in the channel, thus allowing the burden of sediment to be dropped in the relatively slack water upon its margin. Because of these natural embankments or levees, tributary streams are often compelled to flow long distances in nearly parallel direction before effecting a junction. Between the trunk stream and its tributaries, likewise bounded by levees, and between streams and the valley walls, there thus exist low basins which are more or less saucer-shaped (Fig. 452). At flood time, when the levees are overflowed or crevassed, water enters these depressions, and an additional supply may be derived from the walls of the valley. Good illustrations of such lakes are furnished by the flood plain of the former river Warren near the banks of the present Minnesota River (Fig. 453).

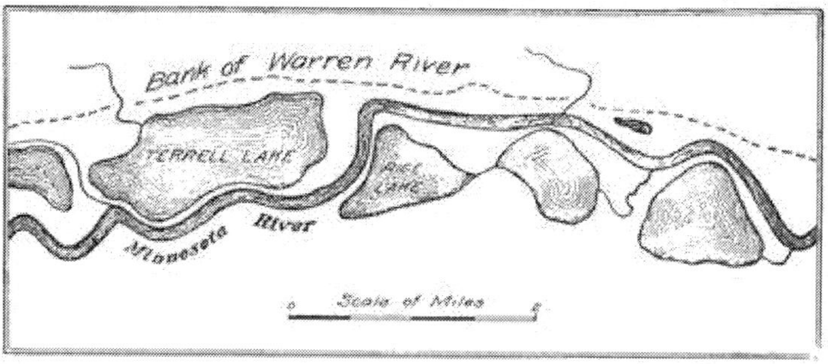

FIG. 453.—Saucer lakes upon the bed of the former river Warren (from the Minneapolis sheet, U. S. G. S.).

Crescentic levee lakes.—As we approach the delta of a river, the size and importance of the levee increases, and here a new type of levee lake may develop in series (Fig. 454). At flood time the levee is breached near the point of sharpest curvature on the convex side (Fig. 454 *a*). When the waters are subsiding, the current is kept away from the old channel by the rising grade of the levee as well as by the inertia of the current, and an entrance to the old channel is first found below the next change in curvature of the meander, since here scour becomes effective in cutting through the levee. The new channel is thus established in the form of a loop inclosing the old one, and the process of levee building now erects a wall about the territory newly acquired by the meander. This territory has the form of a crescent, and when occupied by water produces a crescentic levee lake often joined to its neighbors in series. The abandoned channel now closed at both ends by levees may be occupied by water to produce a subordinate ribbon type of curving trench (Fig. 454 *b*, *c*).

The importance of levees in obstructing drainage to form lakes is only beginning to be appreciated. It has quite recently been shown that when trunk streams are greatly swollen and burdened with sediment while flowing from a receding continental glacier, they may build such high levees as to aggrade their tributary streams above the junctions, even producing reversed grades and so impounding the waters to form extensive lakes. During the "ice age" lakes of this type were formed in Illinois and Kentucky rivers just above their junctions with the Ohio. The old lake floor with its eastern shore line and its protruding islands is easily made out upon the new topographic maps of Kentucky.

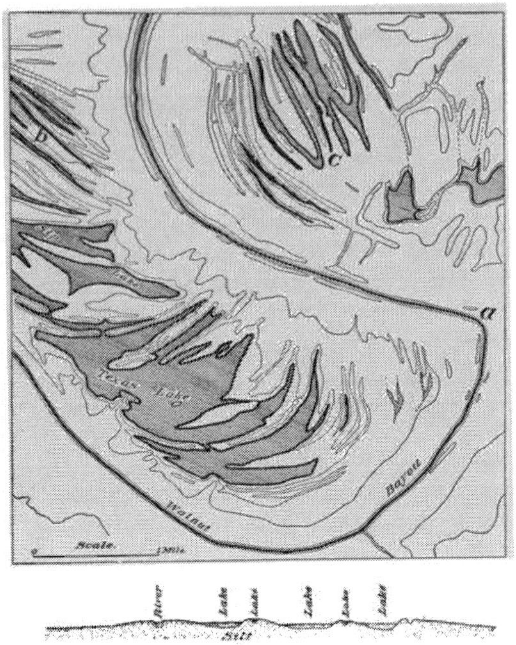

FIG. 454.—Levee lakes developed concentrically in series within meanders of a stream tributary to the Mississippi and flowing upon its delta plain. *b* and *c* are examples of the ribbon type of levee lake due to occupation of the abandoned river channel. The larger number of lakes, of which Sip Lake and Texas Lake are examples, have the form of crescents and lie between abandoned levees (from recent map of U. S. G. S.).

Raft lakes.—Within humid regions the flood plains of our larger rivers are generally forested, and as the river swings from side to side in its perpetual meanderings, the timber which grows upon the convex side of each meander is progressively undermined by the river and felled upon its bank. The prostrate trees remain upon the banks during the low water of the summer season, to be gathered up at the time of flood in the next spring season. It is log jams thus acquired which so generally block the main channel of a river and turn the current across the neck of the meander when cut-offs occur with the formation of ox-bow lakes. When the mass of timber thus gathered up by the river is excessive, as, for example, within the flood plain of the Red River of Arkansas and Louisiana, huge log rafts are produced which dam up the river so effectively as to produce temporary lakes. The impounded waters soon find an outlet over the levee at some point higher up the river, and the waters flowing off through the timbered bottom lands, other logs are caught by the standing timber as in a weir. A second dam is thus formed which is separated from the initial one by open water, and in this way the driftwood dam acquires enormous proportions as it gradually moves up the river. After a period of perhaps a century or more, the lower sections of the jam become decayed and dislodged so as to float down the river.

FIG. 455.—Raft lakes along the banks of the Red River in Arkansas and Louisiana at their fullest recorded development (after A. C. Veatch, U. S. G. S.).

In the lower Red River a great raft of alternating jams and open water reached a length of about one hundred and sixty miles and moved up the river at the average rate of something less than a mile per year. Within the limits of the dam all tributary streams were blocked, so that secondary lakes were formed in a double fringe about the main river (Fig. 455). The great raft which formed here in the latter part of the fifteenth century has now at the beginning of the twentieth been largely removed and measures have been adopted to prevent its re-formation.

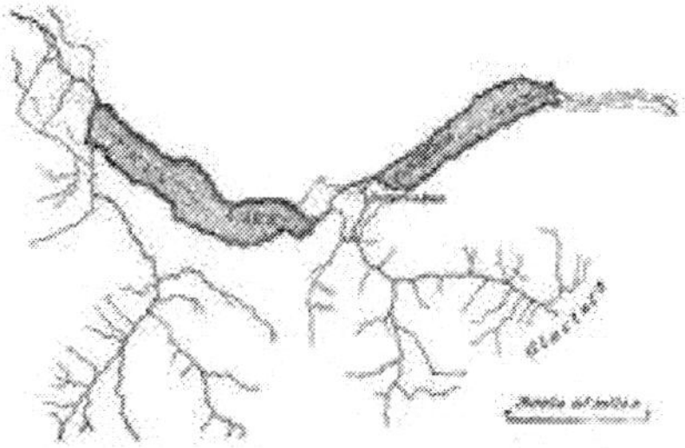

FIG. 456.—The Swiss lakes Thun and Brienz, formed by deltas at the junction of streams tributary to a steep-walled valley.

Side-delta lakes.—It is characteristic of river drainage that the tributary streams enter the main valley on steeper gradients than the trunk stream at the point of junction. Wherever the difference in velocity of the two streams at the junction is large, and the side stream is charged with sediment, a delta will be formed at the mouth of the tributary stream. Such deltas push out from the shore and may eventually block the main channel so as to form a more or less sausage-shaped expansion of the river—a side-delta lake. Traverse and Big Stone Lakes in the valley of the Warren River in Minnesota have been formed in this way (Fig. 354, p. 326). Lakes Thun and Brienz in the Swiss Alps are of similar origin, the beautiful city of Interlaken being built upon the delta plain over the valley of the earlier river (Fig. 456). The Mississippi has similarly been expanded to form Lake Pepin above the delta at the mouth of the Chippewa River.

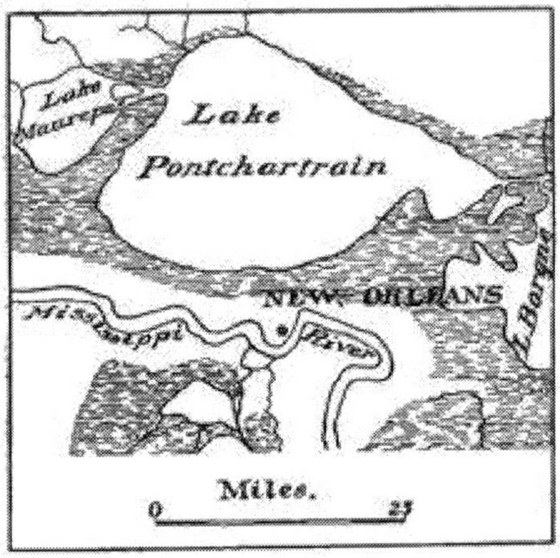

FIG. 457.—Delta lakes formed at the mouth of the Mississippi through the junction of the levees of radiating distributaries with the shore of the estuary (after Berghaus).

Delta lakes.—A somewhat different type of delta lake has been formed in Louisiana, where the "father of waters" discharges into the gulf. Here the various distributaries radiate from the main channel to produce the "bird-foot" delta type and the toes in this foot by their junction with the banks which outline the ancient estuary, have separated in succession a series of basins that before were in direct connection with the sea (Fig. 457). Lake Pontchartrain is the largest of this series, while the so-called Lake Borgne is in process of separation.

Where large deltas push out from the shore into the open sea, the levees which border the individual distributaries are attacked by the waves and their materials are transported by the shore currents and built into barriers. These barriers cut off the re-entrants between neighboring distributaries so as to produce lagoons or lakes (Fig. 458).

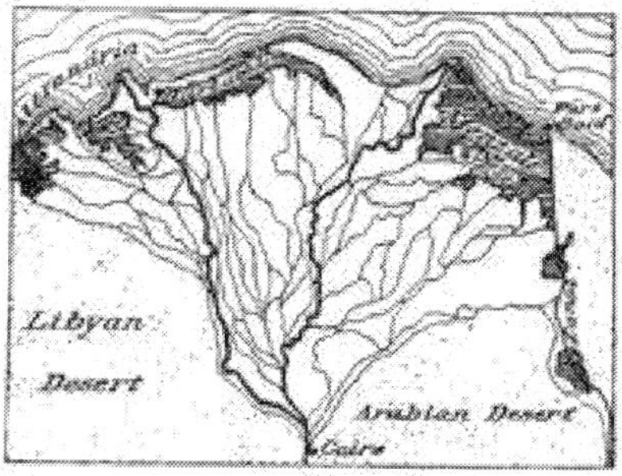

FIG. 458.—A type of delta lakes formed by levees in part destroyed and built into barriers on the margin of the delta of the Nile (after Supan).

A type of delta lake, which more resembles the side-delta lake above described, has formed at the mouth of the Colorado River, where it enters the Gulf of Lower California. The Imperial Valley lying to the north of this delta is the desiccated floor of the earlier Gulf of Lower California which has been captured from the sea by the delta of the Colorado. The rampart of mountains, by which this valley is surrounded, has cut it off from any water supply derived from clouds, and its waters being no longer renewed from the sea, the region has passed through a period of desiccation which has left the Salton Sink as the only existing remnant of the earlier lagoon. It will be remembered that careless operations in diverting distributaries of the Colorado recently reversed this process so that the waters rose in the valley, and expensive emergency operations were necessary in order to again turn the waters of the Colorado into their accustomed channels.

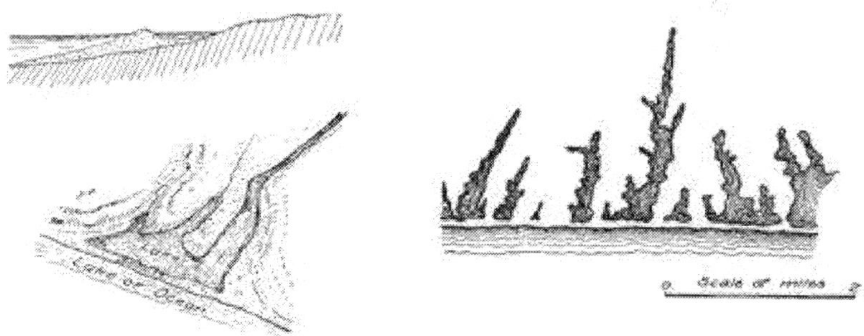

FIG. 459.—Diagrams to illustrate the characteristics of barrier lakes, with an example from the southern coast of the Island of Nantucket.

Barrier lakes.—The Salton Sink illustrates a type of lake which is formed at the border of the sea through the erection of some kind of barrier which captures a small area of the ocean's surface. Though such lakes may be properly described as strand lakes, it is usually at the mouth of a river that the process becomes effective. The common type of *barrier lakes* is found developed on most ragged coast lines where the shore currents have formed first bars and later barriers at the mouths of the estuaries (Fig. 459). Such embankments are usually gently curving or crescent shaped and are composed of sand or shingle which presents a steep landward and a gradual seaward slope.

FIG. 460.—Dune lakes on the coast of France (after Berghaus).

Dune lakes.—Within the narrow strips of shore in which all the fine soil that could be available for plant life has been washed away by the waves, beach sand is exposed to the direct action of the winds. In time of storm the sand is picked up and after drifting in the wind is collected in long ridges parallel to the shore. Constantly traveling along shore, these dunes block the mouths of rivers and thus produce a series of lakes such as are indicated in Fig. 460.

Sink lakes.—Another class of lakes are due either directly or indirectly to the work of underground waters. In districts which are underlain by limestone, the surface water descending along the joints of the limestone may widen these passageways through solution of the rock and at lower levels flow on the floors of caverns eaten out by the same process on bedding planes of the formation. At the intersections of joints, more or less circular shafts known as "swallow-holes" go down to the caves from the surface.

Locally, also the cavern roofs give way so as to choke the galleries with rubble and leave a basin at the surface which has an irregular but generally a more or less oval outline. If sufficiently clogged at the bottom by finer rock débris, these basins become occupied by small lakes which are known as sinks, and constitute one of the best surface indications of a limestone country.

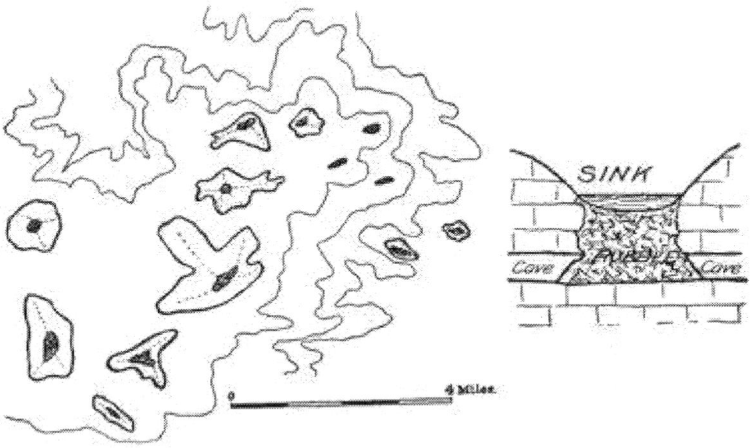

FIG. 461.—Sink lakes in Florida, with a schematic diagram to illustrate the manner of their formation (map from U. S. G. S.).

Karst lakes—poljen.—In the limestone country to the north and east of the Adriatic Sea—the so-called Karst region—there are many interesting features which are directly traceable to the solution of the country rock. Here all the surface water descends in certain districts along the widened joint planes so that the drainage is largely subterranean. The so-called *dolines* or sinks of very regular and symmetrical forms resembling deep bowls cover a large part of the surface.

The entire country is, moreover, faulted in the most intricate fashion into many rift valleys. The drainage being so largely subterranean, these downthrown blocks of crust, the so-called *poljen*, become flooded at certain seasons of the year when the subterranean passages become choked or are too small to carry away all the water. A seasonal lake of this character is the Zirknitz Lake (p. 189).

Playa lakes.—It is the law of the desert that the arid region be walled in by mountains. This encircling rampart forces the clouds to rise, and by robbing them of their moisture leaves the desert dry and barren. Those waters which fall upon the inner margin of the ranges drain toward the interior of this pan-like depression and are not returned to the sea—the desert is without an outlet. Infrequent though they be, the desert rains are of the cloudburst type and in the hills develop torrents whose waters, emerging upon the desert floor, develop lakes in the space of a few minutes or at most hours. In the hot and dry atmosphere the waters of these shallow basins may be sucked up in the space of a few hours but reappear in the same basins at the time of the next succeeding cloudburst. Such ephemeral lakes are known as playas.

Salines.—Desert lakes more favored in their supply of water may be relatively long lived and persist for periods measured in years or centuries. Such lakes are, however, extremely sensitive to climatic changes (see p. 198).

For the reason that they have no outlet the waters of desert lakes become salt through continued evaporation. They are, therefore, spoken of as *salines*. Lake Bonneville, so long as it discharged its waters over the sill of the Red Rock Pass, must have remained fresh; but when the level of its waters had fallen below this outlet, its waters became salt and the content increased as the volume diminished.

The shallow basins upon the floors of desert lakes may have come into existence in various ways; but it would appear that the irregular removal of the soil by the winds, modified as this is by differences in composition of the rock materials and by vegetable growth, and the deposition of sand by the same agent, are by far the most important. Many of the types of tectonic and volcanic lakes which have been described are characteristic of humid and arid regions alike.

Alluvial-dam lakes.—Within the mountains upon the desert borders, the alluvial fans which form at the mouths of valleys, because of the characteristic cloudburst, sometimes obstruct a main valley at the junction with its tributaries. By this process the waters of the main river are impounded in essentially the same manner as are the rivers of humid regions by the deltas of their tributaries.

Résumé.—The types of lakes which we have now considered are arranged below in tabular form so as to show their relationship to important geological processes. While not complete, the list includes the more important classes, as well as others which, while not of common occurrence, are yet of interest in giving further illustration to the processes which have been treated in earlier chapters.

By giving careful attention to criteria which have been above suggested, it should be possible in the greater number of instances at least to determine whether any lake which is visited has had its origin in one or another of the processes described.

CLASSIFICATION OF LAKES

Tectonic Lakes	*Volcanic Lakes*
Newland lakes	Crater lakes
Basin-range lakes	Coulée lakes
Rift-valley lakes	
Earthquake lakes	

Continental Glaciation Lakes	*Mountain Glaciation Lakes*
Morainal lakes	Rock-basin lakes
Pit lakes	Valley moraine lakes
Glint or colk lakes	Landslide lakes
Ice-dam lakes	Border lakes
Glacier-lobe lakes	

River Lakes	*Strand Lakes*
Ox-bow lakes	Barrier lakes
Saucer lakes	Dune lakes
Crescentic levee lakes	
Raft lakes	
Side-delta lakes	
Delta lakes	

Ground Water Lakes	*Desert Lakes*
Sink lakes	Playa lakes
Karst lakes—*poljen*	Salines
	Alluvial dam lakes.

READING REFERENCES FOR CHAPTER XXIX

General:—

I. C. RUSSELL. Lakes of North America. Boston, 1895, pp. 125, pls. 23.

A. P. BRIGHAM. Lakes, A Study for Teachers, Jour. Sch. Geogr., vol. 1, 1897, pp. 65-72.

N. M. FENNEMAN. The Lakes of Southeastern Wisconsin, Bul. 8, Wis. Geol. and Nat. Hist. Surv., 1902 (Rev. Ed., 1910), pp. 188, pls. 37.

A. DELEBECQUE. Les Lacs Français (with Atlas). Paris, 1898. (Work crowned by the Society of Geology of Paris.)

H. R. MILL. Bathymetrical Survey of the English Lakes, Geogr. Jour., vol. 6, 1895, pp. 46-73, 135-166.

A. SUPAN. Grundzüge der Physischen Erdkunde. Leipzig, 1896, pp. 531-548.

H. BERGHAUS. Atlas der Hydrographie. Gotha, 1891, pl. 3.

R. D. SALISBURY. Physiography. 1907, pp. 292-327.

CHARLES RABOT. Revue de limnologie, La Géographie, Vol. 4, 1901, pp. 110-119, 172, 189.

I. C. RUSSELL. A Geological Reconnaissance in Southern Oregon, 4th Ann. Rept. U. S. Geol. Surv., 1884, pp. 442-447. (Basin range lakes.)

ED. SUESS. The Face of the Earth, vol. 4, 1909, pp. 268-286. (Rift valley lakes.)

J. S. DILLER. Crater Lake, Nat. Geogr. Mag., vol. 8, 1897, pp. 33-48, pl. 1; Geology of Lassen Peak Quadrangle, California, Geol. Fol. 15, U. S. Geol. Surv., 1895. (Coulée lakes.)

N. M. FENNEMAN. Lakes of Southeastern Wisconsin, *l.c.*, pp. 4-6. (Pit lakes.)

ED. SUESS. The Face of the Earth, vol. 2, 1906, pp. 340-346, pl. 7. (Glint lakes.)

I. C. RUSSELL. A Preliminary Paper on the Geology of the Cascade Mountains in Northern Washington, 20th Ann. Rept. U. S. Geol. Surv. Pt. ii, 1900, pl. 14. (View of a rock-basin lake.)

E. W. SHAW. Preliminary Statement concerning a New System of Quaternary Lakes in the Mississippi Basin, Jour. Geol., 1911, pp. 481-491. (New type of levee lakes.)

A. C. VEATCH. Formation and Destruction of the Lakes of the Red River Valley, Prof. Pap. No. 46, U. S. Geol. Surv., pp. 60-62, pls. 29-33. (Raft lakes.)

M. NEUMEYER. Erdgeschichte, vol. 1, pp. 595-596. (Poljen.)

CHAPTER XXX

THE EPHEMERAL EXISTENCE OF LAKES

Lakes as settling basins.—Of all the processes which conspire to blot out the lakes with which our northern landscapes are dotted, the one of greatest importance is in most cases a process of filling by the sediments brought in by tributary streams. The carrying of sediment in suspension depends, as we know, upon the velocity of the current, and as this is checked where it reaches the lake margin, all coarser material is at once deposited to form a delta, while the finer sediments are held longer in suspension and finally settle in thin layers over the entire bottom of the lake. Clay deposits surrounded by coarser sediments are thus characteristic of filled lake basins.

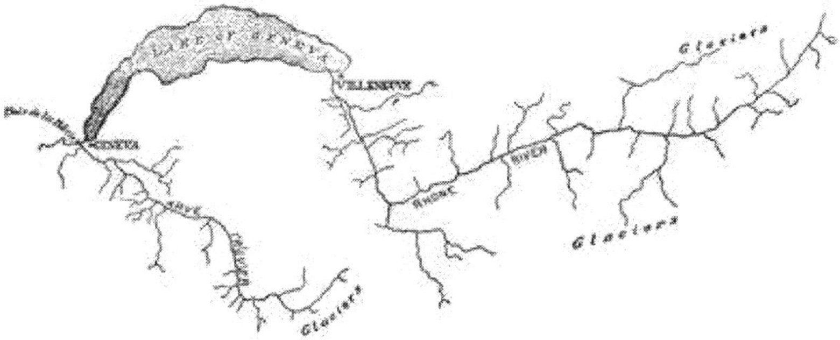

FIG. 462.—Map of the Arve and the upper Rhone to show the importance of Lake Geneva as a settling basin of the larger stream.

How waters are clarified by their passage through a lake is indicated by a comparison of a river system such as the St. Lawrence, with a river like the Missouri and Mississippi. Not only are the lower stretches of the St. Lawrence in striking contrast with the muddy floods of the Missouri and Mississippi; but the delta, which is so remarkable a feature in the Mississippi, has no counterpart in the northern river.

The most noteworthy examples of settling are, however, furnished by the lakes of Switzerland, for the reason that SWISS rivers are heavily charged with rock flour produced beneath the numerous glaciers at the valley heads, and, further, because these rivers descend with turbulent currents to near the borders of the larger lakes. To look out upon the murky waters of the upper Rhone, where they enter Lake Geneva near Villeneuve, and then to watch the flood of crystal water which issues from the lake and passes under the bridge at Geneva, is an object lesson which no traveling student should miss (Fig. 462). Yet even more instructive is a visit to the *Bois de la Bâtie* at the junction of this clear stream with the Arve, a half hour's walk only below Geneva. The waters of the Arve have come on a steep descent directly from the glaciers of the Mont Blanc district, and as they meet the cleared waters of the Rhone, they flow beside them down the common valley without mingling. Dull and opaque, the Arve waters can be discerned for a long distance as a white belt against the left bank of the river, sharply defined against the blue reflecting surface of the Rhone waters (Fig. 463). Upon the banks of the Arve, just above its junction, a cement manufactory has been established to utilize the clays which are here deposited.

FIG. 463.—View looking upstream across the opaque waters of the Arve to the clear reflecting surface of the Rhone. To the right across the Arve is seen the cement works for recovering the Arve sediments.

Wherever lakes are contained in long and narrow valleys, the greater part of the tributary drainage enters at the upper end, and the delta which there forms extends from bank to bank. As it continues to advance into the lake, the earlier water basin is gradually transformed into a level plain of delta deposit, a feature so common as to be deserving of a special name. The Scottish lochs, which are lakes of this type, are each extended in a longer or shorter delta plain described as a *strath*, and this local term may well be given a general application (frontispiece). The city of Ithaca, the seat of Cornell

University, is built upon a strath at the head of Lake Cayuga, and numberless Scottish and Swiss hamlets have been located upon such fertile plains (Fig. 464).

FIG. 464.—The village of Poschiavo in eastern Switzerland, built upon a strath at the head of Lake Poschiavo.

Drawing off of water by erosion of outlet.—Next in importance to the filling up of lake basins as a factor in their early extinction is the cutting down of their channels of outflow. Whenever the walls of the outlet are cut in rock, this draining process is apt to be slow, for the reason that the outlet stream is of filtered water and so lacks the necessary cutting tools. By far the larger number of lakes are, however, held back by dams of loose drift deposits laid down by the earlier continental glaciers; and so the very clarity of the water promotes the erosion of the outlet by allowing the stream's full burden of sediment to be lifted and then removed from the channel.

The pulling in of headlands and the cutting off of bays.—The removal of projecting headlands by wave action, though it increases the area of the lake, yet it decreases directly the volume of lake water through formation of the built terrace, and indirectly in far larger measure through the transformation of bays into quiet lagoons within which the extinguishing process of peat growth is set in operation.

Lake extinction by peat growth.—The first condition for the growth of lake vegetation is quiet water. Within small lakes, such as the kettle basins upon moraines, aquatic vegetation develops rapidly, and bogs of peat might almost be included among the most important distinguishing marks of a glaciated country. Within larger lakes it is only after barrier beaches have been thrown across the mouths of the bays to form natural breakwaters for the waves that this process of lake extinction by peat growth can become effective.

FIG. 465.—View of the floating bog and surrounding zones of vegetation in a small glacial lake of the Yellowstone National Park (after a photograph by Fairbanks).

Many erroneous notions are still held concerning the prime importance of sphagnum in peat formation, owing to the peculiar local conditions under which the early studies were made. Within the glaciated districts of the United States, the formation of peat involves the successive growths of a number of zones of vegetation and the formation of a floating bog which advances into the lake from the shores, followed in turn by belts of low shrubs, tamaracks, and lastly deciduous trees (Fig. 465).

In most cases the first plants to develop in a quiet lake are the water lilies, though these are sometimes preceded by chara and floating bladderwort. Next behind the water lilies come the sedges, which form a mat of floating bog by their grasslike stems sinking down in the water and being there interwoven with the rhizomes below. This mat of sedge is often so firm that cattle may advance upon it to the water's edge, but it is separated by a layer of water from the bed of growing peat at the bottom of the lake (Fig. 466). This bed of peat appears to grow upward toward the surface and become joined to the shore end of the floating bog by decaying vegetation which is dropped from the bottom of the mat above.

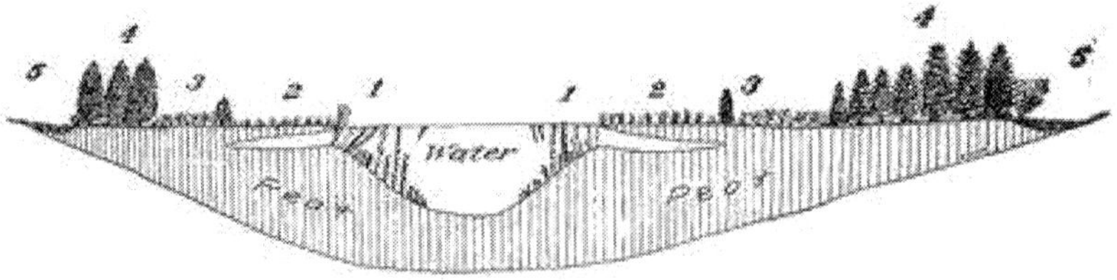

FIG. 466.—Diagram to show how small lakes are transformed into peat bogs (after C. A. Davis).

In order behind the floating bog come the advanced plants of the conifer group, with sphagnum and low shrub here upon a peat base extending to the lake bottom. Behind the belt of shrubs arise the tamaracks and spruces, and lastly, toward the shore, come the deciduous trees and especially poplars, maples, and marginal willows. Upon the margin of the basin there is usually a low trench, or "fosse", filled with water during wet seasons, as a result, no doubt, of seasonal inwash that does not reach the residual lake toward the center of the basin.

Extinction of lakes in desert regions.—In arid regions there are special causes of lake extinction. Thus the blowing in of sand and dust carried for long distances in the air, a by no means negligible factor even in humid regions, here assumes large importance. The now exposed basins of extinct desert lakes afford the evidence, however, of an even greater factor of extinction, in climatic change. The clouds, which at one time found their way into the drainage basin of a lake, may later through the rise of a mountain barrier be cut off, and so with reduced water supply a period of lake desiccation is begun. When, in this process of drying up, the lake level has fallen below that of the outlet, the saline content of the waters begins to increase, and later a stage is reached, as in Great Salt Lake, when the sodium salts are precipitated. When the lake has become extinct, these deposits remain as a witness to the changed climatic condition.

The rôle of lakes in the economy of nature.—It is natural, in considering the extinction of lakes, to give some attention to the rôle which they play in the economy of nature. That lakes filter the water of rivers, and prevent the formation of important delta deposits, has already been noticed. A curious exception to this general rule is furnished by the great delta at the head of Lake St. Clair, just below the outlet of Lake Huron. This anomaly is, however, explained by the peculiar currents of Lake Huron, which are so directed as to sweep the beach sand into the swift current of the outlet, to be deposited in the quiet waters of Lake St. Clair (Fig. 467).

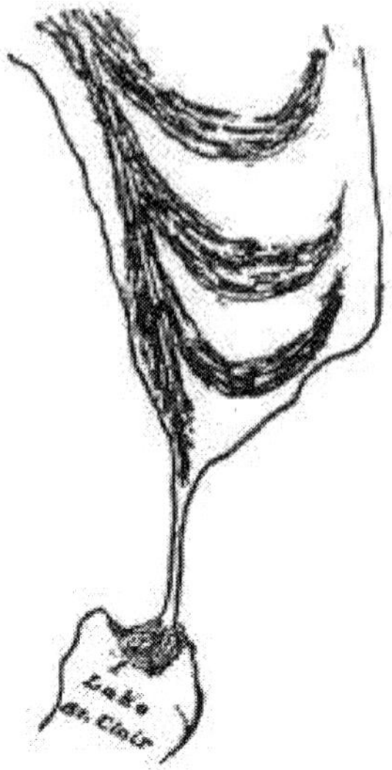

FIG. 467.—Map to show anomalous position of the delta in Lake St. Clair, due to the peculiar currents in Lake Huron (after maps by Cole).

As regulators of the flow of rivers, lakes perform an important function. Such disastrous floods as are characteristic of the spring season within the basin of the lower Mississippi could not occur in the lower St. Lawrence, for the reason that the great basins of the lakes serve as distributing reservoirs. The annual floods, upon which the agriculture of Egypt depends, are explained by the flood waters from the high mountains of Abyssinia entering the Nile *below* the lakes of its upper basin.

In one further respect large inland bodies of water have an important function as regulators. It is the property of water to respond but slowly to the variations in the quantity of heat which reaches the earth's surface from the sun. A larger quantity of heat must be added to or abstracted from a body of water, in order to change its temperature by one degree, than would be required for a like change in the same bulk of earth or rock. Thus bodies of water by more slowly acquiring the summer's heat retard the coming spring, and by storing up this energy and carrying it over into the autumn the warm season is prolonged and early frosts prevented. The fruit belts about the lower Great Lakes are thus dependent upon this regulating property of the lake waters. The discomfort of the long spring of raw weather is thus compensated by an unusually salubrious harvest season.

Ice ramparts on lake shores.—Small ridges known as ice ramparts are formed upon lake shores by the action of lake ice, though subject to so many qualifying conditions that the range of their occurrence is somewhat limited. Within districts where a winter ice cover of some thickness is formed, the shores of lakes are apt to present ridges of bowlders parallel to and near the water's edge, and such lakes have sometimes become known as "wall lakes" (Fig. 468).

FIG. 468.—A bowlder wall upon the shore of a small lake in the Adirondacks of New York.

In many cases these small ridges have been formed at the time of the spring "break up" of the ice; for the ice cover, when once loosened, is drifted in great rafts first against one shore, and later, with a change of wind direction, against another. Under the impact of such heavy rafts, the half-submerged bowlders near the shore are forced up the beach until they lie in a ridge or bowlder wall.

At other times such bowlder walls, and far more interesting ridges as well, result from a kind of ice shove independent of the wind, but caused by expansion within the ice itself during a sudden rise of temperature of the surrounding air. Such ice ramparts require for their explanation a consideration of the sequence of events from the time the ice cover closes the lakes.

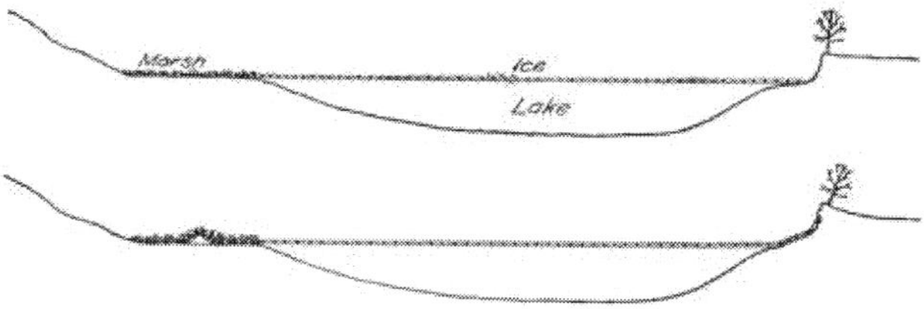

FIG. 469.—Diagrams to show the effect of ice shove in producing ice ramparts upon the shores of lakes (after Buckley with a slight modification).

The first lake ice of early winter forms in most cases with air temperatures a few degrees only below the freezing point of the water. When later a severe "cold wave" arrives, the ice cover is contracted and becomes too small for the lake surface. To this contraction it yields and opens cracks up which the water rises, and in the prevailing low temperature this water is quickly frozen and the lake cover again made complete. Skaters are familiar with the opening of these cracks and the loud "roaring" which accompanies it on cold mornings, the sharp skate runners sometimes starting a crack in the strained ice, as does a light scratch upon glass that is in a similar strained condition.

FIG. 470.—Various forms of ice ramparts (after Buckley).

The original ice cover of the lake, which was formed at near-freezing temperatures, has now received a number of inserted wedges of new ice at a time when its contracted volume has made this possible. If now a "warm wave" succeeds to the "cold wave" in the air, the ice cover expands at a rate corresponding to its rate of contraction, so that a strong pressure is exerted against the shore (Fig. 469). Sliding up the sloping surface of the cut and built terrace, the force of this shove may be deflected upward against the cliff, and if this is of loose materials, the effect may be to ram bowlders into the bank, to push up ramparts or ridges, to overturn trees, etc. (Fig. 470). In marsh land the frozen surface layer may slide over its unfrozen base and be forced up into broken folds (lower diagram of Figs. 469 and 470).

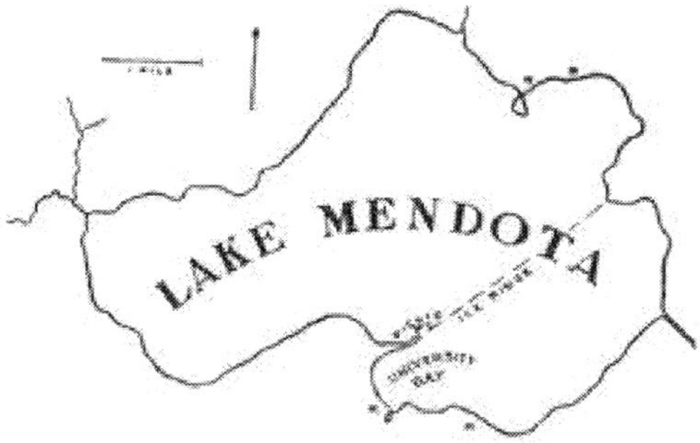

FIG. 471.—Map of Lake Mendota at Madison, Wisconsin, showing the position of the ridge which forms from ice expansion, and the ice ramparts about the shores of the bays (based on Buckley's map).

In order that ice ramparts may be formed, it is necessary that the winter climate of the district be severe and characterized by alternating cold and warm waves, involving considerable range of air temperature below the freezing point. If the lake is small, the push of the ice will be through so small a distance as not to yield appreciable ramparts. If, on the other hand, the lake is too large, the ice cover is not rigid enough to transmit the push to the distant shore, but, like a long beam employed in the same manner to transmit a compressive stress, it is bent out of a straight line and later broken. Thus in a broad lake, with the coming of a "warm wave", the ice cover opens in a crack from shore to shore and finds relief from the stress by pushing up a ridge above the crack. On such lakes ice ramparts are found only about the shores of bays whose expanse does not greatly exceed a mile (Fig. 471).

When there is heavy snowfall, ice ramparts either do not form or are of smaller dimensions, probably in part because the ice is blanketed by the snow and so prevented from sudden elevation of temperature during the "warm wave", but even more because the ice cover is sensibly bowed down under its load and so rendered incompetent to transmit the developed stresses to the shores.

READING REFERENCES FOR CHAPTER XXX

Lake extinction by peat growth:

C. A. DAVIS. Peat, Essays on its Origin, Uses, and Distribution in Michigan, Ann. Rept. Mich. Geol. Surv. for 1906, 1907, pp. 105-182; Peat Deposits as Geological Records, 10th Rept. Mich. Acad. Sci., 1908, pp. 107-112.

G. P. BURNS. Bog Studies. Ann Arbor, 1906, pp. 13.

Ice ramparts:

C. H. HITCHCOCK. Shore Ramparts in Vermont, Proc. Am. Assoc. Adv. Sci., vol. 13, 1869, pp. 335-337.

G. K. GILBERT. Lake Bonneville, Mon. 1, U. S. Geol. Surv., 1890, pp. 71-72.

E. R. BUCKLEY. Ice Ramparts, Trans. Wis. Acad. Sci., etc., vol. 13, 1900, pp. 141-162, pls. 1-18.

WILLIAM H. HOBBS. Requisite Conditions for the Formation of Ice Ramparts, Jour. Geol., vol. 19, 1911, pp. 157-160.

CHAPTER XXXI

THE ORIGIN AND THE FORMS OF MOUNTAINS

A mountain defined.—As ordinarily understood, mountains are elevations upon the earth's surface which rise above the general level of the country. Their summits need not be at great heights above the sea, but it is essential that they project above the average level of the surrounding country by at least a quarter of a mile. Lower elevations are described as hills. On the other hand, the elevation of a plateau like the "High Plains" of the western United States may be as much as a mile, but the vast expanse of nearly level surface precludes the use of the term "mountain." The word is thus applied to a feature of the earth and not merely to an elevated tract.

In a collective sense, though more often in the plural form, the term is properly applied to groups of similar features which have a common origin in local uplift of the land. The origin of mountains used in this sense of mountain complexes is thus connected with some essentially local uplift of the earth's surface. This may take place by the processes of folding and superincumbent fault displacement, by volcanic extravasations or ejections, or by a deeper seated and essentially hydrostatic elevation of rock beds over molten rock material.

The existing *forms* of mountains, as we are to see, are largely shaped by the erosional processes which are set in operation by the uplift itself, though often completed long subsequent to it.

The festoons of mountain arcs.—From our earliest studies of school geographies, we have become familiar with the arrangement of the more important mountains in long chains or systems. Comparatively few persons have given any further attention to the arrangement of the chains, though over large areas of the earth's surface the distribution of mountain ranges is deeply significant. The map of Asia in particular presents a series of great sweeping arcs or crescents which are grouped as though hung upon the map in festoons with knots or vertexes to separate neighboring groups (Fig. 474, p. 438, and Fig. 472).

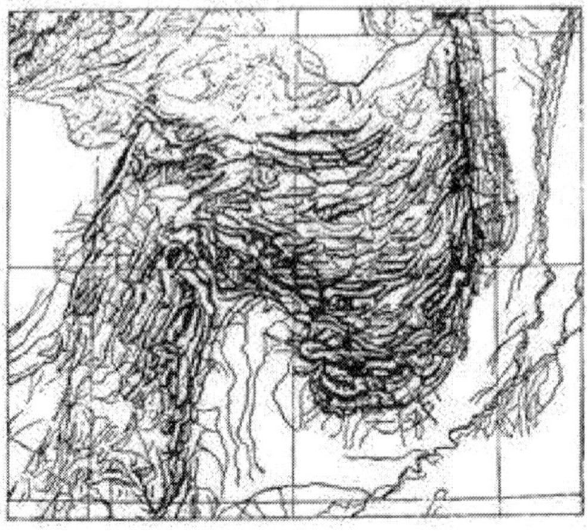

FIG. 472.—The great multiple mountain arc of Sewestan, British India (after de Saint Martin and Schrader).

The significance of these mountain groupings in the evolution of the earth's surface has been pointed out by the great Viennese geologist Suess, to whom we are indebted for focusing upon the plan of the earth an amount of attention which before had been largely given to the preparation of hypothetical sections of strata which were largely buried from sight beneath the earth's surface. Broadly speaking, the mountain arcs may be said to be grouped about those shields of older rock which geological studies have shown to be the oldest land masses upon the globe. Within the northern hemisphere these original continents are represented by the areas of crystalline rock centered over Hudson Bay, the Baltic Sea, and an area in northeastern Siberia known to geologists as Angara Land. In our study of the figure of the earth (Chapter II) it was found that these shields represent the truncated angles of the rounded tetrahedral form toward which the planet is tending (Fig. 3, p. 12).

Theories of origin of the mountain arcs.—The mountain arcs, when studied in detail, are found to be composed of closely folded rock strata, the flexures of which are generally so overturned that their axial planes dip toward the center of the arc (Fig. 473). It was the view of Suess that these arcs are to be explained by a pushing outward of the rock strata from the center of the arc toward its periphery, thus causing a wrinkling of the surface strata and an overriding of the surrounding formations, which upon this hypothesis opposed a greater resistance to the sliding movement. The folding together of the strata due to the sliding naturally involves a very considerable diminution of the surface area presented by the strata (Fig. 22, p. 42). In the case of the Alpine chains it has been estimated that a flat land area, four hundred to eight hundred miles across, has by the folding process been reduced to a width of only about one hundred miles, or from a fourth to an eighth of its former width.

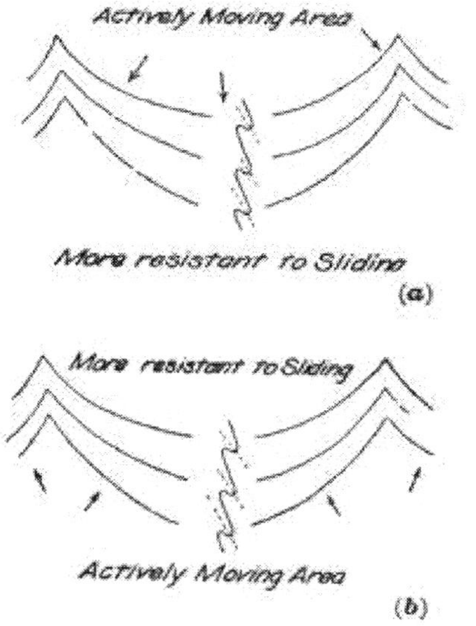

FIG. 473.—*a*, diagram to illustrate the Suess' theory of the origin of mountain arcs; *b*, the author's modification of this view.

The weakness of Professor Suess' theory lies in the fact that such compression as it implies is assumed to be due to an *outward* movement of the relatively small area of the earth's outer shell which is included *within* the arc. It must be obvious that such a movement, being from a center toward three sides at once, would for this circumscribed area involve enormous proportionate reduction in superficial area of the strata and could only result in a hiatus near the center of the arc. No such gap is to be found, and one would, moreover, be difficult to account for upon any plausible hypothesis. On the other hand, the general contraction of the planet as a whole, involving as it does reduction of surface over large areas, is a well-recognized fact; and if it be true that the shields formed by the older continents are less subject to contraction than the remaining portions of the surface, it is easy to understand why the earth's outer skin should be wrinkled by *underfolding* and thrusting about these continental margins. The contrast of this view with that of Professor Suess is expressed in the diagrams of Fig. 473.

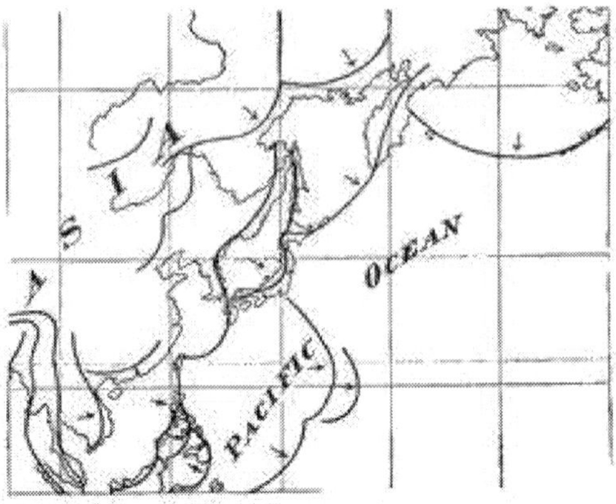

FIG. 474.—Festoons of mountain arcs about the borders of the Pacific Ocean—Pacific type of coast (based upon Suess).

We may illustrate this conception by a stretched sheet of rubber cloth such as is in common use by dentists, upon which a thin layer of hot Canada balsam has been spread. This substance congeals upon cooling to near-normal temperatures, and if a small local area of the balsam layer be chilled and the tension upon the rubber then released, the viscous balsam of the unchilled portion of the layer is thrown into wrinkles about the cooled and more resistant areas. These more resistant portions of the stratum may thus represent the ancient continental shields of our planet.

The Atlantic and Pacific coasts contrasted.—In his studies of mountain arcs in their relation to the plan of the earth, Professor Suess has shown how the arrangements of the mountain chains about the two larger oceans represent two strongly contrasted types. Whereas about the Pacific margin the mountain arcs are, as it were, strung in festoons which trend parallel to and are convex toward the coast, or else lie in fringing garlands of islands in the same attitude (Fig. 474); the mountain chains about the Atlantic become sharply truncated as they reach the coast, and thus indicate that the basin of this ocean has been produced by an inthrow or depression between great marginal displacements in some period subsequent to the formation of the mountains.

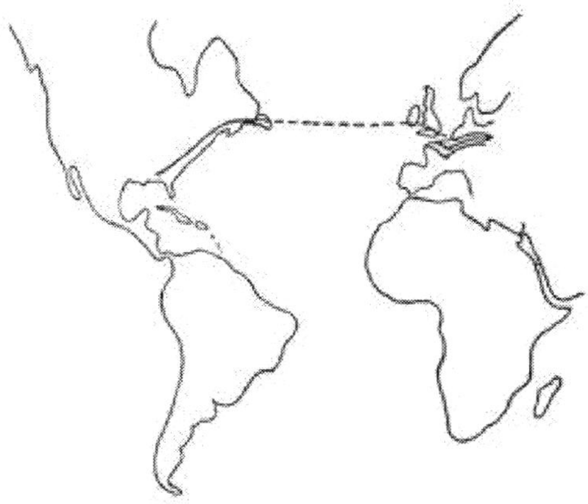

FIG. 475.—The interrupted system of the Armorican Mountains common to western Europe and eastern North America (after Arldt).

Thus the mountain folds of the Appalachian system are in Newfoundland cut off abruptly at the coast line, and the same beds, similarly truncated, are encountered again across the expanse of ocean in the folds at the coast of western Europe (Fig. 475). In discontinuous remnants this ancient mountain chain may be traced in an east and west direction across western and central Europe. We have thus here to do with a single mountain system which extends from central Europe to northern Alabama, out of which a great link has been taken by the subsequent sinking in of the basin of the Atlantic Ocean.

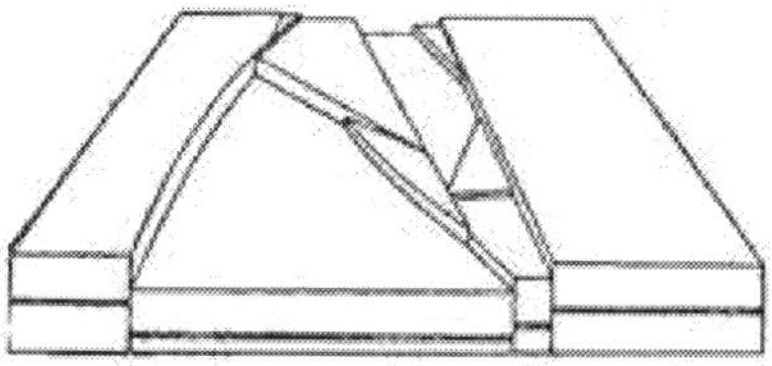

FIG. 476.—Schematic representation of a "zone of diverse displacement" in the Great Basin of the western United States (after Powell).

The block type of mountain.—The inclusion of most elevations in mountain chains and arcs is one of the most obvious facts to any one who has examined world atlases with this subject in mind. Such chains are almost invariably composed of folded rocks, thus indicating that erosion has removed great superincumbent masses of strata since the crustal compression produced the folds at considerable depths below the then surface.

There are, however, large elevated tracts upon the earth's surface which are intersected by deep valleys, but where no arrangement of the elevated portions within chains or ranges is to be detected. In such cases the distribution of mountain and valley may bear a resemblance to a mosaic of disturbed parts which stand at different levels (Fig. 476).

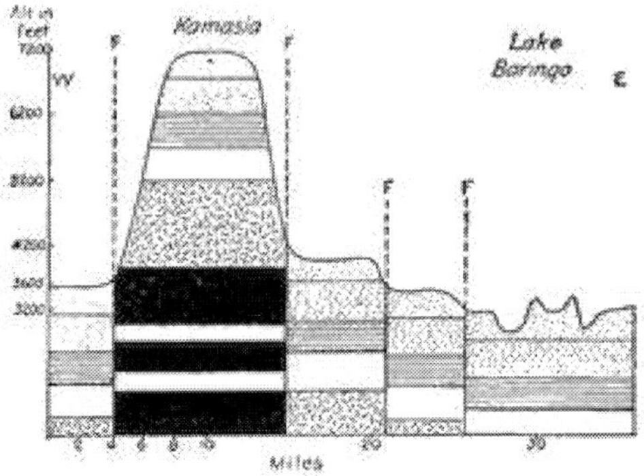

FIG. 477.—Section of an East African block mountain (after J. W. Gregory).

Such block mountain districts are to be found in many parts of the earth's surface, but notably within the Great Basin of the western United States, and in the land area which borders the Indian Ocean upon the west and northwest. In contrast with the mountain arcs, so strikingly exemplified by the continent of Asia as a whole, its extreme southwestern portion is made up of an alternation of plateau and rift valley separated from each other by great displacements. Though modified to some extent by erosion, the elevations seem generally to represent the displaced crust blocks which in mutual adjustments have been left at the highest levels (Fig. 477). The valley of the Jordan, with the mountains of Lebanon rising above it, is near the northern extremity of this faulted mountain region (Fig. 434, p. 404), while the Great Rift valley, crossing east Central Africa, and the many neighboring rifts to the east and west, are graven in lines so deep that an observer upon a neighboring planet might perhaps detect them.

It is not necessary in all cases to assume that the block mountains of a faulted district represent the blocks which in the adjustments were left the highest. Erosion in the course of time accomplishes marvels of transformation, and it may result that heavy masses of more resistant rock eventually project the highest, even though they may represent the downthrown blocks in the fault mosaic (Fig. 43, p. 60).

FIG. 478.—Tilted crust blocks in the Queantoweap valley.

Where in addition to undergoing changes of level the earth blocks have been tilted, the features long since described from our western interior basin as "Basin Range structure" are developed. Here the upper surface of the disturbed earth blocks betrays the evidence of a definite tilt in some one direction (Fig. 478, and Fig. 431, p. 402).

Mountains of outflow or upheap.—An important type of mountain, generally described as volcanic, may be due either to the outflow of lava at the earth's surface, or to accumulations of separated fragments of lava, first thrown into the air, and then deposited by gravity or admixed with water as volcanic mud. Such mountains, both before and after modification by erosion, assume the strikingly characteristic forms which have been fully discussed in Chapters IX and X. The dominant types are the lava dome and the puy, the cinder cone, and the more complex composite cone. Excepting only the surface produced by the few great fissure eruptions and the semivolcanic mesa type, the individual mountains of volcanic origin develop features with notably circular bases.

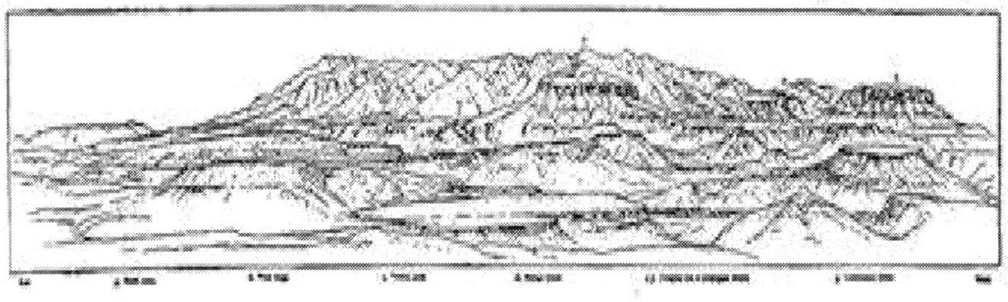

FIG. 479.—Pen drawing of the laccolite of the Carriso Mountain by W. H. Holmes, which shows the jagged surface of the igneous rock core and the sloping tables which still remain of the roof of sedimentary rocks (after Cross).

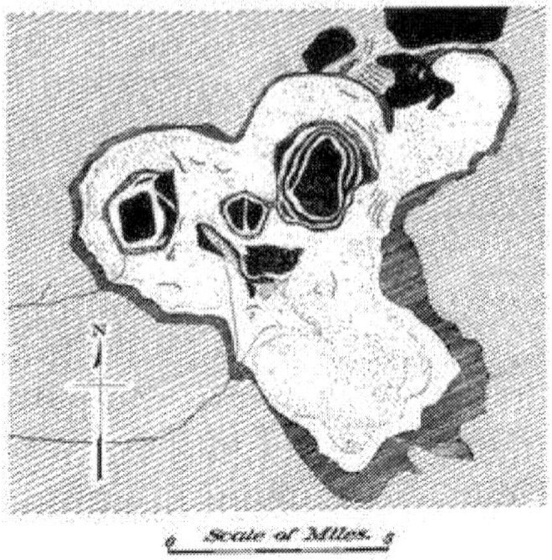

FIG. 480.—Map of laccolitic mountains. A portion of the Judith Mountains, Montana. The intrusive igneous rock is shown in black (after Weed).

Domed mountains of uplift—laccolites.—At a considerable number of widely separated localities upon the earth's surface, mountainous regions are encountered, the central areas or cores of which are composed of intrusive igneous rock such as granite, and about this core the sediments dip away in all directions as though they had once formed a continuous roof above it and had been forced into this dome by hydrostatic pressure of the once viscous material beneath (Fig. 152, p. 143, and Figs. 479 and 480). Examples of such domed mountains of uplift were first described by Gilbert from the Henry Mountains of Utah, but instances are furnished by many elevated tracts, especially within the western United States. Such mountains are known as *laccolites*, but when one margin at least of the igneous core corresponds to a displacement, the mountain is described as a *bysmalite* (Fig. 481).

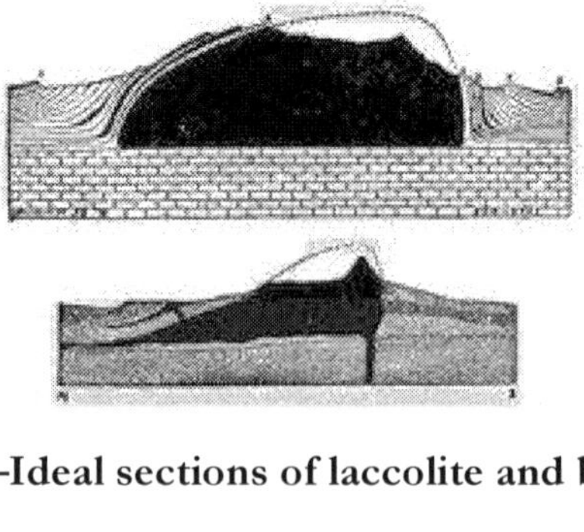

FIG. 481.—Ideal sections of laccolite and bysmalite.

When subjected to long-continued erosion, the generally fissured granitic core of the laccolite weathers in a wholly different manner from the bedded sediments which surround and still in part mount over it. The former usually presents a more or less jagged surface which contrasts sharply with the gently sloping tables of the latter (Fig. 479). About the high granite core of the mountain, the several strata of the uptilted formations present each a steep slope toward this higher land, and a gentler slope in the opposite direction. Such unsymmetrical ridges which surround the mountain area are often referred to as "hog backs" (plate 12 B). The arrangement of the strata in the hog backs thus presents an overlapping series like the shingles upon a roof, except that the overlapping is here from the bottom instead of the top, and the exposed ends thus face toward the crest. Unlike a shingle roof the hog backs do not shed the water which descends to them from the higher levels, but, on the contrary, they cause it to flow in troughs parallel to the base of the slope except where outlets are found through them.

Mountains carved from plateaus.—In the mountain types thus far discussed, the local uplifting of the land has itself developed features which in the aggregate may be referred to as mountains, even though the characters of the original surface are soon destroyed by erosive processes of one sort or the other. Erosive processes are, however, quite competent to produce mountain forms from a featureless plateau, and particularly through the incision by streams of running water, the best studied process of mountain sculpture (see Chapters XI-XIII). This process of throwing valleys about an elevated section of the earth's surface, and so carving out mountains, is sometimes described as *circumvallation*; and if the term "mountain" be applied in its ordinary sense to describe an individual feature, it is clear that most mountains have been formed in this way.

To discuss the characteristic shapes of such mountains would be largely to review the contents of this book, and especially those portions which discuss the character profiles resulting from the action of each sculpturing or molding agent. The work of frost and other weathering agencies, of running water, of mountain and of continental glacier, would all have to be considered in order to evolve the history of each mountain.

In addition to discovering the agents which were chiefly responsible for the shaping of the mountain, we may, further, in many cases determine at what stage the work of one agent has been succeeded by that of another, and at least at what stage of its complete cycle of activity the latest agent is now at work.

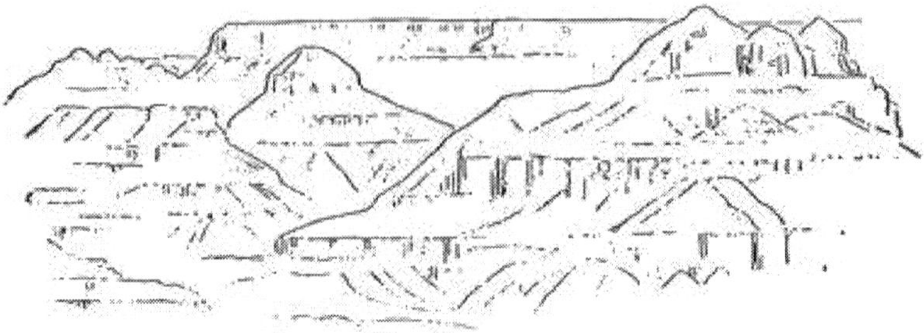

FIG. 482.—The gabled façade so largely developed in desert landscapes and sharply contrasted with the recurring curves in the landscapes of humid districts (from a painting of the Grand Cañon of the Colorado by Moran).

The climatic conditions of the mountain sculpture.—Since the different geological agencies operate either in a different manner or with differences in vigor according to the varying climatic conditions, the mountains of arid regions may in most cases be readily differentiated from those of the more habitable humid sections of country. In broad lines these differences may be summed up in the greater prevalence of the curving line within the landscapes of humid districts. This may be largely ascribed to the influence of the mat of vegetation, which protects the rock surface from more rapid mechanical degeneration, and arrests the sliding movements within the already loosened rock débris. In place of the reversed curves of the lines of beauty, so generally observed in the landscapes of well-watered regions, the desert lands present ever a repetition of the vertical cliff alternating with a sort of many gabled façade which is occasionally due to truncation of mountain spurs by the waves of former lakes, but far more often the outlines of débris cones built up beneath each prominent joint of the cliff walls (Fig. 482).

The effect of the resistant stratum.—In a striking manner mountain landscapes may disclose the influence of the diversified rock materials and of the rock structures as well. After prolonged erosion there is likely to be little correspondence between the positions of the anticlinal folds and the crests of the higher mountains. Such mountains are, in fact, much more likely to rise over synclines than upon the site of anticlines. The traveler who enters the Alps by any of the several railways, or who journeys by steamer over the beautiful lake of Lucerne, has a most favorable opportunity to study the position of the rock folds in the mountain sections that are unrolled in succession before him. Rarely indeed will he find a definite anticline in correspondence with a mountain peak, for the layers which are most resistant have developed the peaks, and it is because the outer layers of the anticlines open by local tension (see Fig. 26, p. 45) that they were first cut away by erosion, so that the hard layers within the synclines are likely to constitute the peaks within the existing surface.

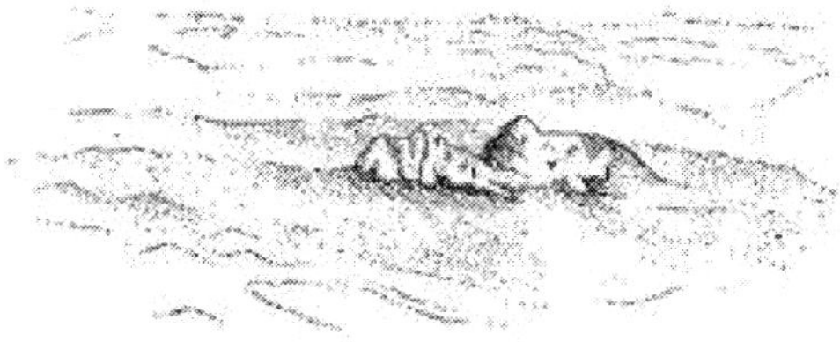

FIG. 483.—The Mythen, composed of Jurassic and Cretaceous sediments, and resting upon softer Tertiary formations. View from a balloon (after a photograph by C. Schmidt).

When, as sometimes happens, an older and likewise more resistant bed has been folded back upon younger and softer formations, an isolated remnant may be found "unrooted" to its base, upon which it appears as though floating within a billowy sea of the softer formations (Fig. 483).

The mark of the rift in the eroded mountains.—Applying the term "mountain" in its collective sense for a circumscribed area of uplifted crust, whether represented to-day by a folded or a faulted complex, a lava mass, or a granite dome; the period of uplift has marked the beginning of the activity of sculpturing agencies. By these the mass is pared down as it is shaped into a more or less intricate design of component and essentially repeating units. In the vernacular the word "mountain" is applied to these units into which the larger mountain mass is subdivided.

FIG. 484.—The battlement type of erosion mountains. Die Drei Zinnen (Three Battlements) in the Dolomites (after Marr).

It has been one of the main objects of this work to point out that the peculiar shapes of these elementary mountains are each characteristic of the erosive agents which produced them, and that each surface has marks which may be recognized in those lines of profile which recur within the landscape—the character profiles. In the subdivision of the larger mass—the *genetical* mountain—to form the numerous smaller masses—the *erosional* or *circumvallational* mountains—there is disclosed a pattern of fractures which has guided the erosional agents in their incisional operations (see Chapter XVII). In high altitudes, where the action of frost is so potent in prying at the wider fractures, this subdivision of the mass may be revealed by the sculpturing of squared towers or battlements (Fig. 484).

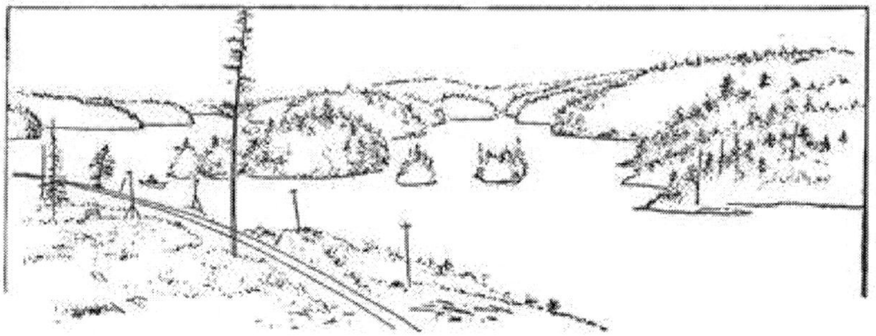

FIG. 485.—Symmetrically formed low islands repeated in ranks upon Temagami Lake, Ontario.

For other examples in which the sculptured surface is largely the handiwork of a single erosional agent, as over vast areas in the Canadian wilderness, the revelation of the fracture design is no less apparent. Here a series of crystalline rocks underlie broad expanses of territory and are without noteworthy variations of hardness and almost bare of surface débris. Sculptured beneath a mantling ice sheet, excavation has naturally been concentrated above the more widely gaping fissures of the joint-fault system, doubtless already marked out in the river network which the glacier overrode. The result has been a division of the surface into a series of low, oval ridges or hummocks, which over vast areas are repeated with monotonous regularity. Wherever the lower levels have been flooded, symmetrical low islands of nearly uniform elevation rise from the expanse of water and may be counted by thousands. Though the smaller islands have notably regular shore lines, the larger ones disclose their composition from smaller units by the breaking of their shores into similar bays spaced with regular intervals (Fig. 485, and Figs. 243 and 245, p. 229).

The ever repeating fracture design of the earth's crust is not restricted to the mountain masses which it has broken up, and the unity of which it has done so much to conceal. It extends far outside the margin of these masses, and is in fact common to whole continents and perhaps even to the planet as a whole. The part played by this design of fractures in the control of the sculpture of landscapes it would be hard to overestimate. Through its influence the striking features molded by one agent have been merged in the contrasted shapes developed by another. It is the great outline blender in the creation of nature's masterpieces of form and color. Thus the lines of this mysterious fracture network, though stamped in indelible characters upon our landscapes, are generally lost in the ensemble effect and may long remain undiscovered. Like a moss-grown inscription upon a slab of marble, though veiled, it may yet be deciphered; and if the veil be withdrawn, the runic characters are disclosed, and one of nature's laws lies open before us.

READING REFERENCES FOR CHAPTER XXXI

Mountain arcs or festoons:—

ED. SUESS. The Face of the Earth, vol. 2, 1906, pp. 201-207; vol. 4, 1909, pp. 498-542.

Block mountains:—

G. K. GILBERT. Surveys West of the 100th Meridian (Wheeler), vol. 3, Geology, Washington, 1875, Pt. 1, pp. 19 *et seq.*, 48.

J. W. POWELL. Report on the Geology of the Eastern Portion of the Uinta Mountains and a Region of Country Adjacent thereto, U. S. Geol. and Geogr. Surv. Ter., II Div. Washington, 1876, pp. 218.

JOHN W. GREGORY. The Great Rift Valley. London, 1896, pp. 422.

Laccolites and bysmalites:—

G. K. GILBERT. Report on the Geology of the Henry Mountains, U. S. Geol. and Geogr. Surv. Ter., 1877, pp. 18-98.

WHITMAN CROSS. The Laccolitic Mountain Groups of Colorado, Utah, and Arizona, 14th Ann. Rept. U. S. Geol. Surv., 1895, pp. 157-241, pls. 7-16.

W. H. WEED and L. V. PIRSSON. Geology and Mineral Resources of the Judith Mountains of Montana, 18th Ann. Rept. U. S. Geol. Surv., Pt. iii, 1898, pp. 485-556, pl. 75.

W. H. WEED. Geology of the Little Belt Mountains, Montana, etc., 20th Ann. Rept. U. S. Geol. Surv., Pt. iii, 1900, pp. 387-400.

VERA DE DERWIES. Recherches géologiques et pétrographiques sur les loccolithes des environs de Piatigorsk (Caucase du Nord). Geneva, 1905, pp. 84, pls. 3.

R. A. DALY. The Mechanics of Igneous Intrusion, Am. Jour. Sci. (4), vol. 15, 1903, pp. 269-278; vol. 16, 1903, pp. 107-126.

JOSEPH BARRELL. Geology of the Marysville Mining District, Montana. A study of Igneous Intrusion and Contact Metamorphism. Prof. Pap. 57, U. S. Geol. Surv., 1907, pp. 151-178.

Climatic condition in relation to land sculpture:—

C. E. DUTTON. Tertiary History of the Grand Canyon District, Mon. 2, U. S. Geol. Surv., 1882, pp. 264, pls. 42.

APPENDIX A

THE QUICK DETERMINATION OF THE COMMON MINERALS

Before one may gain a knowledge of rocks or the architecture of their arrangement within the earth's crust, it is quite essential that some familiarity should be acquired with the appearance and properties of the commonest minerals, and particularly those which enter as essential constituents into the more abundant rocks. To be a competent mineralogist, one must have a rather extended knowledge both of inorganic chemistry and of the science of crystallography, which, fascinating as it is to study, involves some technical knowledge of mathematics and much laboratory experience. Though necessary to any one who contemplates making a career as a geologist, this special study is not essential to a cultural course like the present one. The attempt will here be made to bring together a body of fact, from the study of which the student may quickly learn to recognize the commonest minerals in their usual varieties. The tests he is to apply are mainly physical, and in place of an elaborate discussion of crystal symmetry, pictures only can be supplied.

To the beginner the usual textbook of mineralogy is difficult to read intelligently, for the reason that for each mineral species it sets before him a catalogue of each physical property in its turn, with little indication of those data which in the individual case have special diagnostic value. None the less, however, the student is advised to consider the several properties of each mineral in a definite order, and the following may serve as well as any: crystal or other form, cleavage, fracture, luster, color, streak, transparency, tenacity, hardness, magnetism, and specific gravity. In endeavoring to connect the specific values of these properties with individual mineral species, the chemical composition and the manner of occurrence are not to be forgotten. It is well for the student to be supplied with a small pocket lens and with a pocket knife the blade of which has been magnetized.

Crystal form.—Some mineral species generally occur in more or less definite crystals—are bounded by definite plane surfaces developed when the mineral was formed; others in groups of interfering crystals or aggregates, in which case the mineral is said to be crystalline; while still others are rarely found crystallized at all. Thus in a given case crystal form may, or may not, be important for the diagnosis of the substance. If a mineral species is usually to be found in crystals, the student should be aware of the fact, and if possible should have a mental picture of the common crystal shape or shapes. Without an extended knowledge of crystallography, this must be supplied him by drawings. Since crystals of most species are apt to be distorted, owing to the fact that some planes within the same group appear upon the crystal with a larger development than others, it is convenient to remember that markings, such as lines or etchings upon the crystal faces, are the same throughout the same group of planes, and in the text figures such groups of planes are indicated by the use of a common letter. For crystalline aggregates such terms as fibrous, radiating, massive, or granular have their usual meanings.

Cleavage.—It is characteristic of most crystals that they break or *cleave* along certain directions so as to leave plane or nearly plane surfaces, and the luster of the cleaved surface measures the perfection of the cleavage property. It is important always to note how many such directions of cleavage are present, and, roughly at least, at what angles they intersect—whether they are perpendicular to each other or inclined at some other angle. Further, it should be noted whether a given cleavage is *perfect*, that is, easy, which will be indicated by the thinness of the plates which can be secured. An extremely perfect cleavage is possessed by the mineral mica, whose plates are thinner than the thinnest paper. In the case of imperfect or interrupted cleavage, the fracture surfaces are not plane throughout, but interrupted, the surface "jumping" from one plane to a neighboring parallel one. It is especially important to note whether, in the case of several cleavages possessed by a crystal, all have the same degree of perfection, or whether they exhibit differences.

Fracture.—In minerals with poorly developed cleavage, the fracture surface is described as *fracture*. Fracture is thus perfect in proportion as cleavage is imperfect. The fracture is described as conchoidal when it shows waving spherical surfaces like broken glass. For fine aggregates the fracture is described as even, uneven, earthy, etc., names which are generally intelligible.

Luster.—This term is applied especially to the manner in which light is reflected from mineral surfaces. The most important distinction is made between those minerals which have a *metallic* luster and those which have not, the former being always opaque. Other characteristic lusters are adamantine (like oiled glass), vitreous (glassy), resinous, waxy, etc.

Color.—For minerals which possess metallic luster the color is always practically the same, and hence it becomes a valuable diagnostic property. Of minerals which have nonmetallic luster, the color may be always the same and hence characteristic, but in the case of many minerals it ranges between wide limits and sometimes runs almost the entire gamut of hues, yet without appreciable changes in the chemical composition of the mineral.

Streak.—This term is applied to the color of the mineral powder, and is usually fairly constant, even when the surface color of different specimens may vary within wide limits. In the case of fairly soft minerals the streak is best examined by making a mark on a piece of unglazed porcelain (streak stone).

Transparency (**diaphaneity**).—The terms "transparent", "translucent", "subtranslucent", and "opaque" are used to describe decreasing grades of permeability by light rays. Through transparent bodies print may be read, while translucent bodies allow the light to be transmitted in considerable quantity through them, though without rendering the image of objects.

Tenacity.—This comprehensive term includes such properties as brittleness, flexibility, elasticity, malleability, etc.

Hardness.—Quite erroneous notions are held concerning the meaning of this very common word, which properly implies a resistance offered to abrasion. It is one of the most valuable properties for the quick determination of minerals, since minerals range from diamond upon the one hand—the hardest of substances—to talc and graphite, which are so soft as to be deeply scratched by the thumb nail. For practical purposes it is sufficient to make use of a rough scale of hardness made up from common or well-known minerals. If we exclude the gem minerals, this scale need include but seven numbers, which are: talc, 1; gypsum, 2; calcite, 3; fluor spar, 4; apatite, 5; feldspar, 6; and quartz, 7. A given mineral is softer than a mineral in the scale when it can be visibly scratched by a scale mineral, but will not leave a scratch when the conditions are reversed. If each will scratch the other with equal readiness, the two minerals have the same hardness.

Since it may often be desirable to test mineral hardness when no scale is at hand, the following substitutes may be made use of: 1, greasy feel and easily scratched by the thumb nail; 2, takes a scratch from the thumb nail, but much less readily; 3, scratched by a copper coin and very easily by a pocket knife; 4, scratched without difficulty by a knife; 5, scratched with difficulty by a knife, but easily by window glass; 6, scratched by window glass; 7, scratches window glass with readiness, but a grain of sand may be substituted to represent quartz in the scale.

Magnetism.—Though nearly all minerals which contain important quantities of the elements iron, cobalt, or nickel may be attracted to a strong electromagnet, there are but two common minerals, and these of widely different appearance, whose powder is lifted by a common magnet. Others are, however, lifted after strong heating in the air (*ignition*), and this is a valuable test.

Specific gravity.—Rough tests of relative weight, or specific gravity, may be made by lifting fair-sized specimens in the hand. Better determinations require the use of a spring balance.

Treatment with acid.—The carbonate minerals react with warm and dilute mineral acid so as to give a boiling effect (effervescence), since carbonic acid gas escapes into the air in the process.

PROPERTIES OF THE COMMON MINERALS

The more important common minerals fall into two classes according as they have large economic importance as ores, or enter in an important way into the composition of rocks.

I. The Minerals of Economic Importance

Hematite.—The sesquioxide of iron, Fe_2O_3, and by far the most important ore of iron. Rarely in good crystals, but sometimes in thin opaque scales bearing some resemblance to mica and known as micaceous or specular iron ore. At other times in nodules built up from radial needles (needle ore); in hard masses mixed with fine quartz grains (hard hematite); or in soft reddish brown earth (soft hematite). Color, black to cherry red. The powdered mineral always cherry red or reddish brown, and easily lifted by the magnet after ignition. Hardness 5.5-6.5; specific gravity 5.

Magnetite.—The magnetic oxide of iron, Fe_3O_4, often in crystals like Fig. 486, [1-2]. Black and opaque with a metallic luster. Streak black. Lifted by a magnet and sometimes itself capable of lifting filings of soft iron (lodestone). Hardness 5.5-6.5. Specific gravity 5.

Limonite.—The most abundant and most valuable of the hydrated iron ores, $2\ Fe_2O_3 \cdot 3\ H_2O$. Chemical composition the same as iron rust, with which in the earthy form it is identical. Never in crystals, but often in mammillary or rounded pendant forms resembling icicles, or sometimes clusters of grapes. Its yellow (rust) streak is its best diagnostic property. Ignited it gives off water and becomes magnetic. The streak and its notably lower specific gravity distinguish it from certain forms of hematite which it outwardly resembles. Hardness 5-5.5. Specific gravity 3.6-4.

Pyrite, iron pyrites, or "fool's gold."—The sulphide of iron, FeS_2. The most widely distributed sulphide mineral and now a chief source of the great chemical reagent, sulphuric acid or vitriol. Often, but not always, in crystals (Fig. 486, [3-5]) which have peculiar striæ upon their faces. At other times the mineral is found massive or in radiated needles. Bright metallic luster with the color of new brass, though often tarnished or altered upon the surface to limonite. Hard and brittle, and so distinguished from gold, which is soft and malleable and of the color of the paler old brass (which contained a larger percentage of zinc). Gold is, further, about four times as heavy as pyrite. Hardness 6-6.5. Specific gravity 5.

Chalcopyrite, copper pyrites.—A mixed sulphide of copper and iron. If in crystals, like Fig. 486, [6]; otherwise massive or compact. Luster metallic. Color orange-yellow, often with local blue and green iridescence like a pigeon's throat. Distinguished from pyrite by the deeper color and lower hardness, and from gold, particularly, by its brittleness and lower specific gravity. Hardness 3.5-4. Specific gravity 4.

Galenite, galena.—Sulphide of lead, PbS. The chief ore of lead, and, from admixture of a silver mineral, of silver as well. Usually found in crystals (Fig. 486, [7]). Always cleaves into blocks bounded by six very perfect rectangular faces which, when freshly broken, show a bright silvery luster and quickly tarnish to a peculiarly "leaden" surface. Very heavy. Color and streak lead-gray. Hardness 2.5. Specific gravity 7.5.

Sphalerite, zinc blende.—Sulphide of zinc, ZnS, usually with considerable admixture of sulphide of iron. The great ore of zinc. Not infrequently in crystals (Fig. 486, [8-9]), but more often in cleavable crystalline aggregates. The cleavage in fine aggregates is sometimes difficult to make out, but in coarse-grained masses it is seen to be equally and highly perfect in six different directions, so that a symmetrical twelve-faced form may sometimes be broken out (dodecahedron). Luster like that of rosin (rosin jack), though when with large iron admixture the color may approach black (black jack). The lighter colored varieties are translucent. Hardness 3.5-4. Specific gravity 4.

Malachite.—Hydrated (basic) copper carbonate. The green copper ore and the common surface alteration product of other copper minerals. Usually has a microscopic structure made up of fine needle-like crystals, but generally massive in various imitative shapes not unlike those of the iron ores. Sometimes earthy. Its color is bright green, and it is usually found in association with other characteristic copper ores, such as chalcopyrite and azurite. When relatively pure and in large masses, it is a beautiful ornamental stone. Effervesces with acid. Hardness 3.5-4. Specific gravity 4.

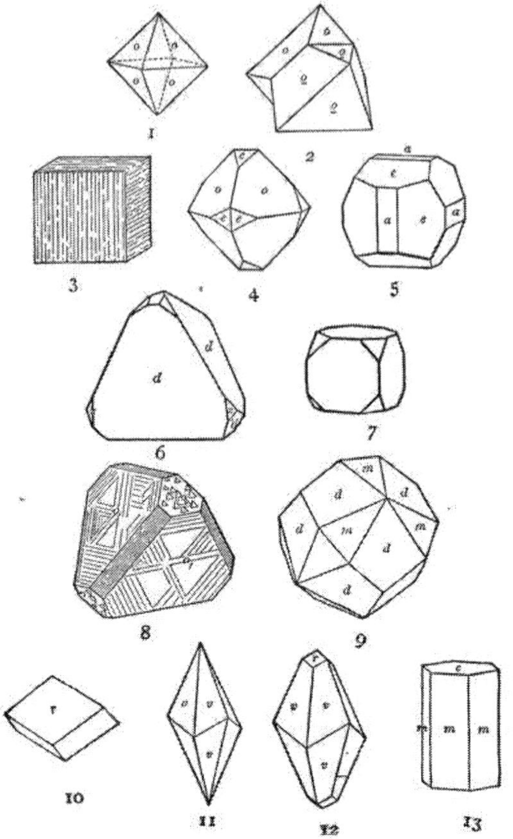

FIG. 486.—Forms of Crystals: 1-2, magnetite; 3-5, pyrite; 6, chalcopyrite; 7, galenite; 8-9, sphalerite; 10-13, calcite.

Azurite.—Hydrated (basic) copper carbonate, less hydrated than malachite, and known as the blue carbonate of copper. Generally in very minute and quite complex crystals, but also in imitative shapes similar to those of malachite, and at other times earthy. Slightly lighter in weight than malachite, from which it is easily distinguished, as from most other minerals, by its bright azure blue color and its somewhat lighter blue streak. Effervesces with nitric acid. Hardness 3.5-4. Specific gravity 3.7-3.8.

Calcite.—Calcium carbonate, $CaCO_3$. Almost always in crystals (Fig. 486, 10-13), or in confused crystal aggregates, though rarely fibrous or dull and earthy. Some of the forms of the crystals are described as "dog-tooth spar", others as "nail-head spar", while still others are modified hexagonal prisms. There is a beautifully perfect cleavage of the mineral along three directions which make angles of about 105° with each other, so that under the hammer the substance breaks into blocks which are shaped like the crystal of Fig. 486, [10]. Usually white or gray, but occasionally faintly tinted. Streak white. Effervesces with cold and dilute mineral acids. An associate of many ores and the chief mineral of limestone. A similar mineral—dolomite—contains in addition magnesium carbonate, has simpler crystals (like the drawing of Fig. 486, [10], but often with rounded faces), and effervesces only when the acid is warmed. Hardness 3. Specific gravity 2.7.

Gypsum.—Hydrated calcium sulphate, $CaSO_4.2\ H_2O$, and the source of plaster of Paris. Often in simple crystals (Fig. 487, [1]) or else "swallow tail", like Fig. 487, [2]; in which case the mineral is generally either transparent or translucent and is described as selenite. Such crystals show a cleavage approaching in perfection that of the micas, but, unlike the mica laminæ, those produced by cleavage in gypsum though flexible are not elastic. There are also fibrous forms of gypsum (satin spar), a fine-grained form (alabaster), and the impure earthy form (rock gypsum). Very soft, light in weight, and difficultly fusible. Color usually white, gray, or pale yellow. Hardness 2. Specific gravity 2.3.

Copper glance.—A sulphide of copper, Cu_2S. Not usually well crystallized, but generally massive and associated or variously admixed with other copper ores such as chalcopyrite, malachite, etc. Fracture conchoidal, luster metallic, color and streak blackish lead-gray, though often tarnished blue or green from surface alterations to the copper carbonates. Softer and heavier than chalcopyrite. Blowpipe or chemical tests are necessary for its identification. Hardness 2.5-3. Specific gravity 5.5-5.8.

Cerussite.—The white or carbonate lead ore, $PbCO_3$, and an important ore of silver as well. Often in crystals of considerable complexity, though Fig. 487, [3-4], shows some common shapes. Often granular, massive, or earthy (gray carbonate ore). Very brittle and with conchoidal fracture. The luster is adamantine or like that of oiled glass. Color generally white or gray. Very heavy, the heaviest of light colored and nonmetallic minerals. Dissolves in nitric acid with effervescence. Hardness 3-3.5. Specific gravity 6.5.

Siderite.—The carbonate or "spathic" ore of iron, $FeCO_3$. Either in crystals resembling in form Fig. 486, [10], but with rounded faces, or cleavable massive to finely granular and earthy. The crystalline varieties cleave easily into smaller blocks of the same form as those of calcite. Color usually gray or brown and streak white. On strongly igniting, the white powder becomes black and magnetic. Lighter in both color and weight than the other iron ores, and unlike them siderite effervesces with acid. Distinguished from calcite by its higher specific gravity and its change upon being ignited. Hardness 3.5-4. Specific gravity 3.9.

Smithsonite.—Carbonate of zinc, $ZnCO_3$, and an important ore of that metal. Seldom found in crystals except as a replacement of calcite crystals, in which case it shows the forms characteristic of the latter mineral. Usually kidney-shaped, stalactitic, or else in incrustations upon other minerals. Sometimes granular or earthy. Brittle. Luster vitreous, color white or greenish gray, though often stained yellow with iron rust. Streak white except when the mineral is stained with iron. Effervesces with warm acid. Hardness 5. Specific gravity 4.4.

Pyrolusite.—Black oxide of manganese, MnO_2, though generally impure from admixture with other manganese oxides. Usually in intricate aggregates which may be columnar, fibrous, mammillary, earthy, etc. Opaque, with color and streak both black. Soft and easily soils the fingers. With hydrochloric acid gives off the choking fumes of chlorine. Hardness 2-2.5. Specific gravity 4.8.

II. The Minerals important as Rock Makers

These minerals are in most cases complex silicates of one or more of a certain number of metals such as aluminium, calcium, magnesium, iron, sodium, potassium, or hydroxyl (OH). For their identification an examination of the physical properties is usually sufficient, whereas of the typical ore minerals already considered, additional chemical tests may be necessary.

Feldspars.—A group of similar alumino-silicates of potassium, sodium, and calcium. The most important of all rock-making minerals. Although with wide variation in chemical composition, the feldspars are yet broadly divided into two classes; the one striated, and the other an unstriated potash or orthoclase variety. The pocket lens is usually necessary in order to make out the striations upon the crystal or cleavage surfaces. When formed in veins, feldspar appears in crystals (Fig. 487, 5-6), but as a rock constituent the mutual interference of crystals prevents the development of bounding faces. Two cleavage directions, nearly or quite perpendicular to each other, are notably different in their perfection. Hard enough to scratch glass, but easily scratched by sand. Color pink (usually orthoclase or microline), white (often albite) to gray. Sometimes with beautiful "pigeon's throat" effect of iridescence (labradorite). Low specific gravity. Hardness 6. Specific gravity 2.5-2.8.

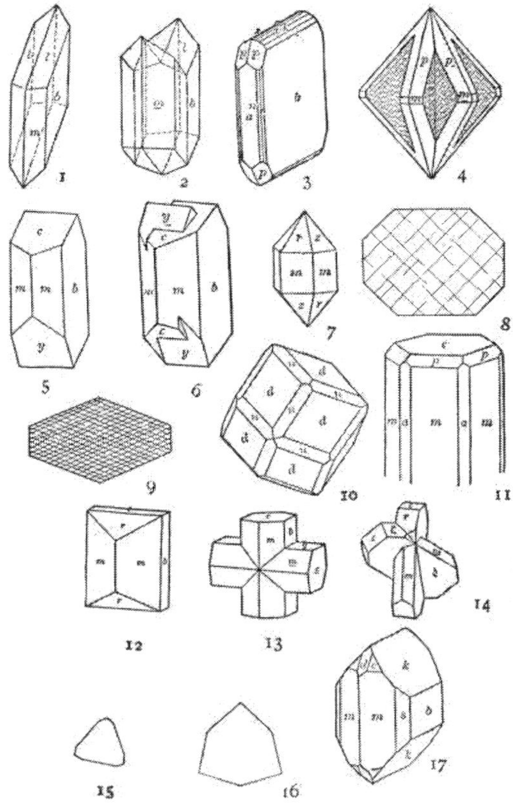

FIG. 487.—Forms of Crystals: 1-2, gypsum; 3-4, cerussite; 5-6, feldspar; 7, quartz; 8, pyroxene (cross section); 9, hornblende (cross section); 10, garnet; 11, nephelite; 12-14, staurolite; 15-16, tourmaline (cross sections); 17, olivine.

Quartz.—Oxide of silicon or silica, SiO_2. Both an important vein mineral associated with the ores and a rock maker. In the former case particularly, often in crystals of notably simple forms (Fig. 487, 7). Few minerals which are not gems are so hard. Remarkable freedom from cleavage so that the mineral breaks much like window glass—conchoidal fracture. Wide range in both transparency and color. Transparent and colorless crystalline variety (rock crystal), brown translucent (smoky quartz), turbid white (milky quartz), and various colored varieties (carnelian, jasper, jet, etc.). Insoluble in acids and infusible. Hardness 7. Specific gravity 2.6.

Micas.—Like the feldspars a group of complex silicates, but here chiefly of potassium, magnesium, iron, and hydroxyl. Abundant as rock makers, the micas are all characterized by the thinnest and toughest of elastic cleavage plates, such as are generally known as isinglass. When a needle is driven sharply through a thin scale of mica, a six-rayed puncture star forms about the needle point. The darker common variety of mica is rich in iron and magnesium and is called biotite, and the lighter colored alkaline variety, muscovite. Hardness 2.5-3.1. Specific gravity 2.7-3.1.

Chlorite.—Generally an intricate mixture of more or less similar microscopic crystals having varying and rather complex chemical compositions and related to the micas, but all characterized by a peculiar leaf green color. These minerals are a common product of hydration weathering in rocks which are rich in magnesium and iron—especially those that contain biotite, pyroxene, or hornblende (see below). Hardness 1-2.5. Specific gravity 2.5-3.

Pyroxenes.—An important group of related rock-making minerals all of which are silicates of the bases magnesium, calcium, aluminium, iron, and manganese. Quite generally developed either in columnar or needle-like crystals which are uniformly shaped in cross section like Fig. 487, [8]. Two rather imperfect cleavages are directed parallel to the longer axis of the crystal and nearly at right angles to each other. The colors of all but the lime varieties are dark and generally green, dark brown, bronze, or black. The lime varieties are white, gray, or pale green. A dark colored and common iron variety is known as augite. Streak generally either white or lightly tinted. Hardness 5-6. Specific gravity 3.2-3.6.

Amphiboles.—A group of minerals of the same chemical composition as the pyroxenes, with which also in most physical properties they agree. The principal distinction is found in the shape of the cross section and in the cleavage (Fig. 487, [9]). Whereas the cross sections of pyroxenes are generally eight sided, those of the amphiboles have six sides, and whereas the cleavage directions of pyroxenes are nearly at right angles to each other (87°), the similar but much more perfect cleavage directions of the amphiboles are inclined at an obtuse angle (124½°). Owing to the obliquity of the amphibole cleavage, fractured surfaces of the mineral appear splintery, which is not in the same measure true of the pyroxenes. A fibrous variety of amphibole, and occasionally other varieties of the mineral, is a not uncommon product of weathering of pyroxenes. Other physical properties of the amphiboles are in the main almost identical with those of the pyroxenes.

Garnet.—Complex alumino-silicates or ferro-silicates of calcium, magnesium, iron, or manganese, or several of these combined. Nearly always in crystals, and usually found in mica schist (see below). The crystals usually have twelve similar faces, each a lozenge (dodecahedron), or else twenty-four similar faces, or the two forms combined (Fig. 487, [10]). Brittle. From any but the gem minerals garnet is easily distinguished by its hardness, which in different varieties ranges from somewhat below to somewhat above that of quartz. The luster is vitreous, and the color runs the gamut of reds, browns, and greens, but with the common hue dark red to black. Streak white. Hardness 6.5-7.5. Specific gravity 3.1-4.3.

Nephelite (**nephelene**).—An alumino-silicate of sodium and potassium. In certain special provinces this mineral is developed in abundance as an essential constituent of igneous rocks, but elsewhere practically unknown. The rare crystals are hexagonal prisms (Fig. 487, [11]), but the mineral is most easily determined by its general resemblance to feldspar, but with the differences of cleavage, luster, and reaction with acid. Whereas the feldspars have two cleavages, either nearly or quite perpendicular to each other and of different degrees of perfection, nephelite has three equal cleavages inclined 60° and 120° to each other and of less perfection than either feldspar cleavage. The luster of nephelite is perhaps the best clew to its identity, since this is greasy and simulated by but few minerals. The fine powder of the mineral treated for some time with strong hydrochloric acid forms a perfect jelly of silicic acid, whereas the feldspars do not. Though itself gray or white and unobtrusive, nephelite is usually associated with brightly colored minerals, which are often the first clew to its presence in a rock. Hardness 5.5-6. Specific gravity 2.5-2.6.

Talc (**soapstone**).—A silicate of magnesium and hydroxyl which is an important alteration product through weathering of certain pyroxene rocks especially. Usually a foliated mass, this product is occasionally fibrous or even granular. Talc is one of the softest of minerals, having a greasy feel and being easily scratched with the thumb nail. The luster of the foliated varieties is apt to be pearly, and the color apple-green to white, though sometimes stained brown from oxide of iron. The streak of the mineral is white except when stained by iron. Although the rocks which are composed mainly of talc (soapstone) are exceedingly soft, they are very tough and remarkably resistant. Hardness 1-1.5. Specific gravity 2.7-2.8.

Serpentine.—Like talc, serpentine is a silicate of magnesium and hydroxyl, and an important product of the breaking down of magnesium minerals in the process of weathering. The mineral is usually found as a fine web of microscopic needle-like fibers, and is best roughly diagnosed by its color and its associated minerals. Like talc it is usually developed within those igneous rocks from which feldspar is lacking, but where either pyroxene or olivine is found in abundance or was previous to alteration. The characteristic color of serpentine is leek-green. The rock largely composed of serpentine is called by the same name, and being exceedingly tough and unchanging is, in spite of its softness, a valuable building and ornamental stone. A red magnesium garnet is apt to be associated with such serpentine masses. Hardness 2.5-4, because of impurities. Specific gravity 2.5-2.6.

Staurolite.—A silicate of aluminium, iron, and hydroxyl. Found in metamorphic rocks usually in association with garnet. Always in crystals bounded by simple forms generally crossed, as shown in Fig. 487, [12-14]. The color is dark reddish brown, and the streak is colorless to grayish. The hardness is exceptional and higher than that of quartz. Hardness 7-7.5. Specific gravity 3.6-3.7.

Tourmaline.—An exceptionally complex silicate of boron and aluminium as well as iron, magnesium, and the alkalies. Found in metamorphic rocks and always crystallized. The crystals are columns or needles whose cross section is the best guide to their identity, since this is a modified triangle unlike that of any other mineral (Fig. 487, [15-16]). Additional diagnostic properties are the characteristic striations which run lengthwise of the crystals upon prism faces, and the lack of any cleavage (difference from hornblende). The hardness is also a valuable property, since this is greater than that of quartz. The mineral is brittle and the fracture subconchoidal. The range in color is as great as, or greater than, that of garnet, though the common forms are jet black. Streak uncolored. Hardness 7-7.5. Specific gravity 3-3.2.

Olivine.—A silicate of magnesium and iron and a rock-making mineral found only in those igneous rocks which have little or no feldspar. It easily suffers alteration by weathering and passes into serpentine, and in fact is seldom found except when at least partially altered to the fibrous webs of that mineral. The form of the unaltered crystals within the rocks is shown in Fig. 487, [17], and, cut in sections, the mineral appears in more or less elongated hexagons. The hardness of the unaltered mineral is about that of quartz. It has rather imperfect cleavages in two rectangular directions, and is usually translucent, with a vitreous luster and a color which is olive-green when not stained brown by oxide of iron. Streak uncolored. Hardness 6.5-7. Specific gravity 3.2-3.3.

APPENDIX B

SHORT DESCRIPTIONS OF SOME COMMON ROCKS

In Chapter IV the classification and the structure of rocks have been briefly discussed. Below are added brief descriptions of the more important common rocks. For rocks as for minerals it is, however, essential that a collection of well-chosen specimens be studied for purposes of comparison. A small pocket lens is a valuable aid in making out the component minerals and the textures of the finer grained rocks.

1. Intrusive Rocks

Granite.—Of granitic texture, though sometimes porphyritic as well. The most abundant mineral constituent is a pink or white feldspar, usually without visible striations, with which there is usually in subordinate quantity a white striated feldspar. Next in importance to the feldspar is quartz, which because of its lack of cleavage shows a peculiar gray surface resembling wet sugar. In addition to feldspar and quartz there is generally, though not universally, a dark colored mineral, either mica or hornblende. The mica is usually biotite, though often associated with muscovite.

Syenite.—Like granite, but without quartz, with more striated feldspar, and generally also the rock has a darker average tint. While biotite is the commonest dark colored constituent of granite, hornblende is more apt to take its place in syenite. Less common than granite, to which it is closely related in origin and in composition.

Gabbro.—A dark colored rock of granitic texture composed of striated feldspar with broad cleavage surfaces and usually an abundance of pyroxene. In contrast to the feldspars of granite, those of gabbroes are often dull and colored grayish yellow or greenish. The pyroxene is often in part changed to fibrous amphibole. Magnetite may be an abundant accessory mineral.

Diabase.—In color dark like gabbro, and of similar constitution. In diabase, however, the feldspar crystals, instead of being broad and of irregularly interrupted outline, are relatively long ("lath-shaped"), and the pyroxene acts as a filler of the residual space between them.

Peridotite.—A heavy and dark colored rock of granitic texture which is nearly or quite devoid of feldspar but contains olivine. When altered, as it generally is, it is largely a mass of serpentine, talc, and chlorite, surrounding cores, it may be, of still unaltered pyroxene and olivine. Magnetite is an abundant constituent, and a red garnet is apt to be present.

2. Extrusive Rocks

Obsidian.—A rock glass rich in silica. It is usually black and breaks with a perfect conchoidal fracture. It often passes over through insensible gradations into pumice, which differs only in its vesicular structure. As regards chemical composition, obsidian and pumice are not notably different from rhyolite (below).

Rhyolite.—A light colored rock of porphyritic texture, often also with fluxion or spherulitic textures, or both combined. The porphyritic appearance is given the rock by large crystals of a glassy, unstriated feldspar and crystals of quartz. Rhyolite is a very siliceous lava containing rather more silica than granite, to which of the intrusive rocks it is most closely related, and from which it differs in its texture and in the manner of its occurrence in nature. Whereas granite is found in great batholites, laccolites, and bysmalites, and consolidated in most cases beneath the earth's surface, rhyolite generally occurs in sheets, flows, or dikes, and consolidated either above or in fissures near to the surface.

Trachyte.—Similar to rhyolite, but usually with a peculiar gray aspect from the greater abundance of feldspar crystals. The rock is less siliceous than rhyolite, contains no quartz crystals, and approaches a feldspar in its average composition.

Andesite.—Similar to rhyolite in appearance and in origin, but more basic and correspondingly dark in color. The porphyritic crystals are of lath-shaped, striated feldspar, with which are associated crystals of either biotite or hornblende or both. A fluxion texture is particularly characteristic of this type of extrusive rock.

Basalt.—A dark colored or black basic rock of porphyritic texture which differs but little from diabase. It may show under the lens fine lath-shaped crystals of striated feldspar associated with crystals of augite, but more frequently the rock is dense and without visible mineral constituents. It is particularly likely to occur divided up into columns six inches to a foot in diameter and known as basaltic columns. Especially fine examples are known from the Giant's Causeway and other localities in the western British Isles.

3. Sedimentary Rocks of Mechanical Origin

Conglomerate (“**pudding stone**”).—A rock made up from pebbles which are cemented together with sand and finer materials. The pebbles are usually worn by work of the waves upon a shore, and may vary in size from a pea to large bowlders. They may consist of almost any hard mineral or rock, though the sand about them is largely quartz.

Sandstone.—A rock composed of sand cemented together either by calcareous, siliceous, or ferruginous materials. Sandstones are described as friable when their surface grains are easily rubbed off, or as compact when they are more firmly cemented. Sandstones are often distinctly banded and are sometimes variously stained with oxide of iron. Those sandstones which have been formed upon a seacoast are known as marine sandstones, while those derived from accumulations collected by the wind in deserts are distinguished as continental deposits. Sandstones form much thicker formations than conglomerates, the latter usually constituting a basal layer only of the sandstone formation (basal conglomerate).

Shale.—A consolidated mud stone which is probably the most abundant rock formation. In large part clay admixed in varying proportions with extremely fine sandy grains.

4. Sedimentary Rocks of Chemical Precipitation

Calcareous tufa (**travertine**).—Not to be confused with tuff, which is a fragmental extrusive or volcanic rock. Calcareous tufa is formed when waters which contain carbonic acid gas and lime carbonate in solution, give off the gas and with it the power to hold the lime in solution. Such a liberation of the gas may occur when the stream is dashed into spray above a cascade, and the lime is then deposited about the site of the falls. Travertine is generally porous and formed of more or less concentric layers or incrustations. A remarkable illustration is furnished by the travertine deposits of Tivoli and other localities near Rome, since here the material supplies a valuable building stone.

Oölitic limestone (**oolite**).—This rock is made up of spherical nodules and so has the appearance of fish roe. Broken apart, each grain reveals in its center a core of siliceous sand about which carbonate of lime has been deposited in concentric layers. It is thought that waters charged with carbonate of lime, in issuing from a river near a sea beach, coat the sand grains of the latter with successive thin films of lime carbonate due to the rhythmic ebb and flow of the tides, evaporation of the adhering water taking place when the sands are exposed at low tide.

5. Sedimentary Rocks of Organic Origin

Limestone.—A generally white or gray rock composed of carbonate of lime with varying proportions of clay, silica, and other impurities. The lime carbonate is usually derived from the hard parts of marine organisms, and the argillaceous and siliceous impurities from the finer land-derived sediments which descend with them to the bottom.

Dolomite (dolomitic or magnesium limestone).—Differs from limestone in containing varying proportions of the mineral dolomite (*ante*, p. 455), which is made up of equal parts of calcium and magnesium carbonates. Difficult to distinguish from limestone unless a chemical test is made for magnesium, though it may be said in general that dolomite is less soluble in cold mineral acids.

Peat.—An accumulation of decomposed vegetable matter within small lakes and in lagoons separated from larger ones (*ante*, p. 429). Peat represents the first stage in the formation of coal from vegetable matter, and differs from the coals by its larger proportion of contained water. Because of this water its fuel value is correspondingly small. It is usually dark brown or black and reveals something of the structure of the plants out of which it was formed.

6. Metamorphic Rocks

Gneiss.—A generally more or less banded (gneissic) metamorphic rock with a mineral constitution similar to granite, and often developed by metamorphic processes from that rock. It may at other times, by processes not essentially different, be derived from sedimentary formations. It usually contains as important constituents unstriated feldspar and quartz, but in addition it may include a striated feldspar, biotite, muscovite, or hornblende, or several of these combined. In proportion as mica or hornblende is abundant, it has a marked banded texture, but it differs from mica schist (see below) not only in the presence of its feldspar, but in the smaller proportion of mica. Biotite gneiss, hornblende gneiss, etc., are terms used to designate varieties in which one or the other of the dark colored constituents predominate.

Mica schist.—A metamorphic rock without feldspar and mainly composed of quartz and light colored mica (muscovite). The abundant mica lends to the rock its characteristic schistose texture, which differs from the usual gneissic texture. In some cases the mica is wrapped about the grains of quartz, but at other times it forms a series of almost continuous membranes separating layers of quartz.

Sericite schist.—A variety of schist which is characterized by an abundance of a peculiar silvery mica rich in the element group hydroxyl. The mica scales are often microscopic and wrought into an intricate web with the quartz constituent.

Talc schist.—A schist made up largely of talc, but with varying proportions of quartz, magnetite, etc. From the abundance of the talc it is usually pale green or white.

Chlorite schist.—A greenish, fine-grained metamorphic rock in which chlorite is the principal mineral, but in which magnetite is a quite characteristic accessory constituent.

Staurolitic garnetiferous mica schist.—A mica schist in which garnet and staurolite are so abundant as to be essential constituents.

Clay slate.—A metamorphosed mud stone or shale. In the process of metamorphism the rock has been hardened, given a slaty cleavage, and innumerable minute scales of mica have developed to produce a silky luster upon the cleavage faces. The color may be gray, green, purple, or black.

Quartzite.—A metamorphosed sandstone in which the sand grains have become enlarged by accretion of silica. Whereas a sandstone fractures about its constituent grains, a break in quartzite is continued through the grains and the cement alike. In contrast to sandstones, the quartzites derived from them are usually lighter in color and often nearly white.

Marble (**crystalline limestone**).—The result of metamorphism upon limestones. Usually white in color but sometimes gray, blue gray, or yellow, and sometimes variously broken or brecciated and stained with iron oxide. Effervesces with cold dilute acid.

Coals.—Under the head of peat the first stage in the formation of coals from vegetable matter has been briefly described. Lignite, or brown coal, represents a further stage and one in which the vegetable structure is still recognizable. It is usually brownish black or black in color and contains a considerable proportion of water. With increased pressure or dynamic metamorphism, further percentages of the volatile constituents are eliminated, and when from seventy-five to ninety per cent of carbon remains, the material burns with a yellow flame and is known as bituminous coal. This is the great fuel for the production of steam. A continuation of the metamorphic processes carries off a further proportion of the volatile matter and leaves a dense, hard, black substance with sometimes as much as ninety-five per cent of carbon. This is the so-called "hard coal" or anthracite generally used for fuel in our houses, for which purpose it is so well adapted because it burns with a production of much heat and almost without smoke.

APPENDIX C

THE PREPARATION OF TOPOGRAPHICAL MAPS

Topographical maps a library of physiography.—For the satisfactory working out in detail of the geology of any region of complex structure, an accurate topographical map is prerequisite. This is so much the more true because nearly all complexly folded or faulted rock masses are to be found in mountainous, or at least in hilly regions. The making of the topographical map must, therefore, precede that of the geological map, and in modern usage the latter is a topographical and a geological map combined in one.

Within certain narrow limits, predictions concerning the geological history of a province may often be made by an expert geologist from examination of an accurate topographical map. Just as in forecasting the weather upon the basis of the usual weather maps, such predictions can sometimes be made with entire confidence in their accuracy, while at other times a guess only may be hazarded. The great value of the modern topographical map is becoming, however, universally acknowledged, and every highly civilized nation has either completed or has in preparation sectional topographical maps of its domain on such a scale as is warranted by its financial condition and its state of development. Thus there is now being accumulated a vast library of geographical and to some extent geological information, of which the student of geology must be prepared to make use.

The nature of a contour map.—More and more the contour map is replacing the earlier and less scientific methods of representing topography on the large scale sectional maps, and hence this type only need here be considered. In the contour map, the relief of the land is represented by a series of curving lines, each the intersection of a particular horizontal plane with the land surface, and the several planes separated by uniform differences of elevation. This altitude interval is known as the contour interval. Its choice is a matter of considerable importance, for though regions of relatively simple topography may be adequately represented upon a map of large contour interval, say one hundred feet, another district may require an interval as short as five feet. A contour map with this interval may be conceived to have been made by flooding the region which it represents and preparing maps of the shore lines for each rise of five feet of the water surface, and superimposing the several maps thus derived with accurate registration one above the other. Wherever the land slopes are steep, the shore lines of the several maps will be crowded closely together and give the effect of a relatively dark local shade; where, upon the other hand, the surface is relatively flat, the several shores will be widely spaced and the effect will be to produce a white area upon the map. Thus in contour maps dark tones indicate steep gradients and pale tones a flatness of surface.

The selection of scale and contour interval.—With the use of the small scale in the contour map, the tones of the map will be correspondingly dark, though the relative differences in tone will remain the same. With the use of a closer contour interval the tones will deepen throughout. The adjustment of scale and contour interval to any given region is a matter requiring experience in topographical mapping, and in addition a knowledge of the geological significance of topographic features. Unfortunately, the element of expense and the special commercial objects held in view, conspire to select scales and contour intervals which are often little adapted to the districts surveyed.

The method of preparing a topographical map.—Having fixed upon the scale and the contour interval which is to be employed, the task of the topographical surveyor is next to fix accurately the positions and the elevations of a sufficient number of points to *control* the map, and then to hang, as it were, upon these points as attachments the design represented by the relief. Were the surface of the ground to be represented by a flexible fabric, the map maker might raise from a flat base a series of stout posts of the heights and in the positions which he has determined, and upon these supports arrange the slopes of the fabric much as drapery is adjusted. The determination of the exact positions and the elevations of his control stations is, therefore, a process coldly precise and formal; whereas in the shaping of the surfaces his attention should be fixed more upon correctly reproducing the shapes than upon fixing accurately the position of every point. As a matter of fact, the position of the average point will be most accurately fixed when the shapes of the features are most clearly comprehended. To some extent, therefore, the topographer should be familiar with the geological significance of the earth features which he is representing.

Laboratory exercises in the preparation of topographical maps.—The principles which underlie the surveyor's method for preparing a topographical map may be learned in the laboratory by the use of models and the simple device shown in plate 24 A and B. To represent the section of country to be mapped a model in plaster of Paris is substituted, and this is placed within a rectangular tank to which locating carriages and altitude gauges are attached that allow the student to fix the position and the elevation of any point upon the surface of the model.

PLATE 24.

A. **Apparatus for exercise in the preparation of topographic maps.**

B. **The same apparatus in use for testing the contours of a map.)**

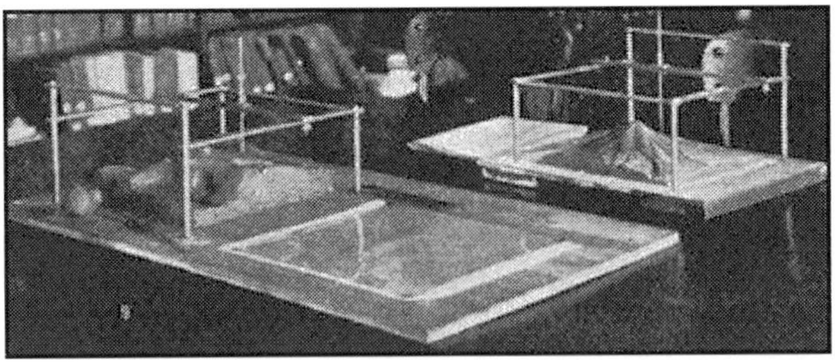

C. **Modeling apparatus in use.**

Upon each model the student "locates", or fixes, the position of a sufficient number of points for the control of his map, entering upon an appropriate map base for each position the altitude which was read from the gauges. Now *with the map always before him* he "sketches in" the forms of the surface by means of contour lines. For this purpose it is often desirable to fix roughly the direction of the steepest slope at a number of places, and noting the differences in elevation between control stations, divide up the distance in accordance with the curves of slope and start the contours at right angles to the slope. Afterwards such sections are connected by sketching in with the model always in view for control (Fig. 488).

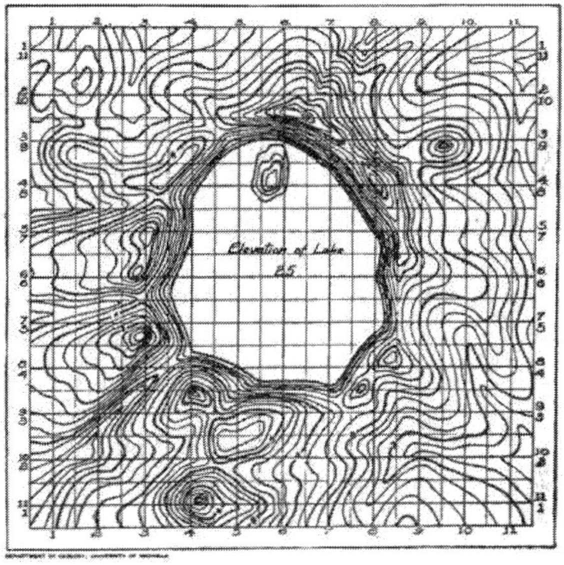

FIG. 488.—A student's map prepared from a model by the use of the contour apparatus represented in plate 24 A.

The verification of the map.—The map prepared, its accuracy may be tested by a simple method which is denied the topographer who has to do with the actual surface of the ground. The locating carriages and altitude gauges are removed from the tank, which is next filled with water and leveled by means of guide marks upon the interior. A few drops of milk or of ordinary clothes blueing are added to the water to render it opaque, and it is then drawn off at the faucet in successive installments, so that the surface drops by layers corresponding in thickness to the contour interval of the map, plate 24 B. As each layer is withdrawn, that contour of the map to which the shore line should correspond is carefully examined and corrected. By such corrections the nature of the first errors made is soon appreciated, and the method of procedure is thus more easily acquired. At the same time the significance of the design of the map is more quickly learned than by a mere examination of the standard government maps.

The work above outlined calls for waterproofed models of suitable form and size, and a series, each of which sets forth some typical feature or series of features, has been designed by Mr. Irving D. Scott.

The preparation of physiographic models.—The apparatus used to prepare the topographic map is adapted also for preparing a physiographic model from a standard topographical map. For this purpose the method is essentially reversed, though the tank is replaced to advantage by a light metal frame elevated upon one side so as to permit a free use of the hands in modeling the clay.

The material used in preparing the model is artists' modeling clay which has a base of beef suet, and hence does not dry out and crack as does ordinary clay. Its form is, therefore, retained indefinitely, and it may be used again and again. Most maps must be enlarged in modeling, and the simplest way is often to photographically or by pantograph enlarge the map to the scale of the model. The map prepared, it is covered by a thin celluloid plate which has cut upon it a series of crossed lines spaced in inches and larger subdivisions to correspond to those of the locating carriages (plate 24 C).

The enlargement of the map is not essential to experienced workers, and the standard map may be covered in similar manner by a transparent plate with "checkerboard" design, the squares of which bear some simple relation in size to the larger divisions of the locating carriages (Plate 24 C, rear).

The method of preparing the model is comparatively simple. Beginning at any point upon the map, the intersection of a heavy contour line with one of the guide lines of the celluloid "position plate" is carefully noted. Both the position and the elevation of this point are fixed by the point of the altitude gauge of the modeling frame, and the clay built up beneath it to that height. With the fingers the clay is now roughly shaped in various directions from this point, the altitude gauge is advanced by the locating carriage so as to correspond in position to the intersection of the next heavy contour line with the same guide line of the position plate, and the elevation for this point similarly adjusted upon the model. As before, the surface of the clay is roughly shaped in advance and upon the sides so as to conform to the indications of the map; and this process is repeated until the work is finished. Corrections for intermediate positions may be carried to any desired degree of refinement which the scale and the accuracy of the map permit. Models which are larger than the area of the modeling frame are prepared by making a square foot at a time by the above described process, and then moving the frame forward and adjusting in a new position by means of the sharp pins in the legs of the apparatus.

READING REFERENCES

WILLIAM H. HOBBS, New Laboratory Methods for Instruction in Geography, Journal of Geography, vol. 7, 1909, pp. 97-104. Also Scot. Geogr. Mag., vol. 24, 1908, pp. 643-652. The Modeling of Physiographic Forms in the Laboratory, *ibid.*, vol. 8, 1910, pp. 225-228.

APPENDIX D

LABORATORY MODELS FOR STUDY IN THE INTERPRETATION OF GEOLOGICAL MAPS

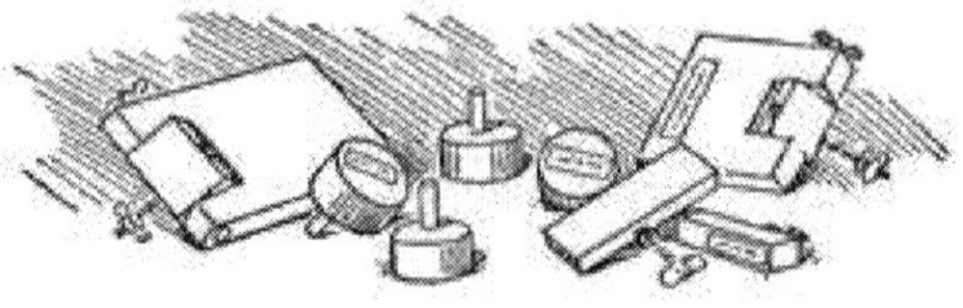

FIG. 489.—Models to represent outcrops of rock.

The laboratory models which have been described on page 63, and are used to represent outcrops in the study of geological maps, are shown in Fig. 489. The drum-shaped blocks serve to represent massive rocks which occur in irregularly shaped masses such as batholites and flows. The long, narrow strips are for intrusive rocks in the form of dikes, while the larger blocks provided with a swivel joint are used for outcrops of sedimentary rocks, and after adjustment they give the dip and strike of the exposure. The wing bolts used in their construction should be of bronze, because of the effect of iron upon the compass. For the same reason tables should not be placed near iron beams or columns. All these blocks can be made by an ordinary carpenter, and should be available in sufficient numbers to arrange problems like those of Figs. 47, 48, and 490. With a view to supplying suggestions for other problems of the same general nature, the three additional field maps of Fig. 491 have been introduced.

FIG. 490.—Special laboratory table set with a problem in geological mapping which is solved in Figs. 47 and 48.

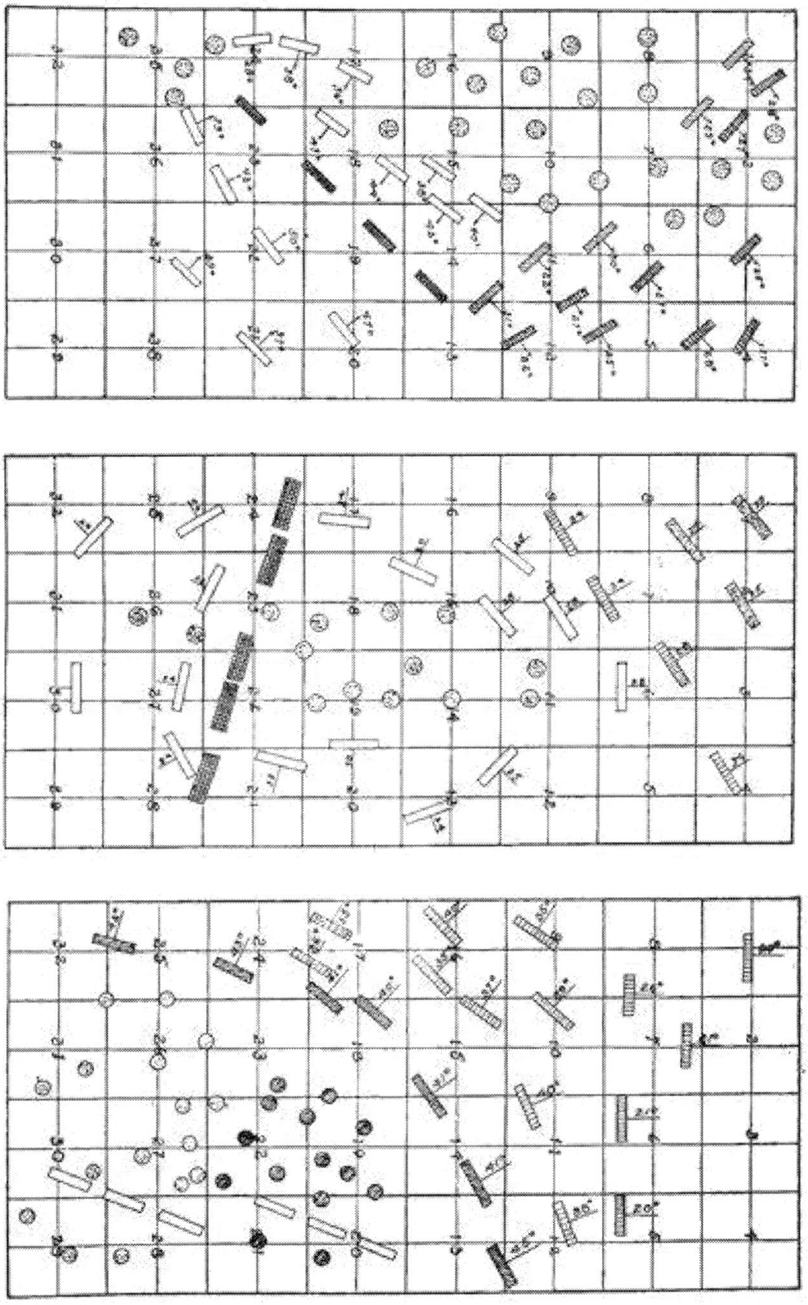

FIG. 491.—Three field maps to be used as suggestions in arranging laboratory tables for problems in the preparation of areal geological maps.

The list of questions given below is intended to indicate the nature of some of the problems which the student should be asked to solve in the preparation of each map. The numbers in parentheses refer to pages in this book where further information is given:—

STRATIGRAPHICAL

1. Of the formations represented what ones are sedimentary and what igneous (Chap. IV, App. B)?

2. Which formations, if any, are separated by unconformities (51-53)?

3. What is the order of age of the sedimentary formations (65)?

4. What are the *exposed* thicknesses of each of these formations (48-49)?

5. Do any of these values represent full thickness of the formation, and if so, which ones?

6. What is the age in terms of the sedimentary formations of each of the igneous rock masses (65)?

7. Which igneous rocks, if any, occur in batholites (143, 441)? Which, if any, in dikes (140)?

STRUCTURAL

8. What formations, if any, have monoclinal dip (42)?

9. Indicate upon the map by dashed lines the crests of all anticlines and the trough lines of synclines.

10. Indicate by arrows the direction of pitch of all plunging anticlines and synclines wherever disclosed by changes of dip and strike (43).

11. Indicate the approximate position of all faults whose position is disclosed (58-61), and, if possible, state which limb is the one downthrown.

12. Prepare suitable geological sections.

READING REFERENCE

WILLIAM H. HOBBS. Apparatus for Instruction in Geography and Structural Geology. III. The Interpretation of Geologic Maps. School Science and Mathematics, vol. 9, 1909, pp. 644-653.

APPENDIX E

SUGGESTED ITINERARIES FOR PILGRIMAGES TO STUDY EARTH FEATURES

The chief value of the laboratory studies discussed in the preceding appendices is as a preparation for observations made in the field—the laboratory *par excellence* of the geologist. The pilgrimages whose itineraries are here suggested have been planned especially for impressing by observation the lessons of this book. Such journeys are best interrupted at a relatively small number of localities which, because already studied in some detail, are specially adapted to serve as centers for local excursions. These localities will in most cases be the great scenic places to which tourists resort, or the seats of universities near which specially detailed explorations have been often made.

Within the United States a few local geological guides have been published, and the Geologic Folios published by the United States Geological Survey are already available for a number of such centers. For one long geological pilgrimage we are fortunate in having a carefully prepared guide, namely, from New York to the Yellowstone National Park and back, with a side trip to the Grand Cañon of the Colorado. Except for the side trip this route, in large measure, corresponds with one here chosen, and for the return journey especially the student is referred to it for information (Geological Guide Book of the Rocky Mountain Excursion, edited by Samuel Franklin Emmons. Comte Rendu de la Congrés Géologique Internationale, 5me Session, Washington, 1891, 1893, pp. 253-487, map and plates 13, figs. 32).

Our journey is begun at New York City, which is built about the deeply submerged channels of an estuary choked with glacial deposits, though the channel may be followed as a deep cañon across the continental shelf to its margin (252, pl. 17 B). New York City is also upon the margin of the glaciated area, the outer terminal moraine of which is well represented on Long Island (298). Across the Hudson in New Jersey is the great Coastal Plain which meets the oldland in a well-defined margin (159, 246, 247). A local geological guide of the vicinity of the metropolis has been written by Gratacap (Geology of the City of New York, Greater New York. Brentanos, New York, 1904, pp. 119, pls. and map).

Traveling by the New York Central Railway, we follow up the Mohawk outlet of the glacial lakes Iroquois and Algonquin (334), first skirting upon the east the great sills of intrusive basalt known as the Palisades, with their markedly columnar jointing and intersections by numerous faults. Above Peekskill we enter the picturesque narrows of the river (174), cut in the hard crystalline rocks of the Highlands. Entering the Mohawk Valley, we pass Syracuse with limestone caverns and well-oriented joints widened by solution through the agency of the descending ground water (181, pl. 6 B). A branch line to the southwest reaches the vicinity of Cayuga Lake and Ithaca, where are well-oriented joints which have controlled the drainage directions, and there is also a typical strath (55, 87, 428).

To Niagara Falls at least a day should be allotted for the "gorge ride" by trolley car, thus making the complete circuit of the brink of the gorge with interruptions and local studies at all important points (352-366, pl. 23 A). From Niagara Falls over the Michigan Central Railway we reach Detroit on the present outlet of the upper Great Lakes as well as of the later Lake Algonquin (334). From this city as a center a trip is made by electric railway to Ypsilanti and Ann Arbor, across the bottoms of the early glacial lakes from the first Maumee to Warren (330-333). The strong Whittlesey beach is encountered at the little station of Ridge Road, and one of the Maumee beaches on Summer Street in Ypsilanti. The city of Ypsilanti is built upon a terrace (165) of the Huron River, and another terrace in the same series is crossed by the electric line. In an excursion of a few miles down the river, passing meanders (164-165) and ox-bow lakes (165, 415), is found an interesting case of stream capture near the little village of Rawsonville (175. See Isaiah Bowman, Jour. Geol., Vol. 12, 1904, pp. 326-334).

Continuing our journey from Ypsilanti over a high moraine (312), Ann Arbor is reached, built upon the level plain of outwash with fosses sometimes separating it from the moraine (281, 314). Upon the campus of the university are great bowlders of jasper conglomerate and jaspilite, which were transported from the north by the continental glacier (305). Across the river from the Michigan Central station and behind the little church is a delta formed in one of the glacial lakes Maumee and here opened in section (168). West of the city is a great valley which was the former course of the Huron River when thus diverted by the continental glacier lying to the eastward of Ann Arbor—border drainage (see Ann Arbor folio by the U. S. G. S., and, further, R. C. Allen and I. D. Scott, An Aid to Geological Field Studies in the Vicinity of Ann Arbor, George Wahr, publisher, Ann Arbor).

Returning to Detroit (M. C. Ry.), the great Sibley quarries in limestone near Trenton may be visited. They display perfect jointing, numerous fossils, and especially well-glaciated surfaces interrupted by deep troughs and showing striæ of several glaciations (304). From Detroit the journey is continued by steamer to Mackinac Island in the strait connecting Lakes Michigan and Huron, passing on the way through the peculiar delta of the St. Clair River (431), and coming in view of the notched headlands, which are a monument to the post-glacial uplift of the glaciated area (250, 341). A day is spent at Mackinac Island and St. Ignace in order to study with some care these uplifted strands of the late glacial lakes (341-344). Chicago may now be reached either by steamer or by rail, and in its vicinity we may see the elevated beaches and the ancient outlet of Lake Chicago (331-332, 347, pl. 22 A. See Chicago Folio, U. S. G. S.). By the Chicago and Northwestern Railway the area of recessional moraines and intermediate outwash plains, and later that of the drumlins, are crossed in journeying to Madison, Wisconsin. By examination of the maps on pages 308 and 317 in connection with the larger scale atlas sheets of the United States Geological Survey (Janesville, Evansville, and Madison sheets), this car journey can be made most instructive in gaining familiarity with the characteristic glacial features, and this study is continued to special advantage in excursions about Madison as a center (316-317, 407). This is the more true since at numerous localities in the vicinity of Madison the well-striated glacier pavement is exposed for comparison of the striæ as regards direction with the axes of the several types of glacial features.

An especially instructive excursion may be made by carriage in a single day to the "driftless area" some twelve miles west of the city. Before reaching it we cross in alternation a series of recessional terminal moraines (pl. 17 C) and outwash plains, and near Cross Plains encounter the partially dissected upland with its arborescent drainage and even sky line (298, 300-301, 312-313, pl. 16 A and B). Typical shore formations (233, 241, 242) are studied to advantage about Lake Mendota in a walking trip to and beyond Picnic Point, where are found the best ice ramparts (431-434. See Buckley, Trans. Wis. Acad. Sci., Vol. 13, pp. 141-162, pls. 18).

Our journey is now continued over the Chicago and Northwestern Railway to Devils Lake near Baraboo, where we cross a salient of the driftless area, within which lies Devils Lake, imprisoned in a former valley of the Wisconsin River, since diverted to another course as a result of the glacial invasion (312-313). The valley here is a former narrows in hard quartzite (466), which towers above the lake in unstable chimneys (300), such as the Devils Tower, but such remnants are not found on the other side of the moraine, being there replaced by rounded rock shoulders. Just north of the lake the marginal moraine which blocks the valley is so characteristic as to merit special study (pl. 17 C). Only a few miles northward along the railway from Devils Lake is Ableman, where, exposed in a high cliff, the hard purple quartzite with beautiful ripple marks to reveal its plane of sedimentation (pl. 11 A) dips vertically, and is overlain by horizontally bedded yellow sandstone. The marked angular unconformity which is thus displayed is further made evident by a basal layer of conglomerate (463) in the sandstone (51-53). Here also are deposits of loess along the river, which display their vertical joint surfaces (207). An excellent geological guide to this interesting district and that of the neighboring "Dalles" of the Wisconsin River has been written by Salisbury and Atwood (The Geography of the Region about Devils Lake and the Dalles of the Wisconsin, etc., Bull. 5, Wis. Geol. and Nat. Hist. Surv., 1900, pp. 151, pls. 38, figs. 47).

If we have taken a conveyance at Devils Lake for Ableman, we may continue in the same manner to Kilbourn, where begin the picturesque Dalles of the Wisconsin River—here a young gorge cut in sandstone, because the Wisconsin was diverted from its old valley to border drainage at the edge of the driftless area (300, 321). The side cañons of the river, through their abrupt zigzags, reveal the control of their courses by the joint system (224). In the journey up the rapids by steamer to inspect the Dalles, we observe many beautiful examples of cross bedding in the sandstone (37).

From Kilbourn we continue our journey to Minneapolis over the Chicago, Milwaukee, and St. Paul Railway, and near Camp Douglas are over a peneplain, out of which rise prominent monadnocks (171). At La Crosse the Mississippi River is reached, flowing beneath bluffs of sandstone which are capped by loess (207). The meanderings and the numerous cut-offs of the Mississippi may be observed to the left (415). Lake Pepin is a side-delta lake blocked by the deposits of the Chippewa River (419).

From Minneapolis an excursion is made to Fort Snelling to view the young gorge of the Mississippi, cut by the Falls of St. Anthony for a distance of about eight miles in manner similar to that of the seven miles of Niagara gorge (354), and to compare this narrow gorge with the broad valley of the Warren River which drained Lake Agassiz (327). Somewhat farther up the Warren River are examples of saucer lakes (416).

From Minneapolis the journey may be continued by the Great Northern Railway to Livingston, Montana, thus crossing between the stations of Muscoda and Buffalo the bed of Lake Agassiz and its marginal beaches (325-328. For local geology of Minnesota consult C. W. Hall, Geology of Minnesota, Vol. 1, Minneapolis, 1903).

The Yellowstone Park is entered from Livingston (Livingston Geological Folio, U. S. G. S.) and departure from it made at the relatively new Union Pacific terminal at the southwest margin. The regular trip through the Park includes visits to the several geyser basins (191-194), Obsidian Cliff (33, 463), the Cañon of the Yellowstone, etc. Good climbers can make a side trip from near the Mammoth Hot Springs to the top of Quadrant Mountain, the remnant of a "biscuit cut" upland (372), and there study the nivation process (368, Yellowstone National Park Folio, U. S. G. S.).

The trip from the Park to Salt Lake City, over the Union Pacific Railway, passes through the Red Rock Pass, the former outlet of Lake Bonneville (423), into the desert of the Great Basin (Chaps. XV and XVI). Great Salt Lake is a saline lake or sink with an interesting record of climatic changes (198, 401). The front of the Wasatch Range, in view and easily reached from Salt Lake City, is deeply scored by the horizontal shore terraces of Lake Bonneville (198, 199), and these terraces are extended at every reëntrant by barrier beaches of great perfection. In the Pleistocene period mountain glaciers in part occupied the valleys of this range, though they did not always extend as far as the mountain front. Big Cottonwood Cañon, which realizes this condition, and the neighboring Little Cottonwood Cañon, from whose front its glacier spread into an expanded foot (264), thus show for comparison in a single view the V and the low U sections respectively (172, 376). Here are also alluvial fans (213) and recent faults which intersect them.

From Salt Lake City the return to New York may be made by the Denver and Rio Grande Railway across deserts and through the Royal Gorge, the cañon of the Arkansas River. A full itinerary of the points of geological interest along this route, and continued to Chicago, Washington, and New York, is supplied in much detail in the guide of the geological excursion to the Rocky Mountains above cited. This the traveling geologist should not fail to study. Some references to points along this journey will be found on preceding pages of this book (219-220, High Plains; 170, Allegheny Plateau in West Virginia; 176, water gap of Harper's Ferry; 176-177, 184-186, side trip up the Shenandoah Valley to Luray Caverns and Snickers Gap; 251, Chesapeake Bay).

Instead of returning directly from Salt Lake City, the traveler, if he has sufficient time at his disposal, may extend his journey southwestward across the Great Basin to Los Angeles. A branch line from this route leaves the Vegas Valley and passes within reach of the famous Death Valley (201) to Tonopah (79) and the Owens Valley (77-78, 92), where are many surface faults dating from the earthquake of 1872 and other less recent disturbances. Returning to the junction point, the route continues across the Colorado and Mohave deserts to Los Angeles. From Los Angeles as a center the exceptionally interesting terraces, caves, and stacks of an uplifted coast are to be seen to best advantage near Pt. Harford (Chap. XIX). The islands of San Clemente and Santa Catalina may also be reached from Los Angeles (239, 248, 249, 250, 256, 257, pls. 5 B, 7 A, 12 A). The return to the East, if made by the Santa Fe Railway, permits of a visit to the Grand Cañon (174, 443) from the station of Williams. From that point eastward the geology of the route is fully covered in Emmons' Guide to the Rocky Mountain Excursion already cited.

For the benefit of those who are privileged to travel in Europe, and the number increases yearly, a pilgrimage is suggested which may easily be made to correspond with plans laid out on the basis of historical, artistic, and scenic points of interest. The only popular guide of a general nature written for geologists traveling abroad appears to be a brief but valuable little paper by Professor Lane (The Geological Tourist in Europe, Popular Science Monthly, Vol. 33, 1888, pp. 216-229). The publishing house of Gebrüder Bornträger in Berlin is now publishing a quite valuable series of geological guides dealing with special districts and written by well-known authorities (Sammlung Geologischer Führer). Of this series some thirteen numbers have already been issued. Many other valuable local guides of a geological nature are the Livrets Guides of the International Geological and Geographical Congresses, and the similar pamphlets supplied in connection with annual meetings of national or provincial geological societies.

Passengers on steamships sailing from the harbor of New York pass out over a deeply submerged cañon (252) largely filled with glacial deposits, through the Narrows (174), and in sight of Sandy Hook, a modified spit (238, 240). To the left are seen the great morainic accumulations at the border of the glaciated area on Long Island (298). In the course of the trans-Atlantic voyage a much-rounded iceberg may be encountered (291), though this is much more apt to occur upon the northern routes from Quebec, and late in the season. Upon entering the English Channel the land on both coasts rises in steep cliffs, where are found all the common shore features well developed (Chap. XVIII). The German steamships pass in sight of Heligoland, that last remnant of wave erosion (236).

While traveling in Europe, the student should consult a map of the glaciated area (299), and so learn to recognize its peculiarities, and carefully mark its marginal moraine (311) and other strongly marked features.

If the British Isles are visited and the more rugged areas are selected, one may study the cirques and other characteristic features due to the presence of mountain glaciers about Snowdon (Chap. XXVI). More mature stages of the same processes are to be found in the Scottish Highlands and the Inner Hebrides, but especially upon the Island of Skye (Fig. 492). A very valuable aid to excursions in this district is Baddeley's Scotland (part I, Dulau, London) and Sir Archibald Geikie's Explanatory Notes to accompany Bartholomew's Geological Map of Scotland (map and notes in cover, Edinburgh, 1892, pp. 23).

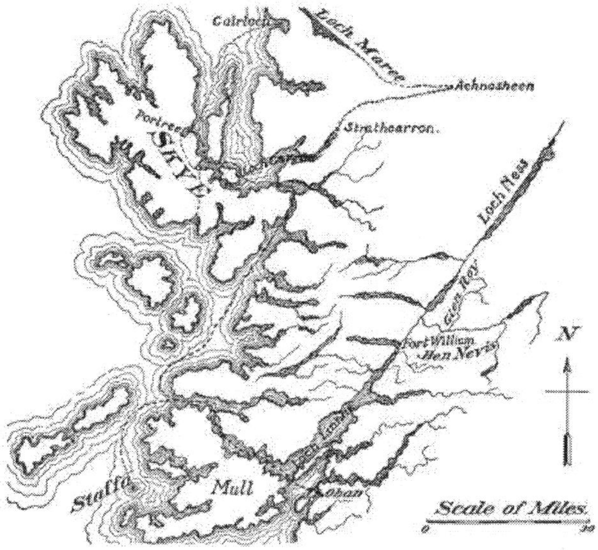

FIG. 492.—Sketch map of Western Scotland and the Inner Hebrides to show location of some points of special geological interest.

It is from Oban, the "Charing Cross of the Highlands", that one should start out upon the summer steamers in order to reach both Skye and Staffa, the latter with fine basaltic columns (463), and Fingal's Cave. In sailing to Skye one passes upon either shore of the narrow fjords many relics left in the dissection of volcanoes (139-143 and Sir A. Geikie, Ancient Volcanoes of Great Britain, Vol. II); also rocky islands and skerries marking submergence (252), and the coast terraces which register a later uplift (250). Skye is a complex of many intrusive and volcanic rocks of such markedly different colors as to appear as tints in the landscape. In the Cuchillin Hills of dark green rises the massive gabbro (462) cut by cirques into the jagged pinnacles of horns and comb ridges (373); while lower down and to the east are rounded domes of rhyolite (463) abraded beneath the glaciers and of a delicate salmon tint. Still lower and to the westward are flat mesas composed of horizontal layers of black basalt under a rich carpeting of the brightest verdure. Eastward across the channel are seen the purplish walls of an ancient sandstone. The jagged gabbro core of the island thus represents a fretted upland (372) and is now the training ground of the Alpinist (Abraham, Rock Climbing in Skye, Longmans, London, 1908), while nestled in one of the bottoms of a U-valley is Loch Coruisk, a typical rock-basin lake (412), its shores of hard rock planed and scored.

From Skye we may go to study the remarkable thrusts (45) on the north shore of Loch Maree, a marked lineament, and one directed at right angles to that on the course of the Caledonian Canal connecting Loch Linne with Loch Ness. This northeast wall of Loch Maree is a strikingly rectilinear fault represented by an escarpment, up which we climb to find at the top the crushed and fluted thrust planes of movement dipping southeastward at a flat angle. Here also are beautiful rock-basin lakes, lying in hollows molded beneath the continental glacier. On our way from Skye we have passed up Loch Carron, a sea loch or fjord (252), and along the strath at its head known as Strathcarron (428).

Returning now to Oban, it is but a short trip by steamer up Loch Linne to Fort William along the striking lineament (226) which continues to Loch Ness and beyond (Fig. 492), and thence by rail to Glen Roy and the neighboring glens of Lochaber (322-325).

From Paris as a starting point, we may visit in a most picturesque region the beautifully preserved craters of extinct volcanoes in the Auvergne of Central France (105, 124, 145), which district is entered from Clermont-Ferrand. Here are found the characteristic puys, steep lava domes of viscous lava (105), which figured largely in the early controversies of geologists concerning the origin of rocks.

FIG. 493.—Outline map of a geological pilgrimage across the continent of Europe.

The rest of our pilgrimage will be so planned as to enter the noble river Rhine at its mouth (Fig. 493), ascend its course to its birthplace in the snows of Switzerland, and after further exploration of the features of this fretted upland, traverse northern and central Italy so as to make our departure for America by the southern route. Entering then upon this course in the Low Countries, we have first the opportunity of observing the characteristics of a great delta with natural levees artificially strengthened as dikes (165-168). Here also are found dunes of beach material which has been raised by the wind into a great rampart near the shore (209-211). Such a wall of dune sand is well displayed at the bathing resort at Scheveningen near the Hague (421). The flood plain of the Rhine (162-165) may be studied in a journey up the river to the university town of Bonn, from whence a day's excursion should be devoted to the relics of volcanoes known as the Seven Mountains (H. von Dechen, Geognostischer Führer in das Siebengebirge, Bonn, 1861). As a preparation for this trip and others in the volcanic Eifel higher up the river, a visit should be made to the mineral and rock collections of the Poppelsdorfer Schloss at the University. In the volcanic Eifel are found some of the most interesting of crater lakes (405), the largest being Lake Laach with its somewhat peculiar volcanic ejectamenta and its picturesque abbey (see von Dechen, Geognostischer Führer zu der Vulkanreihe der Vorder-Eifel, etc., Bonn, 1886. Consult also Lane, A Geological Tourist in Europe, *l.c.*).

Continuing our course up the river from Bonn, we soon enter the gorge of the Rhine cut in an uplifted peneplain (169, 171, 174). From Coblenz, where the Moselle enters the Rhine, a side trip may be made up this tributary river past Zell with its entrenched meanders (173) to the ancient Roman city of Treves. Above Bingen on the Rhine we leave behind us the narrow gorge and rapid current of the river and continue over the broad floor at the bottom of a rift valley (403), lying between the forest of Odin and the Black Forest on the east and the "Blue Alsatian Mountains" far away to the west. At the margins of this plain are beds of loess with their characteristic joint structures and inclusions (207), and in the higher hills on either hand a wealth of intrusive igneous rocks.

At the entrance of the Neckar River to this broad plain is nestled the picturesque castle and university town of Heidelberg, a convenient center for excursions (Julius Ruska, Geologische Streifzüge in Heidelbergs Umgebung, etc., Nägele, Leipzig, 1908, pp. 208, map). At Strassburg (Schwarzwaldstrasse 12) is located the German Chief Station for Earthquake Study, with a particularly large set of modern seismographs. In the university cabinet is also one of the largest and most representative mineral collections in Europe. For excursions in the neighborhood consult Benecke, Sammlung Geognostische Führer, Vol. 5, Elsass, 1900.

From Strassburg we may go by the Black Forest Railway to the Hegau with its volcanic plugs (140), each surmounted by a picturesque castle. We enter next the broadly extended piedmont apron site, above which Lake Constance still remains as a border lake (399). Outwash aprons (314), moraines (311), and drumlins (317) are each in turn encountered. Still continuing our course up the Rhine from Bregenz, we enter the fretted upland (372) of the Alps, mountains composed of great folds and thrusts about a core of intrusive rock (Rothpletz, Sammlung Geologische Führer, Vol. 10, 1902, Thrusts in the Alps between Lake Constance and the Engadine). Some fourteen miles above Chur we pass the terrace produced by successive landslides (414), known far and wide as the Flimser Bergstürz. The further assent of the cascade stairway of this glacier-carved valley brings us to the Furka Pass, from which point magnificent views of the fretted upland are obtained. At the Känzli, a mile from the hotel, one may view the névé of the Rhone Glacier, which may also be easily visited.

We have now followed a great river from its mouth in the sands of Holland to its source in the snows of the higher Alps. Passing over the divide and descending to Gletsch, we may observe the lower end, or foot, of the Rhone glacier and the crevasses and séracs (391) on the steep descent of this radiating glacier (383, 386). The response which glaciers make to climatic changes is here well illustrated by the recession of the glacier front from near the hotel (its position in the '50s of the nineteenth century) to its present position about a mile farther up the valley.

The characteristics of a glaciated mountain valley may be further illustrated by climbing to the Grimsel Pass, which is scratched and striated (377, 385), and then descending the valley of the Aar to Meyringen (377). Near the Grimsel Hospice are the characteristic rock basin lakes (412), and upon the Aar Glacier to our left were carried out the epoch-making researches of Louis Agassiz, the founder of the glacial theory for explaining the drift. We encounter some thirteen rock bars (377). Just before reaching Meyringen we pass the last of these, the Gorge of the Aar, cut by the stream through limestone.

Interlaken (419) may be made the center for additional excursions up the Lauterbrunnen Valley, with its prominent albs (376) and its ribbon fall of the Staubbach (378). By the Jungfrau Mountain railway we may now ascend partly in tunnels of the rock to the Ewigeismeer, and look down upon the névé and bergschrunds of the Great Aletsch Glacier (370, see Baltzer, Sammlung Geologische Führer, Vol. 10, Bernese Oberland, 1906). Returning to Interlaken by way of Grindelwald, one may study the foot of a radiating glacier, the Untergrindelwald glacier, with its tunnel and its milky and braided stream.

Crossing now the Alpine foreland to Villeneuve at the upper end of Lake Geneva and upon a well-developed strath (426, 428), we may look out upon the turbid waters extending far from the shore of the lake. Journeying to Geneva by steamer we note the gradual clearing of the water until at the outlet of the lake it is as clear as crystal. A walking trip from Geneva takes us to the Bois de la Bâtie, where the Arve with turbid waters meets this clear stream (427).

The railroad to Chamonix ascends another cascade stairway (376), affords views of complexly folded sedimentary rocks (43), and at Chamonix itself the mer de glace supplies opportunities for the study of moraines (386, 393) and glacial movement (390-392). To experienced Alpinists the summit of Mount Blanc offers a remarkably extended outlook over the fretted upland of the Alps (pl. 18 A). From the station of LeFayet below Chamonix, one may ascend to the Désert de la Platé, where are Schratten in limestone due to solution (188).

Crossing by one of the passes to the valley of the Rhone at Martigny we may reach Zermatt, to-day the climbing center of the Alps. From the subordinate cirques surrounding this village descend the Gorner, Findelen, St. Theodul, and other components of this radiating glacier. A black tooth of rock, the Matterhorn, towers above the other peaks and shows to greatest advantage this feature of glacial sculpture (374), while the Gorge of the Gorner is a severed rock bar like that of the Aar (377). Either on foot or over the mountain railway we may ascend to the Gorner Grat, a subordinate comb ridge (373) which affords one of the most magnificent and instructive views of radiating glaciers.

From Brig, farther up the Rhone Valley, an excursion is made to the Eggishorn Hotel, a center for study on and about the Great Aletsch Glacier (329, 371, 385, 388, 395, 410). The easy ascent of the Eggishorn is rewarded by a view almost directly downward upon the ice-dammed Márjelen Lake (329, 411).

From Brig one may make his entry into Italy, either over the picturesque Simplon route afoot or by diligence, or else beneath it through the railway tunnel. By an alternation of short steamboat and rail trips the journey is continued in a direction transverse to the longer axes of the border lakes Maggiore, Lugano, and Como, and later southward to Milan. In leaving the village of Como we pass over heavy morainic deposits on the apron borders of the expanded-foot glacier (383, 385) which once occupied the valley above. On the journey from Milan to Venice, over the fertile plains of Lombardy, the similar accumulations about Lake Garda (414) are first encountered at the little station of Lonato and left behind at Somma Campagna (Tornquist, Sammlung Geologische Führer, Vol. 9, Northern Italy, 1902).

The city of Venice is built upon pile foundations in the lagoon behind the barrier beach known as the Lido (242, 428-429). From here we may reach the Karst country by way of Trieste, some of the more interesting and typical features being found near Divača (187-189, 422, pl. 6 A). In a different direction from Venice by way of Belluno we enter the Dolomites with their patterned relief and battlemented towers (228, 445).

Additional centers for geological excursions on the route to our point of departure from Italy are Rome and Naples. At the Italian capitol and in its neighborhood we may study the volcanic Campagna with its beds of tuff (105) and its crater lakes (405. See Sir A. Geikie, The Roman Campagna, Landscape in History and other Essays, Macmillan, 1905, pp. 308-352; also Deecke, Sammlung Geologische Führer, Vol. 8, Campagna, 1901). From Rome it is an easy journey to the cataract of Tivoli with its deposits of travertine (184). In the opposite direction from Rome across the Campagna rise the Alban Hills, ruins of a composite cone with several crater lakes on the sites of former vents. On the summit of the encircling crater rim, like the Monte Somma of the Vesuvian Mountain now a crescent only, is located the chief Italian station for earthquake study.

From Naples we may reach in short excursions and study with some care still active volcanic mountains. To the east is Mount Vesuvius (94, 97, 122, 124, 127-137), which was in grand eruption in April, 1906. Westward from Naples are the Campi Phlegraeii, or burning fields, with many craters. Of these Astroni offers a fine example of a large-cratered cinder cone (105). In the same vicinity are Monte Nuovo (96) and the Solfatara (97), the latter a type of volcano which no longer erupts lava, but in its place emits carbon dioxide and other gaseous emanations (Grotto del Cane). The starting point for excursions in the Phlegræan fields is Pozzuoli with its Temple of Jupiter Serapis (254-255), reached from Naples by an electric line which pierces the wall of an immense crater (Posilippo) composed of fine yellow volcanic ash known as Pozzuolan.

From Naples steamers make short excursions to Sorrento with its deep ash deposits, and to Capri with its blue grotto (257-258). Herculaneum (139) and Pompeii (122), buried during the eruption of 79 A.D., are on the line of the Circum-Vesuvian Railway.

Steamships to New York from Naples call at Gibraltar, the land-tied island *par excellence* (241). Most steamships of the southern route pass through or near the volcanic islands of the Azores, and certain boats touch at Algiers, from which a line of railway gives access to Biskra on the borders of the Desert of Sahara.

Throughout these pilgrimages the traveler should be on the alert to note not only the agent responsible for the features which come under his observation, but, especially where this is the common sculpturing agent of running water, he should not fail to notice the stage of the erosion cycle which is represented (Chapter XIII).

FOOTNOTES

Italian for mouth; plural *bocchi.*

These models and the contouring apparatus are now manufactured for the use of schools and colleges by Eberbach and Son, Ann Arbor, Mich.

This clay is manufactured by the A. H. Abbott Company, art dealers, Wabash Avenue, Chicago.

Numbers in parenthesis refer to pages in this book, where further information is to be found.

Made in United States
Orlando, FL
04 November 2026

90219747R00314